Discrete Choice Models

Discrete Choice Models

Mathematical Methods, Econometrics, and Data Science

Alfred Galichon

Princeton University Press
Princeton and Oxford

Published by Princeton University Press
41 William Street, Princeton, New Jersey 08540
99 Banbury Road, Oxford OX2 6JX

press.princeton.edu.

GPSR Authorized Representative: Easy Access System Europe - Mustamäe tee 50, 10621 Tallinn, Estonia, gpsr.requests@easproject.com

ISBN 978-0-691-22180-9
ISBN (e-book) 978-0-691-22713-9
Library of Congress Control Number: 2025949943

British Library Cataloging-in-Publication Data is available

Editorial: Hannah Paul and Josh Drake
Production Editorial: Kathleen Cioffi
Text and Cover Design: Wanda España
Production: Erin Suydam
Publicity: William Pagdatoon
Copyeditor: Bhisham Bherwani

Jacket image: Kiyoshi Takahase Segundo / Alamy
This book has been composed in Linux Libertine.

Printed in the United States of America

10 9 8 7 6 5 4 3 2 1

MIX
Paper | Supporting responsible forestry
FSC® C008955
FSC www.fsc.org

To Marius, André, and Jacqueline Galichon.

La filosofia è scritta in questo grandissimo libro che continuamente ci sta aperto innanzi agli occhi (io dico l'universo), ma non si può intendere se prima non s'impara a intendere la lingua, e a conoscere i caratteri nei quali è scritto. Egli è scritto in lingua matematica, e i caratteri son triangoli, cerchi e altre figure geometriche, senza i quali mezi è impossibile a intenderne umanamente parola; senza questi è un aggirarsi vanamente per un oscuro laberinto.

—Galileo Galilei (1564–1642), *Il Saggiatore*, cap. 6

Philosophy is written in this grand book that lies continually open before our eyes (I mean the universe), but it cannot be understood unless one first learns to comprehend the language and to read the characters in which it is written. It is written in the language of mathematics, and its characters are triangles, circles, and other geometric figures, without which it is humanly impossible to understand a single word of it; without these one wanders vainly through a dark labyrinth.

—Galileo Galilei (1564–1642), *The Assayer*, chap. 6

Contents

Preface

This book has emerged from several courses and lecture series I gave over the last few years, usually in front of a diverse audience of graduate students and researchers with a background in economics or in mathematics. The goal of this work is to provide a concise treatment of discrete choice models and their applications, leveraging the powerful elegance of mathematics. The mathematical foundation of these models is half optimization theory and half probability and statistics. As we shall explore, there are intricate connections with the theory of optimal transport, which was the subject of my previous book, *Optimal Transport Methods in Economics.* However, in these lectures the framework is broader. The focus is on choice, demand, price formation, and equilibrium, and the statistical estimation of the related models. The exposition is influenced by my own research, and, more broadly, reflects my personal tastes, so there is no attempt at an exhaustive treatment of these models. It is not a survey or a guide to the literature.

Gabriele Buontempo, Enzo Di Pasquale, Alexandre d'Epinay, Alessandro Facchini, Clément Montes, and Jiawei Qian have provided helpful research assistance at various stages of this project. I am grateful to the PhD students of my Applied Microeconometrics class at New York University who have spotted typos and made suggestions on earlier versions of this text, in particular, Diego Cussen, Nilima Pisharody, Caleb Maresca, Vasco Villas-Boas, and Bill Wang. I am hugely indebted to several people who did an exhaustive reading of earlier drafts: Giuseppe Cognata, Facundo Danza, Amine Fahli, Octavia Ghelfi, Leon Guzman, Antoine Jacquet, Isabelle Perrigne, and Maxime Sylvestre. In addition, the editorial team at Princeton University Press, in particular Hannah Paul, Josh Drake, and three anonymous reviewers, provided useful feedback on earlier versions of the manuscript. This project was nourished by insightful conversations with Pierre-André Chiappori, Pauline Corblet, Jeremy Fox, Jean-Marc Robin, Bernard Salanié, Larry Samuelson, Matt Shum, and Jonathan Vogel. I am particularly grateful to Antoine Jacquet, who did not only read and comment, but also suggested improvements in many places. Writing this book while serving as director of New York University's academic center in Paris required considerable multitasking. I could not have focused serenely on this project without the trust placed in me by the University's president, President Linda Mills, and the wonderful collaboration with Professor Eliot Borenstein and Dr. Nancy Morrison. Moreover, I would not have been able to complete the writing in a reasonable amount of time without relying on an exceptional team of highly dedicated administrators at the center, among them, Beth Epstein, Martina Faltova, Chioma Iwuegbu, Valérie Michelin, Marcus Neeld, and Xavier Séguy. Much of my research agenda over the years has been funded by two grants of the European Research Council of the European Union (starting grant No. 313699, and consolidator grant No. 866274). These grants have been transformative, and this book would probably not have come to exist without them.

Discrete Choice Models

Introduction

This book offers a mathematical perspective of discrete choice models and related topics. Discrete choice models, also called multinomial choice models, are used to model situations where the consumer needs to choose one option from among several. One of the first areas where it has been applied has been for the choice of transportation mode, but the methodology has proved useful for understanding many decisions that are in nature economic (related to choice of education, employment, retirement, consumption), demographic (fertility decisions, migration), sociological (marriage, friendship formation), geographic (trade, urban dwelling), and political (voting). While discrete choice models originated in transportation studies and in psychology in the middle of the twentieth century, they have made their way to other disciplines in the last 70 years, to the point that they are now at the core of practically all of the social sciences.

In this book (and in much of the literature), the discrete choice approach is based on the idea of individual rationality: it postulates that individual decision-makers associate a numerical valuation called "utility" to each option that they are faced with. The rational behavior of decision-makers consists of selecting the option that yields the highest utility. Each agent's choice thus results from solving a simple optimization problem: maximizing utility over a finite (i.e., *discrete*) set of options. These choices are then aggregated across individuals in order to account for the share of the population choosing each option, which allows us to express the demand (or "market share") for every option.

Of course, the utilities that individuals associate with each option are most often not observed by the analyst, but they can be inferred based on the data of the choices made. Discrete choice models are thus nonlinear econometric models predicting a categorical outcome—which option has been picked. To train these models, one generally takes a parametric form of the utilities and estimates the value of the parameter vector that leads to fitting as best as possible the observed choices according to some criterion. Many criteria exist, such as maximum likelihood, method of moments, minimax regret, maximum score, and revealed preference. The nature of this criterion very much depends on how exactly the utilities are modeled, computational convenience, desired robustness to misspecification, and so on.

An attractive feature of discrete choice models is the ability to conduct counterfactual analysis: once the utilities have been estimated, it is possible to predict the change in aggregate demand if some policy intervention were to be implemented, such as a tax affecting a transportation mode. Thus, in contrast with less structural methods, such as regression analysis or factor analysis, multinomial choice methods allow us to address in-depth causal questions, as they attempt to capture the core of the decision-making process. Almost built into discrete choice models is the ability to permit welfare analysis; indeed, in discrete choice models the individual utilities are the primary objects of focus, so computing the social welfare is simply a matter of summation. The models will therefore allow us to predict in a very straightforward manner the welfare impact of a policy change.

An important class of discrete choice models relies on the *random utility* paradigm. Take the example of an individual needing to choose a transportation mode. In the random utility framework, one assumes that the utilities that agents associate with each mode have a "systematic" component, which depends deterministically on the agents and an option's observable characteristics, but have in addition an "idiosyncratic" component, which captures the part of the decision that cannot be captured by observable characteristics alone. Even if two agents have the exact same observable characteristics, one may choose the train and one may choose the plane, simply because their idiosyncratic utility terms differ. One therefore speaks of random utility, or stochastic choice, which is a bit of a misnomer because it does not necessarily mean that the idiosyncratic terms are randomly drawn by the agents, or that there is any randomness in the decision-making process. However, as the idiosyncratic term is unknown to the analyst, it is random from their point of view for all practical purposes. And although the analyst does not know the idiosyncratic terms, they typically postulate that the distribution is known, or at least belongs to a parametric family of distributions. From the point of view of the analyst, the idiosyncratic term is a random utility term whose distribution is known.

The random utility paradigm does not allow us to predict the choice that a particular individual will make, but, as chapter 1 will show, it allows the analyst to compute the probability that a particular individual will choose one option or the other. This is closely connected to the classification problem in machine learning—in fact, many of the tools overlap. The random utility framework allows us to compute aggregate quantities, such as the aggregate welfare (the sum of the welfare of individuals in the population); the predicted market shares; and the part of the aggregate welfare that is due to systematic utilities and the part of it that is due to idiosyncratic terms—that part being defined (up to a sign) as the *entropy of choice*, a concept introduced in chapter 1. The analyst also needs to solve the inverse problem of recovering the systematic utilities based on the observation of the market shares, a fundamental problem called the "market share inversion problem." As we shall see in chapter 1, convex analysis is the appropriate mathematical framework to perform these calculations without making any restrictive assumptions about the random utilities. Thanks to convex analysis, we will be able to formulate the basic calculations as a convex optimization problem, which is practically useful and will have important consequences on the analysis of the problem. For instance, it will allow one to deduce results about the existence and uniqueness of a solution to the market share inversion problem.

As we will see in chapter 2, some distributions of random utilities lead to simple formulas for the expression of the welfare, the predicted market shares, and, in some cases, the entropy of the choice and the demand inversion. The most famous case is the logit (or "logistic") framework, which assumes that the random utilities follow an extreme value distribution, more precisely, that of independent and identically distributed (i.i.d.) Gumbel variables. The Gumbel distribution is one of the three max-stable distributions arising in extreme value theory; it is the limiting distribution (after renormalization) of the maximum of a large class of independently distributed random variables. As is probably already familiar to most readers, the logit model allows for simple formulas for the market shares and allows us to solve the market share inversion problem in closed form. It also leads to an entropy of choice that coincides (up to a sign) with the Gibbs entropy.

Yet for all its appeal, the logit paradigm is a very rigid framework that has significant shortcomings. In the transportation mode example, it would specify that the random utilities associated with taking bus, train, and plane are independent, which does not capture the fact that some travelers may dislike air travel in a manner that cannot be predicted by their observable characteristics, thereby introducing a correlation between the random utilities associated with the "bus" and "train" options. Consequently, chapter 2 moves on to exploring distributions of random utility that retain tractability with more flexibility, in particular allowing for this type of correlation. An important class of such distributions presented there is the class of *multivariate extreme value distributions*, discussed in section 2.2. These random variables can be obtained by an ingenious combination of i.i.d. Gumbel variables used as factors, in a way somewhat similar to how any Gaussian vector can be obtained by a linear combination of i.i.d. standard normal factors. Multivariate extreme value distributions, which, following Daniel McFadden, are often called "generalized extreme value distributions" in econometrics, lead to a closed-form expression for the welfare function, the market shares, and sometimes also the entropy of choice, as in the case of the nested logit model, one of the most important representatives of the class.

The logit framework plays a central role in structural estimation, as we begin to see in the subsequent chapter 3 on logistic regression. Consider a stochastic choice problem where the systematic utilities belong to a parametric family and the random utilities are i.i.d. Gumbel. Given a parameter vector, the model predicts the probabilities that each agent will pick the various options, which leads to the specification of a tractable parametric family of choice probability. Assuming that the observations are independently sampled, one can then form the log-likelihood of the sample. In this setting, logistic regression is simply maximum likelihood estimation. It is one of the most important topics of this book, so the entire chapter 3 is dedicated to its study. In addition to being a maximum likelihood estimation problem, logistic regression can be interpreted as a method of moments, and also as a minimax regret procedure. There is a useful link with generalized linear models, an important part of the statistical toolbox that generalizes the Poisson regression: in section 3.4 we recall that connection, known as the "Poisson trick" in the machine learning community, which asserts that logistic regression amounts to a Poisson regression with the addition of a fixed effect associated to each individual. From the computational point of view, logistic regression has many attractive features. It is a convex optimization problem, which leads to easy computational methods as discussed in section 3.3. One can state simple conditions to characterize the existence and the uniqueness of a parameter estimator, as carried out in sections 3.5 and 3.6. When the dimensionality of the parameter is large, the model needs to be regularized by adding a penalty term to prevent overfitting; this can be done for various types of regularizations such as LASSO, with dedicated algorithms such as the proximal gradient descent methods, also recalled in chapter 3.7.

All the models seen up to this point have been based on the logistic framework or its multivariate extreme value generalization (in chapter 2). In contrast, the *characteristics-based approach*, a popular alternative approach to demand estimation, is introduced in chapter 4. It frustrates the hope of getting a closed-form expression for the welfare and other quantities, but it provides a simple and easy-to-interpret geometric description of the interactions between the characteristics of the decision-maker and the characteristics associated with each option. In the most popular version, explored in section 4.1, one assumes that this interaction is a scalar product, or, more generally, a bilinear form. This allows one to make use of Euclidean geometry to describe the choice problem, as one may then locate the characteristics of the consumers who choose a particular option on a polytope in the characteristics space. As we shall see, this problem is closely related to the theory of optimal transport, which was the focus of my previous book (Galichon 2016), and one can leverage the power of computational geometry to efficiently compute the predicted market shares and perform demand inversion. Mixing the characteristics approach and the logistic framework leads to the *random coefficient logit* specification explored in section 4.2, and proposed by Berry, Levinsohn, and Pakes. In this model, the random utility term is the sum of two independent components, one term with a Gumbel distribution and one that is characteristics-based. Here, again, the theory of optimal transport offers insights as discrete choice problems with random coefficient logit random utility can be reformulated in terms of an entropic optimal transport problem, for which many computational tools exist. The random coefficient logit specification is the foundation for Berry, Levinsohn, and Pakes's method for structural estimation, which is presented in section 4.4. This framework deals with endogeneity and allows us to model not only the demand side, but possibly the supply side as well, with imperfect competition among sellers.

Up to this point, and with the exception of section 4.4, the book has regarded the systematic utilities as an exogenous primitive of the model. However, one may want to incorporate random utility specifications into equilibrium models where the systematic utilities depend on the price or other quantities that are adjusted at equilibrium. Chapter 5 offers several examples of such types of models of allocation and equilibrium pricing. The first example is international trade, the primary model of which is the *gravity equation*. In the gravity equation, trade flows are adjusted by supply and demand according to the propensity of pairs of countries to trade with each other, which depends on factors such as geographic distance, trade agreements, cultural proximity, and many other regressors; but they also depend on the sizes of the countries, as measured by their volumes of exports and imports. The trade flows are therefore adjusted at equilibrium by prices, which are materialized by exporter- and importer-fixed effects. Section 5.3 recalls an important reformulation of the gravity equation as a Poisson equation with two-way fixed effects, thus extending the "Poisson trick" to bipartite models.

The incorporation of logistic random utility in various microeconomic frameworks yields variants of the gravity equation. This is the case of the models of matching with transfers pioneered by Gary Becker, and section 5.4 introduces the class of *empirical models of matching*, which are standard models of matching with the addition of a random utility term. The introduction of the random utility term has multiple benefits: accounting for the heterogeneity that is not observed by the analyst; imposing uniqueness of the equilibrium matching in a large population; and providing smoothness, which is desirable for estimation and inference purposes. A pioneering example of an empirical matching framework is the model of Choo and Siow, which has been successfully applied to the analysis of the marriage market, and, to a lesser extent, of the labor market. The Choo and Siow framework is a model of bipartite matching with transferable utility, meaning that prices (wages or other forms of utility transfers via bargaining) adjust at equilibrium in order to clear supply and demand. It incorporates logit heterogeneity, meaning that agents' utilities are the sum of a systematic utility term, which is the outcome of a bargaining process, and a random utility term following an i.i.d. Gumbel distribution. This structure allows the reformulation of the problem of equilibrium matching as a pair of interdependent discrete choice models on each side of the market, and its solution using convex optimization or as a generalized linear model with two-way fixed effects.

The one-to-one, bipartite framework of the Choo and Siow model can be viewed as restrictive in some situations, for example, if one would like to study the same-sex marriage market (which is not bipartite), or the employer-employee matching market (which is not one-to-one). Fortunately, as seen in section 5.5, empirical models of matching à la Choo and Siow can be extended quite directly to more general models of coalition formation. While models of matching do not necessarily put explicit emphasis on prices (although they assume the existence of prices to clear the market), *hedonic models*, covered in section 5.6, directly model their formation. In that paradigm, a good is produced and consumed in different varieties or qualities by heterogeneous producers and consumers. For instance, cars are differentiated on the quality space, and are imperfect substitutes for one another, both from the consumers' and the producers' perspectives. The prices enter the systematic utilities of both consumers and producers, and both sides of the market face a discrete choice problem. At equilibrium, prices adjust so that the demand for each quality from the consumers' side equates to the corresponding supply on the producers' side, and the structural parameters on both sides can be learned in that way.

Our discussion until now has focused on static models, in the sense that it has not taken into account the effect of present decisions on future outcomes. Take maintenance decisions, for example: deciding on the preventive maintenance of a vehicle generates a present-period cost that may exceed the present-period benefit; but performing the maintenance procedure may be justified because it will decrease the cost of operations in the future. Of course, one could incorporate a discounted value in the systematic utilities associated with the maintenance or no-maintenance decision, but this value depends on the various decisions that will be taken in the future, as the decision-maker will be faced with other choices (maintaining the vehicle, operating it, selling it, and so on) at each future period. This is the core of the issue that *dynamic discrete choice models*, covered in chapter 6, are tackling. The chapter starts with a discussion in section 6.1 of these models from a linear optimization point of view that is not typical in most treatments of the topic, but that highlights the connections with the models seen thus far. Logistic random utility is then explicitly introduced in section 6.2, where it is shown that a dynamic discrete choice model is made of a series of static discrete choice problems dynamically linked together by a *Bellman equation.* A distinction must be made between models where a finite number of decisions are made sequentially, which is the finite-horizon case, and models where there is no end to the sequence of decision problems, the infinite-horizon case, where one must discount the values of future period utilities. In the infinite horizon case, the absence of a time horizon induces some stationarity in the value associated to each option in a specific state. Inference is worked out, in section 6.3 for the finite-horizon case, and in section 6.6 for the infinite-horizon case. The chapter concludes with an investigation in section 6.9 of dynamic matching models, which are two-sided dynamic discrete choice models.

While prices, understood as adjustable monetary transfers, have played a prevalent role in our analysis, one should note that there are many situations in the economy when demand is not regulated by monetary prices, but by other adjustment mechanisms. Taxis are a good example: as the price of taxi rides is generally regulated and fixed, the demand for taxis in times of short supply is not regulated by prices, but by waiting lines. If waiting lines are associated with certain overdemanded options, the corresponding utility will decrease—up to the point where demand exactly matches capacity. In that case, waiting times play the role of numéraire, instead of money. While waiting times share some similarities with monetary prices, they are significantly different: they cannot be transferred to the other side of the market. Picking up a passenger who has waited a long time does not make a driver better off. As a result, chapter 7 studies *equilibrium models with nontransferable utility* and builds a specific mathematical machinery for that purpose. The first objective in the chapter is to understand the effect of capacity constraints on systematic utilities through shadow prices. The consequence of rationing on welfare is studied in section 7.2, and monotone comparative statics, which is the study of how utilities respond to changes in capacity, is studied in section 7.3. The machinery developed in this chapter leads to the introduction in section 7.5 of an empirical matching model without prices, which is a nontransferable utility matching model with the addition of random utility terms. The celebrated deferred acceptance algorithm of Gale and Shapley is revisited and adapted to fit into the framework of discrete choice in section 7.6, and its insightful reinterpretation by Hatfield and Milgrom is in turn adapted to the context of discrete choice models and presented in section 7.7.

While the first chapter shows that most of the welfare and demand inversion analysis can be performed without assuming the logit distributional framework, and chapters 2 and 4 cover examples of distributions of random utilities that can be used as alternatives to logit, the subsequent chapters (from chapter 3 to chapter 7) use almost exclusively the logit framework in the interest of simplicity and because of its link with logistic regression. However, it is interesting to note that these chapters could have been written with very general distributions. Chapter 8 goes back to the models of these four chapters and shows how they can be naturally generalized beyond logit random utilities. Some attention needs to be paid, however, to how the adaptation is done. For instance, the natural estimation paradigm is no longer maximum likelihood, as in this context the maximum likelihood estimation problem is not convex anymore, but minimax regret estimation, which coincides with maximum likelihood in the logistic case, but retains convexity and the moment matching interpretation outside of that case. With this empirical strategy in mind, we are able to revisit one-sided discrete choice models (section 8.1), empirical models of two-sided matching (section 8.2), models of coalition formation (section 8.3), and dynamic models (section 8.4).

Positioning. This book touches upon many disciplines from different horizons. As it is hopefully clear from the various examples alluded to above, discrete choice methods span many disciplines in the social sciences. We have mentioned **economics**, **statistics**, **political science**, **marketing**, **psychology**, **operations research**, **geography**, **sociology**, and **transportation studies**; yet the list is not exhaustive. But from a mathematical standpoint, discrete choice models are connected with an exciting mix of tools, whose diversity is evidenced by the range of the topics in the mathematical appendix, appendix A. The general theory and the welfare analysis seen in chapter 1 make heavy use of **optimization theory**, more specifically, **linear programming**, **convex analysis**, and **optimal transport**. The use of special distributions in chapter 2 borrows from **probability theory** and **mathematical statistics**, more specifically, **extreme value distributions**. The characteristic approach seen there uses some **Euclidean geometry**, more specifically, **polyhedral geometry**. The Poisson regression formulation in chapter 3 uses standard **statistical inference theory**. The gravity equation and the empirical models of matching seen in chapter 5 are closely connected with **entropic optimal transport**. The dynamic discrete choice models seen in chapter 6 connect with **dynamic programming**, more specifically **reinforcement learning** and **Markov decision processes**. And the chapter on discrete choice models with limited availability builds on the important theory of **M-functions** and **submodular optimization**. The text makes frequent appeals to **tensor algebra** with **vectorization** and **Kronecker products**, and to **numerical optimization algorithms**. A particular emphasis has been placed on coding. Finally, the Python code demos in appendix B underscore that, far from being abstract constructions, the concepts introduced in this book are practical and implementable.

Scope. In writing this book, choices were made, and the book does not address some topics that are related with the book's subject. The book says little, for example, about nonparametric welfare analysis. Topics such as maximum score estimation, partial identification, counterfactual analysis, and assortment optimization are not covered. The random utility paradigm is prevalent. Other points of view, such as that of the Bayesian approach, are not represented. Also, some valid critiques of discrete choice models—such as the centrality of the assumption of individual rationality, the reliance on distributional assumptions about unobserved heterogeneity, or the inherently individualistic perspective—receive little or no discussion. The reader should not expect an encyclopedic treatment in this text. The selection of topics and points of view offered here is obviously biased toward the author's own tastes and inclinations.

Audience. Given the range of topics, it is hard to predict whether this book will appeal to a very narrow or to a very wide audience; but while writing it I have bet on the latter. Graduate students and researchers in one of the social sciences disciplines listed above and with a good command of college-level mathematics should find it useful to gain a deeper understanding of the mathematical tools on which the models rest. Mathematicians and data scientists may also find the book useful for understanding what types of applications and models economists are working on. My previous book, *Optimal Transport Methods in Economics* (Galichon 2016), has been fortunate to attract this dual audience, and therefore create a bridge between two communities. My hope is that the same will hold true for the present project.

Prerequisites. This book features a number of mathematical appendices to make it as self-contained as possible, but it assumes a minimal level of mathematical background. The reader is expected to possess knowledge of college-level mathematical notions, such as basics of linear algebra, real analysis, and probability and statistics. For example, the book assumes prior knowledge of the matrix product, but not of the Kronecker product; of an exponential distribution, but not of an exponential family; of maximum likelihood estimation, but not of M-estimation; of ordinary least squares, but not of generalized linear models. Notions such as "almost surely," "absolutely continuous distribution," "converging subsequence," "full rank matrix," and so on will be assumed part of common knowledge. Some command of optimization theory will be helpful but is not strictly required, as all the notions are introduced in several appendices. The book also assumes a basic familiarity with Python, as the code demonstrations form an important part of the learning experience. Readers without this background may find it difficult to fully engage with those examples. The appendices do not include a Python tutorial, as numerous high-quality resources are readily available online. Instead, they focus on more advanced topics, such as vectorization and automatic differentiation.

Organization of this book

This book is organized into eight chapters and two appendices. Chapter 1 introduces the basics of random utility models. Chapter 2 focuses on their primary example—the logit model—and its generalizations. Chapter 3 explores the connection between random utility models, logistic regression, and generalized linear models. Chapter 4 examines characteristics-based random utility models. Chapter 5 studies applications to trade, matching, and equilibrium pricing models. Chapter 6 covers dynamic discrete choice models. Chapter 7 deals with discrete choice models under capacity constraints. Chapter 8 extends all the models introduced in the book to general distributions of random utility. Appendix A provides the mathematical toolbox needed for this book. Appendix B provides code demonstrations that illustrate the concepts introduced throughout the chapters of this book.

Notations and conventions

Terminology. Our use of mathematical terminology aims to strike a balance between precision and convenience. By *probability measure* we shall mean a Borel probability measure; by a *set*, a measurable set. A *continuous probability* or *continuous distribution* will mean a probability measure that is absolutely continuous with respect to the Lebesgue measure; a *convex* function will mean a convex function in the usual sense, that is, one that can take any real value or $+\infty$, is lower semicontinuous, and is not identically equal to $+\infty$. A *rectangle* of $\mathbb{R}^d$ is a Cartesian product of d intervals of the real line (which may be either closed or open, either bounded or unbounded).

Abbreviations. We will use a limited number of classical abbreviations: *c.d.f.* for "cumulative distribution function"; *p.d.f.* for "probability density function"; *a.s.* for "almost surely"; *s.t.* for "such that" or "subject to"; *l.s.c.* for "lower semicontinuous," *u.s.c.* for "upper semicontinuous"; and *w.l.o.g.* for "without loss of generality." A *poset* will mean a "partially ordered set."

Standard notations. The following notations, adopted throughout the book, are more or less standard in the literature. The scalar product of two (column) vectors x and y of the same dimensions is denoted by $x^\top y$. The Euclidean norm of x is denoted by $\|x\|$. Given a function $f : \mathbb{R}^d \to \mathbb{R}$, the *gradient* of f at x, denoted by $\nabla f(x)$, is the vector of partial derivatives $(\partial f(x)/\partial x_1, \ldots, \partial f(x)/\partial x_d)^\top$, and $D^2 f(x)$ is the *Hessian matrix* at x, which is the matrix of second derivatives $(\partial^2 f(x)/\partial x^i \partial x^j)$, $1 \leq i, j \leq d$. The set of $y \in \mathbb{R}^d$ such that $f(z) \geq (z-x)^\top y + f(x)$ for all $z \in \mathbb{R}^d$ defines the *subdifferential* of f at x. The *Legendre–Fenchel* transform of f, denoted by f^*, is defined as $f^*(y) = \max_{x \in \mathbb{R}^d} \{x^\top y - f(x)\}$. Given a function $f : \mathbb{R}^k \to \mathbb{R}^l$, the *Jacobian matrix* of f at x, denoted by $Df(x)$, is the matrix of partial derivatives $(\partial f^i(x)/\partial x^j)$, $1 \leq i \leq l$, $1 \leq j \leq k$. For X a subset of $\mathbb{R}^n$, the notation $conv(X)$ stands for the convex hull of X; these are the points of the form $\sum_{i=1}^I \lambda_i x_i$, $x_i \in X$, $\lambda_i \geq 0$, $\sum_{i=1}^I \lambda_i = 1$, for any family (x_i) of elements of X. The notation *cone* (X) stands for the convex cone generated by X, which has the same definition without the requirement $\sum_{i=1}^I \lambda_i = 1$. The *Dirac mass* at x_0, denoted by δ_{x_0}, is the probability distribution that gives unit mass to x_0. Given a compact set C of $\mathbb{R}^d$, $\mathcal{U}(C)$ is the uniform probability distribution on C. The set $L^1(\mathcal{P})$ is the set of functions that are integrable with respect to the probability $\mathcal{P}$. The set $\mathcal{M}(\mathcal{P}, \mathcal{Q})$ is the set of probability measures π such that if $(X, Y) \sim \pi$, then $X \sim \mathcal{P}$ and $Y \sim \mathcal{Q}$. If $\mathcal{P}$ is a probability distribution over $\mathbb{R}^d$ and $T : \mathbb{R}^d \to \mathbb{R}^{d'}$, then the *push-forward* of $\mathcal{P}$ by T, denoted by $T\#\mathcal{P}$, is a probability distribution on $\mathbb{R}^{d'}$ that is the distribution of $T(X)$ where $X \sim \mathcal{P}$. The expectation of a random vector $X \sim \mathcal{P}$ is denoted by $\mathbb{E}_\mathcal{P} X$, and the variance-covariance matrix of X is denoted by $\mathbb{V}_\mathcal{P}(X) := \mathbb{E}_\mathcal{P}[XX^\top] - (\mathbb{E}_\mathcal{P}[X])(\mathbb{E}_\mathcal{P}[X])^\top$. The notation $X \perp\!\!\!\perp Y$ denotes that the random variables X and Y are independent. Also, the c.d.f. associated with $X \sim \mathcal{P}$ is denoted indifferently by F_X or $F_\mathcal{P}$. The *quantile* of that distribution, denoted by Q_X or $Q_\mathcal{P}$, is defined as the right-continuous pseudoinverse of the c.d.f., namely, $Q_X(t) = \inf\{x : F_X(x) > t\}$. Given a set E, the *power set* of E, denoted by 2^E, is the set of subsets of E. If $(E, \geq)$ is a poset and $x \in E$, the notation $[x, \infty)$ will denote the set $\{y \in E : x \leq y\}$; similarly, the set $(-\infty, x]$ will denote the set $\{y \in E : x \geq y\}$. A *correspondence* Γ from E to F, denoted by $\Gamma : E \rightrightarrows F$, is a map $\Gamma : E \to 2^F$. Given a lattice L, the set of sublattices of L is denoted by $\mathcal{L}(L)$. Given x and y in $\mathbb{R}^d$, the notation $x \ll y$ means $x_i < y_i$ for all $1 \leq i \leq d$. Given $x \in \mathbb{R}^d$ and $B \subseteq \{1, \ldots, d\}$, we denote by x_B the subvector of x whose indices are in B, and we identify x with (x_B, x_{B^c}), where B^c is the complement of B in $\{1, \ldots, d\}$. Given a subset E of $\mathbb{R}^d$, the *interior* of E, denoted by E^{int}, is defined as the complement of the closure of the complement of E. The *extended real line*, denoted by $\overline{\mathbb{R}}$, is defined as the set $\mathbb{R} \cup \{-\infty\} \cup \{+\infty\}$. A *permutation* of size n is a bijection of $\{1, \ldots, n\}$ onto itself. If n is a natural number, $[n]$ denotes the set $\{1, \ldots, n\}$. The set of permutations of size n is denoted by $\mathfrak{S}_n$, and called the *symmetric group*. Given a finite set E, the *cardinality* or number of elements of E is denoted by $|E|$. Following standard conventions in economics, given a vector $x \in \mathbb{R}^d$ and an index $i \in \{1, \ldots, d\}$, the notation x_{-i} will represent the subvector of x in $\mathbb{R}^{d-1}$ obtained by removing the ith entry. Given a set X, $cl(X)$ is the closure of X, $conv(X)$ is its convex hull, and $cch(X)$ is its convex closure. The *convex indicator function* of X, denoted by ι_X, is equal to 0 on X, and to $+\infty$ on the complement of X. The *binary indicator function*, denoted by $\mathbf{1}_X$, is equal to 1 on X and to 0 on the complement of X. Whenever there is no ambiguity, we will call either function an *indicator function*. For $X \in \mathbb{N}$ and $x \in [X]$,

$\mathbf{e}_X^x$ denotes the xth vector of the canonical basis of $\mathbb{R}^X$; it is the xth column of the identity matrix of order X. Given two vectors z and z' in $\mathbb{R}^n$, we denote by $z \wedge z'$ the vector $(\min(z_i, z_i'))_i$, and by $z \vee z'$ the vector $(\max(z_i, z_i'))_i$. We denote by z^+ the vector $z \vee 0$ and by z^- the vector $(-z) \vee 0$. If v is a vector of $\mathbb{R}^n$, we denote by $diag(v)$ or $\boldsymbol{\Delta}_v$ the square diagonal matrix of size $n \times n$ with v on the diagonal.

Notations introduced in this book. This book will also introduce some original notations that are not standard in the literature. Given a probability distribution $\mathcal{P}$ over $\mathbb{R}^{\mathcal{Y}}$, the *welfare function*, a.k.a. *Emax function*, $G_{\mathcal{P}}$ is denoted by $G_{\mathcal{P}}(U) = \mathbb{E}_{\mathcal{P}}\left[\max_{y \in \mathcal{Y}} \left\{U_y + \varepsilon_y\right\}\right]$, where $\varepsilon \sim \mathcal{P}$. When there is no ambiguity, the subscript $\mathcal{P}$ will be omitted. When $G_{\mathcal{P}}$ is differentiable, its gradient is denoted by $\boldsymbol{\pi}_{\mathcal{P}}(U) = \nabla G_{\mathcal{P}}(U)$ and called the *market share map*. The Legendre–Fenchel transform of $G_{\mathcal{P}}$, denoted in a standard way by $G_{\mathcal{P}}^*$, will be referred to as the *generalized entropy of choice* associated with $\mathcal{P}$. If A is a matrix of term A_{ij}, and $f: \mathbb{R} \to \mathbb{R}$, $f(A)$ denotes the matrix of term $f(A_{ij})$. The notation $[0{:}n]$ denotes the set of integers $\{0, 1, \ldots, n\}$. If I and J are integers, the notation $[I \times \mathrm{J}]$ denotes the list of pairs ij for $1 \leq i \leq I$ and $1 \leq j \leq \mathrm{J}$, ordered in the lexicographic ordering: $11, 12, \ldots, 1\mathrm{J}, 21, 22, \ldots, 2\mathrm{J}, \ldots, I1, I2, \ldots, I\mathrm{J}$. This notation extends to lists of triples $[I \times \mathrm{J} \times K]$ and to any tuples.

CHAPTER 1

Theory of Random Utility Models

Discrete choice models represent a crucial advancement in understanding and predicting consumer behavior, particularly when it comes to making choices from among a finite set of alternatives. Unlike other demand models that might focus on the quantity of goods consumed, discrete choice models aim at capturing the qualitative aspects of consumer decisions—why consumers prefer one product or service over another. From a historical perspective, the roots of discrete choice modeling can be traced back to the pioneering work of Daniel McFadden in the 1970s, for which he was awarded the Nobel Prize in Economics in 2000. His work laid down the foundation for the conditional logit model, enabling economists to analyze and predict choices made by individuals, taking into account various factors including personal preferences, product attributes, and socioeconomic variables.

The emphasis on quality rather than quantity introduced by discrete choice modeling permits a more nuanced analysis of demand than an approach focused purely on quantity, and allows the study of substitution patterns, capturing how consumers shift between options when parameters of the choice are altered. This approach is particularly useful, for instance, in transportation planning, marketing strategy, and policy evaluation, where understanding the factors that influence the selection of one alternative over others can lead to better product design, more targeted marketing campaigns, and more effective public policies. Furthermore, because of their versatility, discrete choice models are a crucial building block of equilibrium models; in this book, we shall build on them to construct models of international trade, matching models, or hedonic models, to name just a few. But to get there, we need to cover the basics. Let's get started.

1.1 Setting

Consider a population of consumers who need to make a choice from among a finite number Y of alternatives $y \in [Y] = \{1, \ldots, Y\}$, or choose none of these alternatives, which is an option in its own right, denoted by $y = 0$ and called the default option. For instance, consumers may consider alternative transportation modes to reach a certain destination, or by default they may choose not to travel. The full set of options (default option and alternatives) will be denoted by $[0 : Y] = \{0\} \cup [Y]$.

We assume that a consumer i associates an amount of utility $U_y + \varepsilon_{iy}$ with each option $y \in [0 : Y]$, and we impose the normalization $U_0 = 0$. All consumers agree over the value U_y, which is called the *systematic utility* associated with y; furthermore, each consumer draws an additional amount of *idiosyncratic utility*, or *utility shock* ε_{iy}. Given that the idiosyncratic utility is heterogeneous and distributed across the population, we will treat $\varepsilon_i = (\varepsilon_{iy})_y$ as a random vector of which each entry is associated with an alternative. For this reason, ε is often called "random utility," although it is not random from the point of view of the consumer i, who is aware of their own utility shock. The distribution $\mathcal{P}$ of ε_i over the population of consumers i is typically known, or postulated, by the analyst. This model is called the *random utility model*, also called *multinomial choice model* when $Y > 1$, and *binomial choice model* when $Y = 1$, in which case the set of options has two elements, $[0 : Y] = \{0, 1\}$.

The consumer's problem is choosing the utility-maximizing alternative. Given a vector of utility shocks ε (we have dropped the subscript), the indirect utility of the consumer, defined as the utility associated with the optimal alternative, is

$$u(\varepsilon) = \max_{y \in [0:Y]} \left\{ U_y + \varepsilon_y \right\} = \max_{y \in [Y]} \left\{ U_y + \varepsilon_y, \varepsilon_0 \right\}, \tag{1.1.1}$$

which is a random quantity, as it depends on the realization of the vector of utility shocks. Note that the normalization $U_0 = 0$ is without loss of generality, since replacing U_y by $U_y - U_0$ yields the same chosen option. By replacing ε_{iy} by $\varepsilon_{iy} - \varepsilon_{i0}$, it would also be possible to impose a similar normalization, $\varepsilon_{i0} = 0$, on the random utility associated with the default option, but in most cases there will be little to gain in doing so.

We denote by $G(U)$ the expected indirect utility of a consumer drawn from the population, that is,

$$G(U) = \mathbb{E}_{\mathcal{P}}\Big[\max_{y \in [Y]} \{U_y + \varepsilon_y, \varepsilon_0\} \Big], \tag{1.1.2}$$

which is the average welfare of a consumer randomly drawn from the population. The function G is called *welfare function*, or (for obvious reasons)

Emax function. It is also sometimes known as *McFadden's social surplus function.* This function does not have a simple closed-form expression in general, but in the case of the binomial model it can be expressed as an integral over the real line.

Example 1.1 (Binomial choice). In the case of the binomial choice model, $Y=1$ and we can assume without loss of generality that $\varepsilon_0=0$, so that the vector of utility shocks has only one dimension, ε_1. Letting f and F be respectively its p.d.f. and c.d.f., we have $G(U_1)=\mathbb{E}_{\mathcal{P}}[\max\{U_1+\varepsilon_1,0\}]=\int_{-U_1}^{\infty}(U_1+e)f(e)\,de$. It follows by integration by parts that

$$G(U_1)=\int_{-\infty}^{U_1}(1-F(-e))\,de. \tag{1.1.3}$$

1.2 Demand Map and Welfare Function

Note that, without additional assumptions, the optimal alternative resulting from problem (1.1.1) may not be unique. For instance, if $U_y=0$ for all $y\in[0:Y]$ and if the random vector ε were also identically equal to zero, then the consumer would be indifferent among all options $y\in[0:Y]$. To prevent that situation from occurring with a positive probability, we will assume that the conditional distribution $\mathcal{P}$ of ε has a density with respect to the Lebesgue measure, which implies that the set of optimal alternatives in the consumer's problem (1.1.1) is almost surely unique, as then the probability of a tie $\mathcal{P}(U_y+\varepsilon_y=U_z+\varepsilon_z)$ for $y\neq z$ is zero. We state:

Assumption 1.1 (Continuity). *The distribution $\mathcal{P}$ of the vector of utility shocks ε has a density with respect to the Lebesgue measure on $\mathbb{R}^{Y+1}$.*

This assumption will allow us to define the *market shares*, which are the probabilities that a consumer randomly drawn from the population will choose each option $y\in[0:Y]$. The expression of the market share of option y as a function of U is given by

$$\boldsymbol{\pi}_y(U):=\mathcal{P}(U_y+\varepsilon_y\geq U_z+\varepsilon_z,\forall z\in[0:Y]), \tag{1.2.1}$$

where we can indifferently replace the weak inequalities by strict ones. Equation (1.2.1) defines a *market share map*, also called *demand map* or *conditional choice probability map.* Note that the bold notation $\boldsymbol{\pi}(U)$ denotes a function that maps the vector U of systematic utilities to the numerical vector π of market shares π_y, which is denoted using a non-bold font. If each π_y is the market share of option $y \in [0:Y]$, the entries of the vector π sum to 1, and thus the market share of the default option can be deduced from the other market shares by $\pi_0 = 1 - \sum_{y \in [Y]} \pi_y$.

Because we hold fixed $U_0 = 0$, and because π_0 may be deduced from the other entries of π, we may view the market share map $\boldsymbol{\pi}$ as mapping the vector U of systematic utilities associated with nondefault alternatives to the vector π of choice probabilities of nondefault alternatives, and thus in the sequel we shall see the market share map $\boldsymbol{\pi}$ as a function from $\mathbb{R}^Y$ to $\mathbb{R}^Y$. Again, there is a simple expression of $\boldsymbol{\pi}$ in the binomial case.

Example 1.1 (Binomial choice, continued). Back to the binomial choice model: we have $\boldsymbol{\pi}_1(U_1) = \mathcal{P}(U_1 + \varepsilon_1 \geq 0)$, that is, after denoting by F the c.d.f. of ε_1,

$$\boldsymbol{\pi}_1(U_1) = 1 - F(-U_1). \tag{1.2.2}$$

A comparison of expressions (1.1.3) and (1.2.2) hints at a connection between the welfare function G and the market share map $\boldsymbol{\pi}$, and, indeed, there is a general relation between these objects in the multinomial case. To see why, let's run a thought experiment. Imagine that we increase the utility U_y associated with one alternative $y \in [Y]$ by some small amount δ so that U_y is replaced by $U_y + \delta$, but we keep unchanged all the other entries U_z for $z \neq y$. Let's ask how the social welfare $G(U)$ responds. Intuitively, this should raise the welfare of some consumers, but not all of them. In fact, if δ is small enough, the preferred option of most consumers will not change. So, the consumers who choose y will benefit from the change; their indirect utility will be increased by δ. The welfare of the other consumers, who pursue other options, will be unchanged. As a result, we may expect the welfare to be increased by $\delta \times \boldsymbol{\pi}_y(U)$, which is the increase in utility times the fraction of consumers who choose y. This suggests that $G\left(U + \delta \mathbf{e}_Y^y\right) \simeq G(U) + \delta \boldsymbol{\pi}_y(U)$ if δ is small, where $\mathbf{e}_Y^y$ denotes the yth vector of the canonical basis of $\mathbb{R}^Y$, hence $\boldsymbol{\pi}_y(U) = \partial G(U) / \partial U_y$, or in words: *the market share of an option equals the partial derivative of the welfare function with respect to the systematic utility of that option.* This discussion is not completely rigorous because some consumers will switch choices as the systematic utilities move; but one can show that the mass of consumers who switch has only a second-order effect in δ. As a result, the above intuition holds as a rigorous theorem, known as the Daly–Zachary–Williams theorem:

Theorem 1.2.1 (Daly–Zachary–Williams). *Under assumption 1.1, the welfare function G is C^1, and its gradient is the market share map $\boldsymbol{\pi}$, that is,*

$$\boldsymbol{\pi}(U) = \nabla G(U). \tag{1.2.3}$$

A first example of a random utility model where G and $\boldsymbol{\pi}$ can be computed in closed form is provided by the logit specification:

Example 1.2 (Logit model). In the logit model, the vector ε has independent components that have the centered type-I extreme value distribution with scale parameter 1, a.k.a. the centered Gumbel distribution, denoted by $\mathcal{G}_0$. As one shall see in chapter 2, section 2.1, in that case $\max_y \left\{U_y + \varepsilon_y\right\}$ is distributed as $\log(\sum_y \exp U_y) + \eta$, where $\eta \sim \mathcal{G}_0$, and thus, given that $U_0 = 0$, one has

$$G(U) = \log\left(1 + \sum_{y \in [Y]} \exp U_y\right) \text{ and } \boldsymbol{\pi}_y(U) = \frac{\exp U_y}{1 + \sum_{z \in [Y]} \exp(U_z)}. \tag{1.2.4}$$

In particular, when $Y = 1$, the *binomial logit model* is such that

$$\boldsymbol{\pi}_1(U_1) = \frac{\exp U_1}{1 + \exp U_1} = \frac{1}{1 + \exp(-U_1)}, \tag{1.2.5}$$

and the map $\boldsymbol{\pi}_1$ is called a *logistic function.*

Note that the logit model just introduced satisfies the property of *independence of irrelevant alternatives* (IIA). This property states that, for any pair of alternatives $y, y' \in [0 : Y]$, the odds ratio $\boldsymbol{\pi}_y(U)/\boldsymbol{\pi}_{y'}(U)$—that is, the ratio of the probability of choosing y to that of choosing y'—depends only on the corresponding utilities U_y and $U_{y'}$. In exercise 2.4, the reader is asked to show that it is the only random utility model to have that property. In this sense, the logit model can be seen as the simplest among all random utility models.

Theorem 1.2.1 suggests that the function G plays an important role in the analysis of demand, which calls for an investigation of some of its properties. First, we see from its definition in expression (1.1.2) that $G(U)$ is the expectation over ε of the functions $U \mapsto \max_y\{U_y + \varepsilon_y, \varepsilon_0\}$ that are convex in U. As a result, $G(U)$ is convex, and this fact will play a major role in this chapter and subsequent ones. Next, if alternative $z \neq y$ becomes more attractive, the market share of y cannot increase, so $\boldsymbol{\pi}_y(U)$ is nonincreasing in U_z for $z \neq y$, as can also be verified directly from (1.2.1). Equivalently, $\partial G(U)/\partial U_y$ is nonincreasing in U_z for $z \neq y$, which means that the function G is submodular. The submodularity of G also plays an important role, in particular in the theory we will develop in chapter 7. We summarize the discussion above into:

Proposition 1.2.1. *The welfare function G is convex and submodular.*

Example 1.2 (Logit model, continued). In the logit model, a straightfoward calculation shows that $D^2 G(U) = diag(\pi) - \pi\pi^\top$, where $\pi = \nabla G(U)$. Hence,

$D^2G(U)$ is symmetric positive definite, and its off-diagonal entries are negative (it is called a *Stieltjes matrix*), which is consistent with our knowledge that G is convex and submodular.

1.3 Demand Inversion and Entropy of Choice

So far, we have solved the direct problem of demand: we have considered systematic utilities U_y as data and gone on to predict the market shares π_y. This is the "theorist's problem," which seeks to predict the behavior of agents based on structural parameters. In contrast, the "econometrician's problem" consists of solving the inverse problem: taking market shares as data, how to deduce the corresponding vector of systematic utilities. In this section, we assume throughout that there are $Y+1$ options, one default one $y=0$ with $U_0=0$, and Y alternatives $y\in[Y]$. Under assumption 1.1, given a vector π with nonnegative entries satisfying $\sum_{y\in[Y]}\pi_y<1$, we would like to characterize and compute the set

$$\boldsymbol{\pi}^{-1}(\pi)=\left\{U\in\mathbb{R}^Y:\boldsymbol{\pi}(U)=\pi\right\}. \tag{1.3.1}$$

This is the *market share inversion* problem, pioneered by the work of Hotz, Miller, and Berry, which is also called "demand inversion," "conditional choice probability inversion," or the "identification problem." It plays a central role in econometrics and industrial organization and will be a key building block for the analysis of structural models.

Reformulation as an optimization problem. We start by reformulating the demand inversion problem (1.3.1) in terms of an optimization problem. Given that $\boldsymbol{\pi}(U)=\nabla G(U)$, we may view the equation $\boldsymbol{\pi}(U)=\pi$ as the first-order conditions associated with the maximization of the concave function $U\mapsto\pi^\top U-G(U)$. Hence, we may state:

Proposition 1.3.1. *Under assumption 1.1, $U\in\boldsymbol{\pi}^{-1}(\pi)$ if and only if U is a solution to*

$$\max_{\tilde{U}\in\mathbb{R}^Y}\{\pi^\top\tilde{U}-G(\tilde{U})\}, \tag{1.3.2}$$

or, in other words, $\boldsymbol{\pi}^{-1}(\pi)=\arg\max_{\tilde{U}}\{\pi^\top\tilde{U}-G(\tilde{U})\}$.

This reformulation is useful for computational purposes: indeed, problem (1.3.2) is a convex optimization problem, and therefore can be computed using the convex optimization toolbox recalled in appendix A.1.2. But the reformulation is also useful for understanding properties of the inverse

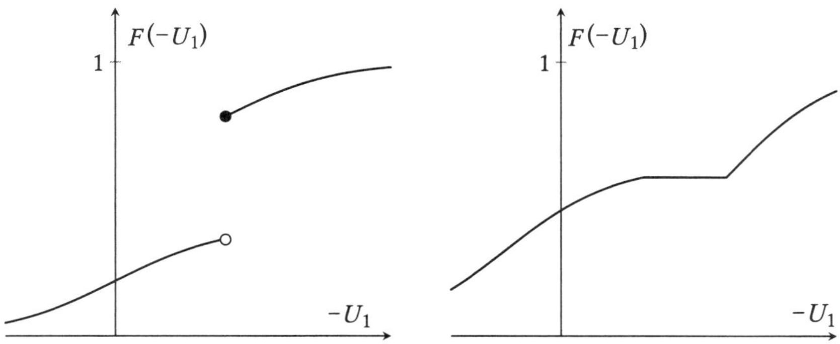

(a) Distribution has a mass point.

(b) Distribution does not have full support.

Figure 1.1: Examples of c.d.f. with jumps and flats.

demand. In exercise 1.4 this result, in combination with the fact that G is convex and submodular, is used to show that the set $\boldsymbol{\pi}^{-1}(\pi)$ is a convex sublattice of $\mathbb{R}^Y$ (i.e., a set that is stable by the $\vee$ and $\wedge$ operations).

Existence and uniqueness of inverse demand. Two important questions in the market share inversion problem are the **existence** and **uniqueness** of elements U of $\boldsymbol{\pi}^{-1}(\pi)$. To give some intuition of what drives the answers, it is useful to consider the binomial choice model of example 1.1. Recall that in this case $\boldsymbol{\pi}_1(U_1) = 1 - F(-U_1)$, where F is the c.d.f. of ε_1, so $U \in \boldsymbol{\pi}^{-1}(\pi)$ holds in the binomial model if and only $F(-U_1) = 1 - \pi_1$, so the problem when $Y = 1$ essentially consists of inverting the c.d.f. F. Nonexistence of U_1 for $\pi_1 \in (0, 1)$ may arise if the graph of F has a jump, as illustrated in the left part of figure 1.1; but the continuity of F_1, which follows from assumption 1.1, rules that out. However, nonuniqueness may still arise if the graph of F has horizontal parts, as seen on the right part of figure 1.1; in that case, there will be two distinct values U_1 and $U_1^\star$ such that $F(-U_1) = F(-U_1^\star)$. This suggests that uniqueness issues may arise if the distribution of ε has incomplete support. Based on this intuition, in addition to assumption 1.1, we therefore add an important full support assumption:

Assumption 1.2 (Full support). *The probability distribution $\mathcal{P}$ of the vector of utility shocks ε has full support.*

The following result, which the reader is asked to prove in problems 1.2 and 1.3 following arguments in Galichon and Salanié and in Chiong, Galichon, and Shum, answers the existence and uniqueness questions.

Theorem 1.3.1. *Let $\pi \in \mathbb{R}^Y$ be such that $\pi_y > 0$ for all $y \in [Y]$ and $\sum_y \pi_y < 1$. Then problem* (1.3.2) *admits:*

(i) ***at least*** *one solution U under assumption 1.1 (continuity).*
(ii) ***at most*** *one solution U under assumption 1.2 (full support).*

As we saw in section 1.2, the map $\boldsymbol{\pi}$ is defined under the assumption of the continuity assumption 1.1. It follows from theorem 1.3.1 that under both assumptions 1.1 and 1.2, the map $\boldsymbol{\pi}$ is one-to-one from $\mathbb{R}^Y$ to the set of vectors $\pi \in \mathbb{R}^Y$ such that $\pi_y > 0$ for all $y \in [Y]$ and $\sum_y \pi_y < 1$. That set is an open set; it is the interior of the $(Y-1)$-dimensional standard simplex, whose definition is recalled in appendix A.2.1.

Generalized entropy of choice. Note that when $\boldsymbol{\pi}^{-1}(\pi)$ has only one element, there is a simple way to represent the solution vector U as a function of π. Denote by $G^*(\pi)$ the value of problem (1.3.2), which, in convex analysis, is known as the Legendre–Fenchel transform of G. We have by the envelope theorem that the vector U that is optimal in (1.3.2) is the vector of partial derivatives of $G^*(\pi)$ with respect to π. Hence, we shall see that $U \in \boldsymbol{\pi}^{-1}(\pi)$ if and only if $U = \nabla G^*(\pi)$, and therefore $\boldsymbol{\pi}^{-1} = \nabla G^*$. This suggests that G^* has a useful role to play. Galichon and Salanié defined this function as:

Definition 1.3.1 (Generalized entropy of choice). *Let G be defined as in (1.1.2) above. The* generalized entropy of choice *associated with $\mathcal{P}$ is defined as G^*, the Legendre–Fenchel transform of G:*

$$G^*(\pi) = \max_{U \in \mathbb{R}^Y} \left\{ \sum_{y \in [Y]} \pi_y U_y - G(U) \right\}. \tag{1.3.3}$$

As usual, we have an explicit expression for G^* in the case of binomial random models.

Example 1.1 (Binomial choice, continued). In the case of the binomial model, we have $U_1 = -F^{-1}(1-\pi_1)$, and therefore by integration, if F is the c.d.f. of ε_1, one has

$$G^*(\pi_1) = -\int_0^{\pi_1} F^{-1}(1-m)\, dm. \tag{1.3.4}$$

Note that, by definition of G^*, we have for every $U \in \mathbb{R}^Y$ and $\pi \in \mathbb{R}^Y$ such that π is in the domain of G^* that $G^*(\pi) \geq \sum_y \pi_y U_y - G(U)$. As a result, we have $G(U) \geq \sum_y \pi_y U_y - G^*(\pi)$, with equality if $\pi = \boldsymbol{\pi}(U)$, as we just saw.

Also note that $G^*(\pi)$ will take value $+\infty$ as soon as one entry of π is negative, or if the sum of its entries exceeds 1. As a result, we may state:

Proposition 1.3.2. *We have*

$$G(U) = \max_{\pi \in \mathbb{R}^Y} \left\{ \sum_{y \in [Y]} \pi_y U_y - G^*(\pi) \right\}, \tag{1.3.5}$$

or equivalently by making the domain of G^* explicit,

$$G(U) = \max_{\pi \in \mathbb{R}^Y} \left\{ \sum_{y \in [Y]} \pi_y U_y - G^*(\pi) : \pi_y > 0, \sum_{y \in [Y]} \pi_y < 1 \right\}. \tag{1.3.6}$$

The result in proposition 1.3.2 is familiar in convex analysis, where it is well known that a convex function is the Legendre–Fenchel transform of its own Legendre–Fenchel transform; see appendix A.1.1. However, its interpretation in our context is particularly interesting. It means that the social aggregation of the choices of individual decision-makers with random utility consists in maximizing an objective function that is the sum of the individual welfare $\sum_y \pi_y U_y$, plus a perturbation term equal to $-G^*(\pi)$, which is minus the entropy of choice. That term may be seen as the effect of the random utility terms on the choice problem.

Our choice of the reference to "entropy" in the name of G^* comes from the fact that in an important special case, the logit case, the function G^* boils down to the usual entropy function:

Example 1.2 (Logit model, continued). In the logit model, $G^*(\pi)$ is defined for π such that $\pi_y > 0$ and $\sum_{y \in [Y]} \pi_y < 1$, and we have

$$G^*(\pi) = \sum_{y \in [Y]} \pi_y \log \pi_y + \pi_0 \log \pi_0 = \sum_{y \in [0:Y]} \pi_y \log \pi_y, \tag{1.3.7}$$

where π_0 is defined as $\pi_0 = 1 - \sum_{y \in [Y]} \pi_y$. This is (up to a sign flip) the usual definition of the Gibbs entropy associated with the probability distribution (π_y). One can see what proposition 1.3.2 becomes in the logit case: if the vector of systematic utilities is (U_y), the chosen distribution π is the solution to

$$\max_{\pi \in \mathbb{R}^Y} \left\{ \sum_{y \in [Y]} \pi_y U_y - \sum_{y \in [Y]} \pi_y \log \pi_y - \pi_0 \log \pi_0 \right\}. \tag{1.3.8}$$

We state the following result, which combines results by Galichon and Salanié, and by Chiong, Galichon, and Shum:

Theorem 1.3.2. *Let $\pi \in \mathbb{R}^Y$ be such that $\pi_y > 0$ and $\sum_{y\in[Y]} \pi_y < 1$. Then under both assumptions 1.1 and 1.2, $G^\star$ is differentiable, and one has $\boldsymbol{\pi}^{-1}(\pi) = \nabla G^*(\pi)$, which is the vector U solution to problem* (1.3.2).

We can provide an interpretation of the meaning of $G^*(\pi)$ defined in expression (1.3.3). If $\pi = \nabla G(U)$, then U attains the maximum in that expression, and we have

$$G^*(\pi) = \sum_{y\in[Y]} \pi_y U_y - G(U). \tag{1.3.9}$$

We let $y^\star(\varepsilon) \in \arg\max_{y\in[0:Y]} \{U_y + \varepsilon_y\}$ be the random variable valued in $[0:Y]$ that is the optimal choice of an agent with a vector of utility shocks ε. This choice is almost surely unique; the random nature of $y^\star(\varepsilon)$ is driven by the randomness of ε. We have

$$G(U) = \mathbb{E}\left[U_{y^\star(\varepsilon)}\right] + \mathbb{E}\left[\varepsilon_{y^\star(\varepsilon)}\right] = \sum_{y\in[Y]} \pi_y U_y + \mathbb{E}\left[\varepsilon_{y^\star(\varepsilon)}\right], \tag{1.3.10}$$

and substituting the expression of $G(U)$ from (1.3.10) into (1.3.9) yields

$$G^*(\pi) = -\mathbb{E}\left[\varepsilon_{y^\star(\varepsilon)}\right]. \tag{1.3.11}$$

Therefore, the entropy of choice $G^*(\pi)$ is interpreted as minus the expected amount of heterogeneity needed to rationalize the vector of market shares π. The binomial example developed in example 1.1 provides a good illustration for this:

Example 1.1 (Binomial choice, continued). Going back to the binomial example, one has $G(U) = \mathbb{E}_{\mathcal{P}}[(U_1 + \varepsilon_1)\mathbf{1}\{U_1 + \varepsilon_1 \geq 0\}] = U_1\pi_1 + \mathbb{E}_{\mathcal{P}}[\varepsilon_1 \mathbf{1}\{\varepsilon_1 \geq -U_1\}]$, so we have $-G^*(\pi_1) = \mathbb{E}_{\mathcal{P}}[\varepsilon_1\mathbf{1}\{\varepsilon_1 \geq -U_1\}]$, which is consistent with the expression obtained in (1.3.4). As $\varepsilon_1\mathbf{1}\{\varepsilon_1 \geq -U_1\} = \varepsilon_{y^\star(\varepsilon)}$, one is led to expression (1.3.11).

Inverse demand and optimal transport. Expression (1.3.11) provides the intuition for a characterization of G^* based on optimal transport, which we explain next. Denote by $\mathcal{M}(\mathcal{P}, \pi)$ the set of joint distributions λ of random pairs $(\varepsilon, \tilde{y})$, where $\varepsilon \sim \mathcal{P}$ and $\tilde{y} \sim \pi$. Given $(\varepsilon, \tilde{y}) \sim \lambda$, where $\lambda \in \mathcal{M}(\mathcal{P}, \pi)$, the expression $\varepsilon_{\tilde{y}}$ is understood as the entry $\tilde{y}$ of vector ε. We have, by a result of Galichon and Salanié:

Theorem 1.3.3 (Inversion theorem). *Let G^* be the entropy of choice associated to a distribution $\mathcal{P}$. Then:*

(i) $-G^*(\pi)$ *is the value of a Monge–Kantorovich problem, whose primal formulation is*

$$- G^*(\pi) = \max_{\lambda \in \mathcal{M}(\mathcal{P},\pi)} \mathbb{E}_\lambda \left[\varepsilon_{\tilde{y}}\right], \tag{1.3.12}$$

and whose dual formulation is

$$-G^*(\pi) = \min_{u,U} \int u(\varepsilon)\, d\mathcal{P}(\varepsilon) - \sum_{y \in [0:Y]} \pi_y U_y \tag{1.3.13}$$

$$s.t.\ u(\varepsilon) - U_y \geq \varepsilon_y \quad \forall \varepsilon \in \mathbb{R}^{Y+1},\ \forall y \in [0:Y].$$

(ii) Moreover, under assumption 1.1, then $U \in \boldsymbol{\pi}^{-1}(\pi)$ *if and only if there exists* u *such that* (u, U) *is the unique solution to* (1.3.13) *such that* $U_0 = 0$.

Note that part (i) of this theorem requires neither assumption 1.1 nor assumption 1.2, while part (ii) requires only assumption 1.1.

The logit model offers a closed-form expression for the entropy of choice—which is simply the opposite of the usual Gibbs entropy.

Example 1.2 (Logit model, continued). In the logit case, we have

$$G^*(\pi) = \begin{cases} \sum_{y\in[Y]} \pi_y \log \pi_y + \pi_0 \log \pi_0 & \text{if } \pi_y \geq 0 \text{ for all } y \text{ and } \sum_{y\in[Y]} \pi_y \leq 1 \\ +\infty & \text{otherwise,} \end{cases} \tag{1.3.14}$$

where $\pi_0 = 1 - \sum_{y\in[Y]} \pi_y$, and in the case when $G^*(\pi) < +\infty$ with the inequalities above holding strictly, we have

$$(\boldsymbol{\pi}^{-1})_y(\pi) = \frac{\partial G^*(\pi)}{\partial \pi_y} = \log \frac{\pi_y}{\pi_0}. \tag{1.3.15}$$

1.4 Sampling

We now consider a sample of consumers $i \in [I]$ whose utility shocks $\varepsilon_i \in \mathbb{R}^{Y+1}$ are independently drawn from the distribution $\mathcal{P}$. Let $\mathcal{P}_I$ be the empirical distribution of the sample, which assigns weight $1/I$ to each observation i. The objects that we introduced in sections 1.2 and 1.3, namely the welfare function, the entropy of choice, and the market share map and its inverse, which are all defined with reference to the population distribution $\mathcal{P}$, all have empirical analogs obtained by replacing the population distribution by its empirical counterpart $\mathcal{P}_I$. We can define the empirical welfare function $\hat{G}$ and the empirical market share map $\hat{\pi}$ as

$$\begin{cases} \hat{G}(U) = \frac{1}{I} \sum_{i \in [I]} \max_{y \in [Y]} \left\{ U_y + \varepsilon_{iy}, \varepsilon_{i0} \right\}, \\ \hat{\pi}_y(U) = \frac{1}{I} \sum_{i \in [I]} 1 \left\{ U_y + \varepsilon_{iy} \geq U_z + \varepsilon_{iz}, \forall z \in [0:Y] \right\}. \end{cases} \tag{1.4.1}$$

The code in appendix B.1.1 computes these quantities on an example. Although the empirical distribution $\mathcal{P}_I$ satisfies neither assumption 1.1 (it does not have a density) nor assumption 1.2 (its support is a finite set), we can still define the empirical entropy of choice $\hat{G}^*(\pi)$ using expression (1.3.3). Without surprise, the quantity $\hat{G}^*(\pi)$ is the opposite of the value of a finite-dimensional optimal transport problem, and we have the following result due to Chiong, Galichon, and Shum:

Theorem 1.4.1. *Let $\hat{G}^*(\pi)$ be the entropy of choice associated with the empirical distribution $\mathcal{P}_I$. Then:*

(i) The quantity $-\hat{G}^*(\pi)$ is the opposite of the value of the finite-dimensional optimal transport problem

$$\max_{\lambda_{iy} \geq 0} \sum_{iy \in [I \times 0:Y]} \lambda_{iy} \varepsilon_{iy} \tag{1.4.2}$$

$$\text{s.t.} \sum_{y \in [0:Y]} \lambda_{iy} = \frac{1}{I},$$

$$\sum_{i \in [I]} \lambda_{iy} = \pi_y,$$

whose dual expression is

$$\min_{u_i, U_y} \frac{1}{I} \sum_{i \in [I]} u_i - \sum_{y \in [Y]} \pi_y U_y \tag{1.4.3}$$

$$\text{s.t. } u_i \geq U_y + \varepsilon_{iy}, \quad U_0 = 0.$$

(ii) The convergence $\hat{G}^*(\pi) \to G^*(\pi)$ holds almost surely as $I \to +\infty$.

(iii) Under assumption 1.1, consider $\pi = \boldsymbol{\pi}(U)$, and let $\left(u^I, U^I\right)$ be any solution to (1.4.3). Then the convergence $U^I \to U$ holds almost surely as $I \to +\infty$.

It is useful to provide an algebraic formulation of expressions (1.4.2) and (1.4.3). To do this, we need to understand a fundamental concept, *vectorization*, which is covered in appendix A.3.1. Vectorization consists of representing (ε_{iy}) not as a matrix, but as a vector, which we do using the so-called row-major order, which is the lexicographic order on indices. For instance, if $[I]=\{i_1,i_2\}$ and $[0:Y]=\{y_0,y_1\}$, the vectorized representation of (ε_{iy}) will be $(\varepsilon_{i_1y_0},\varepsilon_{i_1y_1},\varepsilon_{i_2y_0},\varepsilon_{i_2y_1})$. We see that it corresponds to listing the pairs of indices iy as elements of the list $[I\times Y]$. In that case, formula (A.3.4) in appendix A.3.1 on Kronecker products shows that the constraint $u_i - U_y \geq \varepsilon_{iy}$ can be vectorized as $(\mathbf{I}_I\otimes \mathbf{1}_{Y+1})u-(\mathbf{1}_I\otimes \mathbf{I}_{Y+1})U\geq\varepsilon$, and, therefore, problem (1.4.3) finds a compact expression as

$$\begin{aligned}\min_{u\in\mathbb{R}^I,U\in\mathbb{R}^{Y+1}} \quad & \frac{1}{I}\mathbf{1}_I^\top u-\pi^\top U \\ s.t.\quad & (\mathbf{I}_I\otimes \mathbf{1}_{Y+1})\,u-(\mathbf{1}_I\otimes \mathbf{I}_{Y+1})\,U\geq\varepsilon, \\ & U_0=0.\end{aligned} \tag{1.4.4}$$

The code in appendix B.1.2 provides an example for computing problem 1.4.4 using the linear programming solver Gurobi.

1.5 Asymptotic Theory

In this section, we shall investigate how the sample uncertainty on the measurement of the vector of choice probability π affects the estimation of the systematic utility U. In order to do this, we need to say a bit more about the data-generating process. Let us assume that the sample size is I, and that individual i in the sample made a choice $y_i\in[0:Y]$ after observing the realization of their vector $(\varepsilon_{iy})_{y\in[0:Y]}$. The empirical market share associated with option $y\in[0:Y]$ is then simply the fraction of the individuals in the sample who have chosen y, that is,

$$\hat{\pi}_y=\frac{1}{I}\sum_{i\in[I]}\mathbf{1}\{y_i=y\}. \tag{1.5.1}$$

This is a parametric model whose vector of parameters is $U\in\mathbb{R}^Y$. The probability that individual i chooses y is then $\boldsymbol{\pi}_y(U)=\partial G(U)/\partial U_y$, so the log-likelihood of the sample is

$$\ell(U)=\sum_{i\in[I]}\log\boldsymbol{\pi}_{y_i}(U)=I\sum_{y\in[0:Y]}\hat{\pi}_y\log\boldsymbol{\pi}_y(U),$$

with the understanding that $\boldsymbol{\pi}_0(U) = 1 - \sum_{y \in [Y]} \boldsymbol{\pi}_y(U)$. The gradient of the likelihood is the vector of the partial derivatives with respect to U_z, $z \in [Y]$, which have expression

$$\frac{1}{I}\frac{\partial \ell(U)}{\partial U_z} = \sum_{y \in [Y]} \frac{\partial^2 G(U)}{\partial U_y \partial U_z} \left(\frac{\hat{\pi}_y}{\boldsymbol{\pi}_y(U)} - \frac{\hat{\pi}_0}{\boldsymbol{\pi}_0(U)} \right),$$

leading to

$$\frac{1}{I} \nabla \ell(U) = D^2 G(U)\, \delta\left(\hat{\pi}, U\right) \text{ where } \delta_y\left(\hat{\pi}, U\right) := \frac{\hat{\pi}_y}{\boldsymbol{\pi}_y(U)} - \frac{\hat{\pi}_0}{\boldsymbol{\pi}_0(U)},\ y \in [Y].$$

The maximum likelihood estimator is such that $\nabla \ell(U) = 0$, which is equivalent to $\delta\left(\hat{\pi}, U\right) = 0$, which in turn is equivalent to $\boldsymbol{\pi}_y(U) = \hat{\pi}_y$. Consequently, we have:

Theorem 1.5.1. *The maximum likelihood estimator in the model above is given by*

$$\hat{U} = \boldsymbol{\pi}^{-1}\left(\hat{\pi}\right) = \nabla G^*\left(\hat{\pi}\right). \tag{1.5.2}$$

Thanks to theorem 1.5.1, one can work out the asymptotic distribution of the estimated utilities $\hat{U}$ based on the asymptotic distribution by the delta method,[1] which is a simple linearization of (1.5.2) around the true value of the parameter U when the sample size I gets large. We start by working out the asymptotic distribution of the multinomial distribution $\hat{\pi}$. A calculation based on (1.5.1) shows that for y and y' in $[Y]$, $cov\left(\hat{\pi}_y, \hat{\pi}_{y'}\right) = \left(\mathbf{1}\left\{y = y'\right\}\pi_y - \pi_y \pi_{y'}\right)/I$, which shows that the variance-covariance matrix of $\hat{\pi}$ is V_π / I, where, letting $\boldsymbol{\Delta}_\pi = diag(\pi)$,

$$V_\pi = \boldsymbol{\Delta}_\pi - \pi \pi^\top. \tag{1.5.3}$$

Hence, as $I \to +\infty$, the central limit theorem implies that $\hat{\pi} = \pi + I^{-1/2} Z + o_P(I^{-1/2})$, where $Z \sim \mathcal{N}(0, V_\pi)$. As a result, one has $\hat{U} = \nabla G^*(\hat{\pi}) = \nabla G^*(\pi) + I^{-1/2} D^2 G^*(\pi)\, Z + o_P(I^{-1/2})$, and therefore we can state:

Theorem 1.5.2. *Under assumptions 1.1 and 1.2, one has the convergence of $\hat{U}$ to U at rate $I^{1/2}$ as $I \to +\infty$, and the following convergence in distribution holds:*

$$I^{1/2}\left(\hat{U} - U\right) \Rightarrow \mathcal{N}\left(0, D^2 G^*(\pi)\, V_\pi D^2 G^*(\pi)\right). \tag{1.5.4}$$

[1]See, e.g., van der Vaart (1998), chapter 3.

Theorem 1.5.2 highlights the role played by the Hessian of the entropy of choice D^2G^*. Given that the map ∇G^* is the inverse map of ∇G, it follows that

$$D^2G^*(\pi) = \left(D^2G(U)\right)^{-1}, \text{ where } \pi = \nabla G(U),$$

and both these matrices are symmetric positive definite. Further, D^2G has positive diagonal terms and nonpositive off-diagonal terms, while all the entries of D^2G^* are nonnegative.

Without surprise, $D^2G^*(\pi)$ has a closed-form expression in the logit model.

Example 1.2 (Logit model, continued). In the logit model, for $\pi \in \mathbb{R}^Y$ such that $\pi_y > 0$ for all $y \in [Y]$ and $\sum_{y\in[Y]} \pi_y < 1$, we have

$$\begin{cases} D^2G(U) = \Delta_\pi - \pi\pi^\top, \text{ where } \pi = \nabla G(U), \\ D^2G^*(\pi) = \Delta_\pi^{-1} + \mathbf{1}_Y\mathbf{1}_Y^\top/\pi_0, \end{cases} \tag{1.5.5}$$

where $\mathbf{1}_Y \in \mathbb{R}^Y$ denotes the column vector of 1s, of size Y, and $\pi_0 = 1 - \sum_{y\in[Y]} \pi_y$. This shows that in the logit model, one has $D^2G(U) = V_\pi$; therefore, expression (1.5.2) becomes $I^{1/2}(\hat{U} - U) \Rightarrow \mathcal{N}(0, V_\pi^{-1})$, where $V_\pi^{-1} = \Delta_\pi^{-1} + \mathbf{1}_Y\mathbf{1}_Y^\top/\pi_0$.

The previous results allow a discussion on the precision of simulation methods for discrete choice models. One can see from equation (1.5.3) that the variance of $\hat{\pi}_y$ is $(\pi_y - \pi_y^2)/I$. If π_y is small (with respect to 1), this is at first order equal to π_y/I, and thus the standard deviation is of order $\sqrt{\pi_y/I}$. A crude reasoning goes as follows: if the number of options is Y, then some π_y will necessarily be smaller than or equal to $1/Y$. Say π_y is equal to $1/Y$. Then the standard deviation of its simulator will be $1/\sqrt{YI}$; and, therefore, in order to provide a relative precision (defined as the ratio of the two standard deviations of $\hat{\pi}_y$ divided by π_y) of the order of ρ, we need to set I to the order of at least $4Y/\rho^2$, which means that if we want a relative precision of 0.05 and if we have 20 options, we need to set the number of draws I equal to over 40,000.

1.6 Comparative Statics

In this section, we are interested in *comparative statics*, which is the study of how the market share map and its inverse respond to an increase or decrease in one of their arguments. First, we study the impact of a change in the systematic utility U_y on the vector of market shares $\boldsymbol{\pi}(U)$. Intuition suggests that if U_y increases, then option y becomes more attractive than any alternative $z \neq y$ and some individuals may decide to switch from z to y. Therefore the market share of y will increase, and the market share of z will decrease, or at least will not increase. As theorem 1.2.1 implies that $\pi_y(U) = \partial G(U)/\partial U_y$, it follows that $\partial \pi_y(U)/\partial U_z = \partial^2 G(U)/\partial U_y \partial U_z$, and as we know from proposition 1.2.1 that G is submodular, it follows that $\partial \pi_y(U)/\partial U_z \leq 0$, and we can state:

Proposition 1.6.1. *Under assumption 1.1, for each $y \in [Y]$, the market share of y, $\pi_y(U)$ is increasing in U_y, and for each $y \in [0:Y]$, $\pi_y(U)$ is nonincreasing in U_z for all $z \in [Y]$, $z \neq y$. That property is called* weak substitutes.

Note, however, that proposition 1.6.1 only states that the market share of z is nonincreasing with respect to U_y, not that it is decreasing; indeed, z may not be a strong substitute for y, which means no individual may be indifferent between y and z for their first choice. This is the case in the model depicted in figure 1.2, where $\varepsilon_y = \epsilon^\top \xi_y$, ϵ is a two-dimensional random vector with a continuous distribution, and for each option y, ξ_y are two-dimensional vectors. In that case (when the continuity assumption 1.1 is met but not the full support assumption 1.2), we see that $y=1$ is not a strong substitute for $y=4$ as the demand zones for these two options are not adjacent, so increasing U_4 does not affect the demand area for $y=1$.

However, if we assume that the distribution of the random utility vector ε has full support, in addition to assuming it is continuous, then all the options become strong substitutes to one another, and an increase in the systematic utility of one option will strictly decrease the market shares of all the alternatives. We state:

Proposition 1.6.2. *Under assumptions 1.1 and 1.2, the* strong substitutes property *holds, which means that for each $y \in [0:Y]$, $\pi_y(U)$ is decreasing in U_z for all $z \in [Y]$, $z \neq y$.*

If π is continuously differentiable, then the weak substitutes property holds if and only if the Jacobian matrix $D\pi$—that is, the matrix $(\partial \pi_y(U)/\partial U_z)_{yz}$—has nonpositive off-diagonal terms. A matrix with nonpositive off-diagonal terms is called a Z-matrix; this explains why the functions with the weak substitutes property are called Z-functions; see definition A.3.7 in appendix A.3.6. If one imposes further that $\sum_y \pi_y(U)$ is increasing in U_z for all z, then the matrix $D\pi_y(U)$ is called *diagonally dominating*, which means that the increasing behavior of $\pi_y(U)$ with respect to U_y offsets its decreasing behavior with respect to U_z for $z \neq y$, the other entries of U. It is well known that diagonally dominating matrices are invertible and that their inverses have nonnegative entries. A slightly more general class of matrices beyond

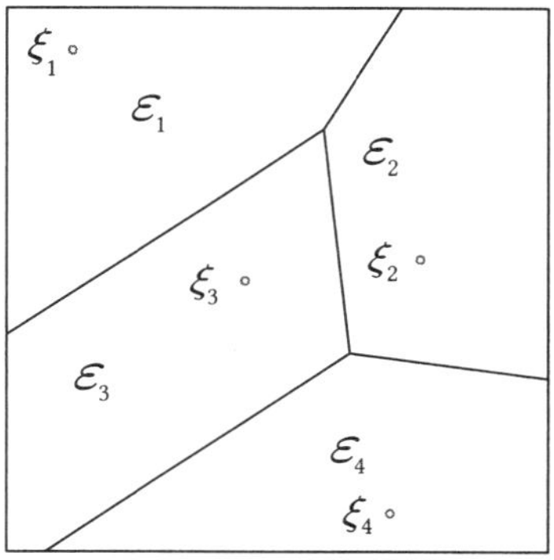

Figure 1.2: This model will be introduced in section 4.1 under the name of the pure characteristics model. There, $\varepsilon_y = \epsilon^\top \xi_y$, where $\epsilon = (\epsilon_1, \epsilon_2) \sim \mathcal{U}([0,1])^2$ is uniformly distributed on the square, and ξ_y are vectors in $[0,1]^2$ for each $y \in [4]$. Options $y=1$ and $y=2$ are strong substitutes, as their demand areas are adjacent. Options $y=1$ and $y=4$ are not strong substitutes since they do not share a common frontier; however, they are connected substitutes, since they have joint neighbors, $y=2$ and $y=3$.

diagonally dominating matrices is the class of M-matrices, which are precisely Z-matrices that are invertible and whose inverses have nonnegative entries. This notion extends beyond matrices to nonlinear functions, and definition A.3.11 in appendix A.3.6 introduces the class of M-functions, which are Z-functions that are inverse isotone.

A function π is inverse isotone if whenever the ordering relation $\pi(U) \leq \pi(U')$ holds for two vectors U and U', then $U \leq U'$. Inverse isotonicity implies injectivity: indeed, if π is inverse isotone, then $\pi(U) = \pi(U')$ implies $\pi(U) \leq \pi(U')$ and $\pi(U) \geq \pi(U')$, which respectively imply $U \leq U'$ and $U \geq U'$, which allows us to conclude $U = U'$.

We have seen that under the usual assumptions 1.1 and 1.2, the map π satisfied the strong substitutes property, and hence was a Z-function. One can show that under the same set of assumptions, it is an M-function, and therefore it is inverse isotone.

Theorem 1.6.1 (Isotonicity of the inverse demand map). *Under assumptions 1.1 and 1.2, the map π is inverse isotone.*

By definition of an M-function, the theorem implies that under the two assumptions of continuity and full support, the inverse demand map is isotone. A proof of this result can be found in appendix A.3.6, where the result follows from theorem A.3.6. In fact, Berry, Gandhi, and Haile have proven a stronger result closely related to theorem A.3.9 in appendix A.3.6. Without making the full support assumption 1.2, but assuming *connected strong substitutes* instead—meaning that one cannot partition the alternatives into two groups that are not substitutable one for the other—they arrive at the same conclusion as that in theorem 1.6.1. This case is illustrated in figure 1.2, where strong substitutes are not satisfied, but connected strong substitutes are.

1.7 Exercises

Exercise 1.1. *Show that if $\mathcal{P}$ is a distribution on $\mathbb{R}^Y$ that satisfies assumptions 1.1 and 1.2, then $\boldsymbol{\pi}_y(U) > 0$ for all $y \in [Y]$.*

Exercise 1.2. *Consider $Y = 2$ and no default option and assume $\varepsilon_1 = \varepsilon_2$ and that these coinciding terms have an $\mathcal{N}(0,1)$ distribution. Does the distribution $\mathcal{P}$ of this random vector satisfy assumption 1.1? And assumption 1.2?*

Exercise 1.3. *There is no default option in this exercise. Assume the distribution $\mathcal{P}$ of $(\varepsilon_y)_{y\in[Y]}$ satisfies $\mathbb{E}_{\mathcal{P}}[\varepsilon_y] = 0$ for all $y \in [Y]$. Letting $\pi \in \mathbb{R}^Y$ be such that $\pi_y \geq 0$ for all $y \in [Y]$ and $\sum_{y\in[Y]} \pi_y = 1$, show that $G^*(\pi) \leq 0$.*

Exercise 1.4. *Under assumption 1.1:*

(i) Show that the set $\boldsymbol{\pi}^{-1}(\pi)$ is a convex subset of $\mathbb{R}^Y$. Hint: *One may use the results from appendix A.1.1.*

(ii) Show that it is a sublattice of $\mathbb{R}^Y$, meaning that it is stable by $\vee$ and $\wedge$. Hint: *One may use the results from appendix A.3.5.*

Exercise 1.5. *There is no default option in this exercise. Assume the distribution $\mathcal{P}$ of $(\varepsilon_y)_{y\in[Y]}$ satisfies $\mathbb{E}_{\mathcal{P}}[\varepsilon_y] = 0$ for all $y \in [Y]$. Show that*

$$G(0) \geq G(U) - \max_{y\in[Y]} U_y \geq 0.$$

Deduce that when the Y-dimensional vector of random utility shocks ε follows an i.i.d., centered Gumbel distribution, then

$$\sigma \log Y \geq \sigma \log\Big(\sum_{y\in[Y]} \exp(U_y/\sigma)\Big) - \max_{y\in[Y]} U_y \geq 0.$$

Exercise 1.6. *Consider the binomial probit model, i.e., $Y=1$, and assume the distribution of $(\varepsilon_0, \varepsilon_1)$ is Gaussian with zero mean, variances σ_0^2 and σ_1^2, and correlation $\rho \in (-1,1)$. Using theorem 1.5.2, give an explicit formula for the 95% confidence interval of the estimator of U_1.*

Exercise 1.7. *Show that the value of problem*

$$\max_{\pi \in \mathbb{R}_+^Y} \left\{ \sum_{y \in [Y]} \pi_y a_y - \sum_{y \in [Y]} \pi_y \log \frac{\pi_y}{\sum_{\tilde{y} \in [Y]} \pi_{\tilde{y}}} \right\}$$

*equals zero if $\sum_y \exp a_y \leq 1$, and $+\infty$ otherwise. (*Hint: *One can use the expression of G^* in the logit model without a default option.)*

Exercise 1.8. *We consider a random utility model with Y options and no default option. For $y \in [Y]$, we define ε_y as the random utility associated with option y, where the distribution of the vector ε satisfies the continuity assumption 1.1. Further, we assume that the welfare function G is Y times differentiable.*

(i) *Relate $\boldsymbol{\pi}_y(U)$ for $y \in [Y]$ to the cumulative distribution function of a random vector (to be made explicit) evaluated at $(U_y - U_z)$, $z \neq y$.*

(ii) *Deduce that*

$$\operatorname{sign}\left(\frac{\partial^Y}{\partial U_1 \partial U_2 \dots \partial U_Y} G(U)\right) = (-1)^{Y-1}. \tag{1.7.1}$$

Exercise 1.9 (Assessing simulation accuracy). *Assume three options ($Y=3$) and no default option, and let $\mathcal{P}$ be the distribution $\mathcal{N}(0, \mathbf{I}_Y)$, which means that the ε_y are i.i.d. standard Gaussian variables.*

(i) *Use simulations to approximate the vector of market shares $\boldsymbol{\pi}_y(U)$ where $U = (0.1, 0, 0.2)$, as well as the approximate variance-covariance matrix of that approximation.*

(ii) *Use simulations to approximate the vector $\boldsymbol{\pi}_y^{-1}(\pi)$ of systematic utilities associated with the vector of market shares $\pi = (0.2, 0.5, 0.3)$ (normalized by $U_1 = 0$), as well as the approximate variance-covariance matrix of that approximation.*

1.8 Problems

Problem 1.1 (Defining a discrete choice map without the continuity assumption). *The goal of this problem is to investigate an analog of the notions presented in chapter 1 when assumption 1.1 is no longer satisfied by the distribution $\mathcal{P}$ of the random utility terms. Assume that there are Y options and no default option.*

(i) *To get started, consider the binomial case when there are two options $y \in \{1,2\}$, $\varepsilon_1 = 0$, and $\varepsilon_2 = -1$ or $\varepsilon_2 = 1$ with probability 1/2 each. If $U_1 = 0$ and $U_2 = 1$, show that π_2, the possible market share associated with $y = 2$, can take any possible value between $1/2$ and 1.*

(ii) *Show that the welfare function G and the entropy of choice G^* are defined even in the absence of assumption 1.1.*

(iii) *Show that G is not differentiable if assumption 1.1 does not hold.*

(iv) *Show that a market share vector π is associated to a vector of systematic utility U if and only if there is a joint distribution $\lambda \in \mathcal{M}(\mathcal{P}, \pi)$ (where that notation is defined just before the statement of theorem 1.3.3) such that for $(\varepsilon, \tilde{y}) \sim \lambda$, one has*

$$\Pr_\lambda\left(\tilde{y} \in \arg\max_{y \in [Y]} \left\{U_y + \varepsilon_y\right\}\right) = 1.$$

(v) *Show that this is equivalent with the fact that for all $B \subseteq Y$, the following inequalities should hold*

$$\Pr_{\mathcal{P}}\left(\max_{z \in B}\{U_z + \varepsilon_z\} > \max_{z \in Y \setminus B}\{U_z + \varepsilon_z\}\right) \leq \sum_{y \in B} \pi_y$$

$$\leq \Pr_{\mathcal{P}}\left(\max_{z \in B}\{U_z + \varepsilon_z\} \geq \max_{z \in Y \setminus B}\{U_z + \varepsilon_z\}\right).$$

(vi) *Show that the previous conditions are equivalently expressed by $\pi \in \partial G(U)$, where $\partial G(U)$ is the subdifferential of G at U, defined as*

$$G(U) := \left\{\pi \in \mathbb{R}^Y : G(U') \geq G(U) + \pi^\top (U' - U)\right\}.$$

Problem 1.2 (Existence of inverse demand). *The goal of this problem is to provide a proof of part (i) of theorem 1.3.1: if assumption 1.1 holds, and if $\pi \in \mathbb{R}^Y$ is such that $\pi_y > 0$ for all $y \in [Y]$ and $\sum_y \pi_y < 1$, then problem* (1.3.2) *admits at least one solution U.*

(i) *Letting* $\pi_0 = 1 - \sum_{y \in [Y]} \pi_y$, *and denoting for each* $y \in [Y]$ *by* F_y *the c.d.f. of the random variable* $\varepsilon_y - \varepsilon_0$, *show that one has*

$$U_y^{\min} := -F_y^{-1}\left(1 - \pi_y\right) \le U_y \le U_y^{\max} := -F_y^{-1}\left(\pi_0\right).$$

(ii) *For* $t \in \mathbb{N}$, *define* U^{t+1} *recursively such that* $U^0 = U^{\min}$ *and*

$$\boldsymbol{\pi}_y\left(U_y^{t+1}, U_{-y}^t\right) = \pi_y \text{ for } y \in [Y].$$

Show that U^t *is well defined. Show that for all* $y \in [Y]$ *and* $t \in \mathbf{N}$, *(a)* $U_y^t \le U_y^{t+1}$, *(b)* $\boldsymbol{\pi}_y\left(U^t\right) \le \pi_y$, *and (c)* $U_y^{t+1} \le U_y^{\max}$.

(iii) *Deduce that* U^t *converges to* $U^\star$ *such that* $\boldsymbol{\pi}(U^\star) = \pi$, *and hence that* $U^\star$ *is the solution to problem* (1.3.2), *thereby proving part (i) of theorem 1.3.1.*

Problem 1.3 (Uniqueness of inverse demand). *The goal of this problem is to provide a proof of part (ii) of theorem 1.3.1: if assumption 1.2 holds, and if* $\pi \in \mathbb{R}^Y$ *is such that* $\pi_y > 0$ *for all* $y \in [Y]$ *and* $\sum_y \pi_y < 1$, *then problem* (1.3.2) *admits at most one solution* U. *Assume by contradiction that* G *is not strictly convex, in which case there are two distinct vectors* U *and* U' *such that* $G((U+U')/2) = \left(G(U) + G\left(U'\right)\right)/2$.

(i) *Letting* $\tilde{y} \in \arg\max_{y \in [0:Y]} \left\{U_y + \varepsilon_y\right\}$, $\tilde{y}'$ *the same random variable associated with* U', *and* $\tilde{y}''$ *the one associated with* $(U+U')/2$, *show that the inequalities*

$$U_{\tilde{y}} + \varepsilon_{\tilde{y}} \ge U_{\tilde{y}''} + \varepsilon_{\tilde{y}''} \text{ and } U'_{\tilde{y}'} + \varepsilon_{\tilde{y}'} \ge U'_{\tilde{y}''} + \varepsilon_{\tilde{y}''}$$

hold in fact as equalities *almost surely.*

(ii) *Show that the previous conclusion contradicts the full support assumption, which proves part (ii) of theorem 1.3.1.*

1.9 References and Notes

The random utility setting presented in sections 1.1 and 1.2 is generally credited to Marschak and his collaborators in the 1960s; see, e.g., the papers Block and Marschak (1959) and Becker, DeGroot, and Marschak (1963). Its use in econometrics owes much to the pioneering work of Daniel McFadden (1981,

1977), who popularized the concept of welfare function in this setting, with antecedents in Williams (1977) and Small and Rosen (1981). The Daly–Zachary–Williams theorem, theorem 1.2.1 in this text, is due to the work of Daly and Zachary (1979) and of Williams (1977). The logit model, introduced in example 1.2, owes its name ("logit" for "logistic unit") to Berkson (1944). It was popularized by Cox (1958) in the binomial case, in relation with the logistic function, and by Luce (2005), who provided an axiomatization involving the independence of irrelevant alternatives. The monograph by Anderson, De Palma, and Thisse (1992) is a good source for applications to industrial organization. Optimization problems with an entropic regularization, such as problem (1.3.8), are prevalent in many fields of economics. In macroeconomics, they are used to capture model uncertainty in robust control theory; see the book by Hansen and Sargent (2008).

The demand inversion problem studied in section 1.3 has been studied in the pioneering work of Hotz and Miller (1993), who highlighted its importance in dynamic problems, and of Berry (1994) for applications to industrial organization. This author provided a constructive inversion method based on Jacobi iterations. The definition of generalized entropy of choice (definition 1.3.1) and its role in demand inversion are due to Galichon and Salanié (in the initial 2009 preprint and in the version published in 2022, where proposition 1.3.2 is stated). The proof of theorems 1.3.1 and 1.3.2 follows from arguments in Galichon and Salanié (2022), theorem 2 and in Chiong, Galichon, and Shum (2016), theorem 2, while theorem 1.3.3 is due to Galichon and Salanié (2022), theorem 2. The framework of optimal transport, and its applications to economics, are covered in my 2016 monograph, *Optimal Transport Methods in Economics* (Galichon 2016).

A popular general reference on simulation methods for discrete choice models as discussed in section 1.4 is the monograph by Train (2009). The simulation procedure proposed in equation 1.4.1 is called the *accept-reject simulator*. More elaborate simulation techniques exist and will be described in section 4.1 below. Bayesian approaches in a more strict sense are covered in the monograph by Rossi, Allenby, and McCulloch (2005).

The linear programming formulation of the simulated demand inversion problem is due to Chiong, Galichon, and Shum (2016). Part (i) of theorem 1.4.1 follows from theorem 4 there, and parts (ii) and (iii) follow from theorem 5 in Chiong, Galichon, and Shum (2016). Problems 1.2 and 1.3 are broadly based on Galichon and Salanié (2022), theorem 2, and Chiong, Galichon, and Shum (2016), theorem 2.

In statistical physics and thermodynamics, the entropy is a measure of the microscopic disorder of a system; it quantifies the amount of energy of the system that cannot be used to act on macroscopic variables. The welfare function of the logit model $G(U) = \log \sum_y \exp U_y$ is called a *log-partition function*, and the corresponding vector of market shares $\pi_y = \exp(U_y)/(\sum_z \exp U_z)$ is called a *Gibbs distribution.*

1.10 Summary

SUMMARY OF CHAPTER 1

In this chapter we introduce random utility models (RUMs), where a consumer's choice from among a set of alternatives is modeled as a function of both systematic utilities, which are common across all consumers, and idiosyncratic utilities, which vary by individual. Agents choose from among $Y+1$ options $y \in [0:Y]$ yielding indirect utility $u(\varepsilon) = \max_{y\in[Y]} \left\{U_y + \varepsilon_y, \varepsilon_0\right\}$, where U_y is a quantity, called systematic utility, that only depends on y, and ε is a $(Y+1)$-dimensional vector with distribution $\mathcal{P}$. We shall maintain the normalization $U_0 = 0$ throughout.

Section 1.2 delves into the welfare function, which quantifies the average consumer welfare within a population, and the demand map, which defines the probabilities of choosing from among available options based on their utilities. It highlights the Daly–Zachary–Williams theorem, which connects the welfare function's gradient to the market share map. The welfare function, defined as a function of U by $G(U) = \mathbb{E}_{\mathcal{P}}[u(\varepsilon)]$, is such that, by theorem 1.2.1, the quantity defined as $\pi_y(U) = \partial G(U)/\partial U_y$ is the market share of option y. This map has a simple expression in the binomial case, where its expression involves a c.d.f. (see example 1.1), and in the multinomial logit case, where it involves the Gibbs distribution (see example 1.2).

In section 1.3, the focus shifts to the demand inversion problem, a key concern in econometrics and industrial organization, aiming to recover systematic utilities from observed market shares. Given the observation of a vector of market shares (π_y), we solve for the vector of systematic utilities (U_y) such that $\pi(U) = \pi$. Definition 1.3.1 introduces the concept of the generalized entropy of choice and its significance in addressing demand inversion. The generalized entropy of choice, defined by $G^*(\pi) = \max_U \left\{\pi^\top U - G(U)\right\}$, is an extension of the usual entropy, to which it simplifies in the multinomial logit case. Theorem 1.3.2 shows that the vector U solving the demand inversion problem is the gradient of G^*, or, in other words, $U_y = \partial G^*(\pi)/\partial \pi_y$; and theorem 1.3.3 asserts that U can be characterized as the solution to an optimal transport problem.

Section 1.4 covers simulation methods for estimating market shares and the empirical welfare function using population samples, explaining the computation methods and their implications for demand inversion. There we consider a sample of vectors ε_i drawn from $\mathcal{P}$. The computation of market share is implemented in Python in appendix B.1.1. Theorem 1.4.1 reformulates the demand inversion problem in the sample as a linear programming problem, for which code is provided in appendix B.1.2.

Section 1.5 discusses the asymptotic properties of estimators within random utility models, particularly how sample uncertainty impacts the estimation of systematic utilities. Consistency and asymptotic normality results, as the sample size becomes large, are worked out.

Section 1.6 provides comparative statics results that predict how demand responds when utilities of certain alternatives are increased, and how inverse demand responds when all market shares of nondefault alternatives move in the same direction.

CHAPTER 2

The Logit Model and Its Generalizations

In this chapter, we shall go back to the logit model, which featured as an example in the previous chapter, and we shall study its properties in detail as well as an important generalization. The generalization will be based on the class of *multivariate extreme value* (MEV) distributions, somewhat inappropriately called *generalized extreme value* (GEV) in economics.[1] In MEV-based random utility models, one associates to each option y a random utility ε_y, which is a random variable with a very specific distribution, the Gumbel distribution described in section 2.1 below. In the plain logit model to be seen in section 2.1, these random utility terms are independent across alternatives; but in its generalizations based on multivariate extreme value theory, these random utility terms have a rich dependence structure, to be explored in section 2.2. The simplest of these models beyond plain logit is the celebrated *nested logit model*; we shall describe it in section 2.3. We will then cover the two other families of max-stable distributions after the Gumbel, the Fréchet and the Weibull distributions, in section 2.4, where we relate them with the Gumbel distribution, and where we state the fundamental theorem of extreme value distribution, the Fisher–Tippett–Gnedenko theorem. Lastly, we will end the chapter with a continuous extension of the logit model.

2.1 The Logit Model and the Gumbel Distribution

The logit model, introduced in example 1.2 in chapter 1, is based on the *standard Gumbel distribution*, denoted by $\mathcal{G}$, which is the distribution

[1]In the mathematical statistics literature, the term "generalized extreme value" denotes a family that embeds the three max-stable families described in section 2.4. What economists since McFadden have called "generalized extreme value" is called "multivariate extreme value" in statistics, a terminology we will adhere to moving forward.

with c.d.f.

$$F_{\mathcal{G}}(u) = \exp\left(-\exp(-u)\right), u \in \mathbb{R}. \tag{2.1.1}$$

The mean of the standard Gumbel distribution is γ, the Euler–Mascheroni constant, defined by $\gamma = -\int_0^{+\infty} \exp(-t)\log(t)dt$, which is approximately equal to 0.5772, and its variance is $\pi^2/6$. By extension, the *Gumbel family of distributions*, or, more succinctly, *Gumbel distributions*, describes the set of distributions that can be obtained by a location-scale transformation of $\mathcal{G}$. The *centered Gumbel distribution* is the member of the family whose c.d.f. is $F(u) = \exp(-\exp(-u-\gamma))$; it has a mean of zero.

The Gumbel distribution is a particular member of the family of "max-stable" distributions. In general, a distribution $\mathcal{P}$ is called *max-stable* if for $\varepsilon_1, \ldots, \varepsilon_Y$, Y i.i.d. copies of $\mathcal{P}$, there are scalars $a_Y > 0$ and $b_Y \in \mathbb{R}$ such that

$$\max_{y\in[Y]} \left\{\varepsilon_y\right\} =_D a_Y \epsilon + b_Y, \text{ where } \epsilon \sim \mathcal{P}. \tag{2.1.2}$$

In the case where $\mathcal{P}$ is the standard Gumbel distribution $\mathcal{G}$, one can compute the c.d.f. of $\eta = \max_y \varepsilon_y$, which is

$$F_\eta(u) = \Pr(\varepsilon_y \leq u, \forall y \in [Y]) = \exp\left(-Y\exp(-u)\right) = F_{\mathcal{G}}(u - \log Y),$$

and one sees that (2.1.2) holds with $a_Y = 1$ and $b_Y = \log Y$, which shows that the Gumbel distribution is max-stable. Conversely, all max-stable distributions whose support is unbounded are Gumbel distributions, as formalized in the following result, which is a direct consequence of the Fisher–Tippett–Gnedenko theorem (theorem 2.4.1 in section 2.4 below):

Proposition 2.1.1. *The standard Gumbel distribution $\mathcal{G}$ is (up to a location-scale transform) the only max-stable distribution whose support is unbounded above and below.*

Let $\varepsilon_y \sim \mathcal{G}$ be Y i.i.d. standard Gumbel random variables indexed by $y \in [Y]$, and given U a Y-dimensional vector, define a random variable $\tilde{u} = \max_y\{U_y + \varepsilon_y\}$. We saw in (2.1.2) above that if $U_y = 0$, then $\tilde{u}$ is Gumbel. In fact, the distribution of $\tilde{u}$ is Gumbel even when U is different from zero. A simple calculation similar to the one carried out above shows that the c.d.f. of $\tilde{u}$ at $u \in \mathbb{R}$ is $\exp(-\exp(-(u - \log\sum_y \exp U_y)))$, which means that $\tilde{u} - \log\sum_y \exp U_y$ has a standard Gumbel distribution. We state:

Proposition 2.1.2. *If $\varepsilon_1, \ldots, \varepsilon_Y$ are Y i.i.d. standard Gumbel random variables, then*

$$\max_{y\in[Y]} \left\{U_y + \varepsilon_y\right\} =_D \log\left(\sum_{y\in[Y]} \exp U_y\right) + \epsilon \quad \textit{where } \epsilon \sim \mathcal{G}. \tag{2.1.3}$$

From expression (2.1.3), we can easily deduce the welfare function and the market share map in the case of the i.i.d. standard Gumbel distribution:

Theorem 2.1.1. *Assume ε_y are Y i.i.d. standard Gumbel distributions. Then the welfare function $G(U) = \mathbb{E}[\max_y\{U_y + \varepsilon_y\}]$ and the market share map $\pi(U) = \nabla G(U)$ are respectively given by*

$$G(U) = \log\Big(\sum_{y\in[Y]} \exp U_y\Big) + \gamma \quad \textit{and} \quad \pi_y(U) = \frac{\exp U_y}{\sum_{y\in[Y]} \exp U_y}. \tag{2.1.4}$$

Note that the constant γ in the expression of G is an artifact of the fact that we have assumed that the ε_y have a standard Gumbel distribution, and thus have a mean of γ. If annoying, this is innocuous, since the constant disappears when we derive G in order to obtain the market shares. Of course, γ disappears if we replace "standard Gumbel" by "centered Gumbel" in the claim above.

Preference for variety. Because each entry ε_y is independent, the logit model implies a "preference for variety." If there are Y options with the same systematic utility $U_y = U$, the indirect utility is $u(\varepsilon) = U + \log Y + \epsilon$, which increases with Y as the logarithm of Y. Preference for variety—and the underlying idea that more choice makes consumers more happy overall—is a natural idea, and often empirically grounded. However, the rigid structure of the logit model can take this idea a bit too far, and produce unexpected conclusions, as illustrated by a well-known feature of the logit model, the so-called "bus paradox," which we discuss next.

The blue bus, red bus paradox. Suppose that an individual has the choice between two transportation modes, say, the train ($y = 0$, whose systematic utility U_0 is normalized to zero) and the bus ($y = 1$, with systematic utility U). In the logit model, it is very easy to compute the market share of the train; it is $1/(1 + \exp U)$.

Now suppose that we segment one of the options, say, the "bus" option, into two subcategories according to a criterion that is quite irrelevant to the choice: say, the color of the bus. Assume there is a "red bus," which we call option $y = 1a$, and a "blue bus," denoted by $y = 1b$. We assume that the systematic utility associated to the blue bus and the red bus options is the same, and is still U as before. In that case, the market share of the train is $1/(1 + 2\exp U)$. This is not a desirable feature of the model: we would expect the combined market share of both bus options in the new model to coincide with the market share of the bus option in the previous model, and, therefore, the market share of the train to be the same in the two cases. Instead, by partitioning an option using an irrelevant criterion, one has increased its market share, as if the systematic utility for taking the bus had been increased from U to $U + \log 2$.

This paradox is in fact no paradox. We have seen that an inherent feature of the logit model is the preference for variety. Here, by providing more bus options, we have increased the probability that agents will draw a sufficiently large utility shock associated with a bus of some kind, thus increasing the overall market share of buses. We shall see other random utility structures, such as the nested logit model described in section 2.3 below, which do not suffer—or suffer less—from this undesirable feature. There, choice will be sequential: the decision-maker first chooses the transportation mode (train or bus), and then, if relevant, the color of the car (blue or red). By increasing the correlation of the random utility associated with the color within the "nest" (here, the transportation option), one manages to mitigate the effects of the paradox. In another approach, the characteristics approach, to be seen soon in chapter 4, consumers directly care about the characteristics of the products they choose, and the paradox completely disappears.

2.2 Multivariate Extreme Value Generalizations

The logit model imposes a lot of structure to the random utility shocks. In particular, it assumes that the utility shocks across two alternatives ε_y and $\varepsilon_{y'}$ are independent. This may be acceptable in some settings, but there are a number of settings where this is hard to fathom. Consider, for instance, the problem of modeling the demand for housing goods, and assume that y stands for the number of bedrooms. It is rather implausible to assume that the utility shock associated with a two-bedroom apartment is uncorrelated with the utility shock associated with a three-bedroom apartment. Because of these limitations, we move on to describing a natural class of distributions that embeds the logit model while removing some of its drawbacks: the class of multivariate extreme value distributions, which allows for correlation between the random utility terms. As it turns out, multivariate extreme value distributions can be introduced as combinations of i.i.d. Gumbel variables acting as factors in a specific way.

Consider (η_x), a vector of X i.i.d. standard Gumbel variables ("factors") indexed by $x \in [X]$. One seeks to create other standard Gumbel variables using the entries η_x as factors. If U is such that $\sum_{x\in[X]} \exp U_x = 1$, one has by proposition 2.1.2 that

$$\varepsilon = \max_{x\in[X]} \{U_x + \eta_x\} \sim \mathcal{G}. \tag{2.2.1}$$

Now, we can construct another scalar random variable ε' using the same factors η_x, but with a different vector U', still assuming $\sum_x \exp U'_x = 1$; that is, we can define

$$\varepsilon' = \max_{x\in[X]} \left\{U'_x + \eta_x\right\} \sim \mathcal{G}, \tag{2.2.2}$$

and depending on the discrepancy between the vectors U and U', one may expect to see more or less correlation between ε and ε'. Indeed, if $U = U'$, then $\varepsilon = \varepsilon'$, so ε and ε' are perfectly correlated; while if U and U' are such that $U_1 = 0$ and $U_x = -\infty$ elsewhere, and $U'_2 = 0$ and $U'_x = -\infty$ elsewhere, then $\varepsilon = \eta_1$ and $\varepsilon' = \eta_2$, in which case ε and ε' are independent.

Example 2.1. In the coding example in appendix B.2.1, we set for all $x \in [X]$

$$U_x = \ln \frac{x^{-\lambda}}{\sum_{x\in[X]} x^{-\lambda}} \text{ and } U'_x = \ln \frac{(1+X-x)^{-\lambda}}{\sum_{x\in[X]} x^{-\lambda}},$$

and we compute by simulation the correlation of ε and ε' obtained from expressions (2.2.1) and (2.2.2) with this choice of vectors U and U'. In exercise 2.5, it is shown that the random utility vector $(\varepsilon, \varepsilon')$ converges in distribution to a single-nest binomial nested logit model, which, as we shall see in section 2.3 below, implies that the correlation tends to $1-\lambda^2$ as $X \to +\infty$.

The previous discussion suggests a general way of coupling two or more Gumbel random variables using the entries of η as factors. By considering a family U_{xy} of numbers indexed by $x \in [X]$ and $y \in [Y]$ such that $\sum_x \exp U_{xy} = 1$ for all y, we can define

$$\varepsilon_y = \max_{x\in[X]} \left\{U_{xy} + \eta_x\right\}, \tag{2.2.3}$$

and we have $\varepsilon_y \sim \mathcal{G}$ for all $y \in [Y]$, so each ε_y has a standard Gumbel distribution. Hence this construction has allowed us to construct entries of ε that are more or less correlated between complete dependence and complete independence. There is a strong resemblance between this construction and the construction of correlated Gaussian variables using i.i.d. normal factors—the Gaussian distribution is stable for the addition, so the factor representation in the Gaussian case is additive, while the Gumbel distribution is stable for the maximum, which explains that the representation (2.2.3) should involve a maximum.

The c.d.f. of the vector (ε_y) defined by (2.2.3) is $F_\varepsilon(u) = \Pr(\varepsilon_y \le u_y, y \in [Y])$, which, given the construction of ε, is equal to the probability that $U_{xy} + \eta_x \le u_y$ for all x and y, which in turn equals the probability that $\eta_x \le \min_y\{u_y - U_{xy}\}$ for all $x \in [X]$. But from the independence of the entries of η, it follows that $F_\varepsilon(u) = \prod_{x\in[X]} \Pr(\eta_x \le \min_{y\in[Y]}\{u_y - U_{xy}\})$, that is,

$F_\varepsilon(u) = \exp(-\sum_{x\in[X]} \exp(-\min_{y\in[Y]}\{u_y - U_{xy}\}))$; hence

$$F_\varepsilon(u) = \exp\left(-h\left(\exp(-u)\right)\right) \text{ where } h(b) := \sum_{x\in[X]} \max_{y\in[Y]} \left\{b_y e^{U_{xy}}\right\}. \quad (2.2.4)$$

We see from expression (2.2.4) that the function h is positive homogeneous of degree 1. This motivates the following definition:

Definition 2.2.1 (Multivariate extreme value class). *The distribution of a random vector ε in $\mathbb{R}^Y$ with c.d.f. F_ε belongs to the multivariate extreme value class if $h(b) = -\log F_\varepsilon(-\log b_1, \ldots, -\log b_Y)$ is positive homogeneous of degree one. In that case, the function h is called the* dependence function *associated with the distribution of ε.*

According to this definition, if the distribution of ε is multivariate extreme value with c.d.f. F_ε, then $F_\varepsilon(u) = \exp(-h(\exp(-u)))$ where h is positive homogeneous of degree one. This implies that for each $y \in [Y]$, the c.d.f. F_y of the scalar random variable ε_y evaluated at $t \in \mathbb{R}$ is obtained by taking u such that $u_y = t$ and $u_{y'} = +\infty$ for $y' \neq y$, which obtains $F_y(t) = F_\varepsilon(u)$, and thus $F_y(t) = \exp(-h(\exp(-t)\mathbf{e}_Y^y))$, where $\mathbf{e}_Y^y$ is the yth vector of the canonical basis in dimension Y. By positive homogeneity of h, we get $F_y(t) = \exp(-\exp(-(t - \log h(\mathbf{e}_Y^y))))$. As a result, ε_y is a Gumbel random variable with mean $\log h(\mathbf{e}_Y^y) + \gamma$.

Expression (2.2.4) allows us to compute the c.d.f. of the random variable $\tilde{u} = \max_y\{U_y + \varepsilon_y\}$ where the vector (ε_y) has a multivariate extreme value distribution with c.d.f. F_ε. The c.d.f. of $\tilde{u}$ at $u \in \mathbb{R}$ is $F_{\tilde{u}}(u) = \Pr(\varepsilon_y \leq u - U_y, y \in [Y]) = \exp(-h(\exp(U - u\mathbf{1}_Y)))$, and using the positive homogeneity of h, one is led to:

Proposition 2.2.1. *If the distribution F_ε of the random vector ε in $\mathbb{R}^Y$ belongs to the multivariate extreme value class, with associated dependence function $h(b) = -\log F_\varepsilon(-\log b_1, \ldots, -\log b_Y)$, then* $\max_y\{U_y + \varepsilon_y\} = \log h(\exp U) + \epsilon$, *where ϵ is a standard Gumbel variable, and therefore*

$$\mathbb{E}\left[\max_{y\in[Y]} \left\{U_y + \varepsilon_y\right\}\right] = \log h(\exp(U)) + \gamma. \quad (2.2.5)$$

In the case where the random variables ε_y, $y \in [Y]$ are i.i.d. standard Gumbel variables, it is easy to see that $h(b) = \sum_y b_y$. The formula $F(v) = \exp(-h(\exp(v)))$ implies that F_ε is the product of functions of v_y, which means that the random variables ε_y are independent. Hence the additivity on $h(b)$ is directly related to the independence between the random utility terms. More generally, if $h(b) = h_1(b) + h_2(b)$ where $h_1(b)$ does not depend on b_y for

$y \in A$ and $h_2(b)$ does not depend on b_y for $y \in [Y] \backslash A$, then the random vectors $(\varepsilon_y)_{y \in A}$ and $(\varepsilon_y)_{y \notin A}$ are independent.

It follows from proposition 2.2.1 that if ε has a multivariate Gumbel distribution, then the random variable ϵ defined by $\epsilon = \max_y\{\log b_y + \varepsilon_y\} - \log h(b)$ follows a standard Gumbel distribution. One has

$$\max_{y \in [Y]} \left\{ b_y \frac{\exp \varepsilon_y}{\sum_{y' \in [Y]} \exp \varepsilon_{y'}} \right\} = h(b) \frac{\exp \epsilon}{\sum_{y' \in [Y]} \exp \varepsilon_{y'}}$$

so, letting $c = \mathbb{E}[e^{\epsilon}/(\sum_y e^{\varepsilon_y})]$ and $Z_y = e^{\varepsilon_y}/(c \sum_y e^{\varepsilon_y})$ yields $h(b) = \mathbb{E}[\max_y\{b_y Z_y\}]$. Therefore, we have shown that:

Proposition 2.2.2. *If $F(x) = \exp(-h(\exp(-x)))$ is the distribution of a multivariate extreme value distribution, there is a random vector Z defined on $\mathbb{R}_+^Y$ such that the dependence function h takes expression*

$$h(b) = \mathbb{E}\left[\max_{y \in [Y]} \{b_y Z_y\} \right], \tag{2.2.6}$$

which is called the Pickands representation *of the multivariate extreme value distribution.*

For example, if ε_y are i.i.d. standard Gumbel variables, the combination of proposition A.2.4 and proposition A.2.5 implies that a Pickands representation of the i.i.d. Gumbel distribution is provided by $Z_y = 1/D_y/(\sum_y 1/D_y)$, where (D_y) follows a flat Dirichlet distribution of order Y, whose definition is recalled in appendix A.2.1.

To illustrate further, we can come back to the factor model (2.2.2)–(2.2.3) of multivariate extreme value distributions with which we introduced this section, and determine its Pickands representation. In order to interpret the function h as an expectation, we consider $[X] = \{1, \ldots, X\}$ as a probability space endowed with the uniform probability measure which assigns probability $1/X$ to each state $x \in [X]$, and we define the random vector $(Z_y)_y$ by its value in state x: $Z_{xy} = X \exp(U_{xy})$. One has $\mathbb{E}[Z_y] = \sum_x Z_{xy}/X = 1$ for all $y \in [Y]$, and it follows from expression (2.2.4) that the Y-dimensional random vector Z_{xy}, defined on the probability space $[X]$, provides the Pickands representation of the random vector ε.

Two comments are in order. First, it is easy to see that the distribution of the random vector Z in the Pickands representation is not unique: indeed, if a vector (Z_y) satisfies equation (2.2.6), then (Z'_y) defined by $Z'_y = Z_y W$ also does so as soon as W is a positive random variable such that $\mathbb{E}[W|Z] = 1$. Second, note that if $\mathbf{e}_Y^y$ is the yth vector of the canonical basis in dimension Y, then $h(\mathbf{e}_Y^y) = \mathbb{E}[Z_y]$, so if F_ε is such that ε_y is a standard Gumbel variable, then $\log h(\mathbf{e}_Y^y) = 0$ and therefore $\mathbb{E}[Z_y] = 1$.

As a consequence, the following theorem provides the welfare function and the market share map in the case of the multivariate extreme value distribution:

Theorem 2.2.1. *Assume a random vector* (ε_y) *on* $\mathbb{R}^Y$ *has multivariate extreme value distribution* $\mathcal{P}$ *with a dependence function* h. *Then the welfare function and the market share map are respectively given by*

$$G(U) = \log h(e^U) + \gamma \quad \textit{and} \quad \pi_y(U) = \frac{e^{U_y}}{h(e^U)} \frac{\partial h}{\partial b_y} (e^U).$$

2.3 The Nested Logit Model

The nested logit model, pioneered by Ben-Akiva in the 1970s in his PhD thesis, is based on the idea that a decision-maker faces a choice between Y options that can be grouped into *nests*. Options that are in the same nest are assumed to be closer to each other, and, accordingly, their random utility terms will be positively correlated. For instance, the options may be detailed travel options, with the nests being the transportation mode (train, air, bus, car). The nested logit model assumes that agents pick the transportation mode first, and then pick their final option within the transportation mode, but there is positive correlation among the random utilities associated with train options, with air options, with bus options, and so on. Let Y be the total number of options, and for each option $y \in [Y]$, denote by $x_y \in [X]$ the associated transportation mode. In the terminology of nested logit models, the transportation mode is called a *nest*, and $[X]$ is the set of nests. We denote by $\mathcal{Y}_x$ the set of $y \in [Y]$ such that $x_y = x$. In order to describe the mathematical structure of the model, it is useful to state a lemma from probability theory.

Preparation: A decomposition lemma. We shall state a result showing that one can decompose a standard Gumbel random variable $\varepsilon \sim \mathcal{G}$ into the sum:

$$\varepsilon = \alpha\eta + \zeta, \text{ where } \eta \perp\!\!\!\perp \zeta \text{ and } \eta \sim \mathcal{G}, \tag{2.3.1}$$

for some random variable ζ that remains to be determined. This will introduce some correlation between several Gumbel variables. Indeed, setting $\varepsilon_y = \alpha\eta_y + \zeta$, where the terms η_y are independent standard Gumbel variables, but the same random variable ζ appears in each ε_y, one sees that

$$cov(\varepsilon_1, \varepsilon_2) = var(\zeta) = var(\varepsilon) - \alpha^2 var(\eta) = (1 - \alpha^2) var(\varepsilon), \tag{2.3.2}$$

which implies that the correlation coefficient between ε_1 and ε_2 is $1 - \alpha^2$.

The question is therefore whether there exists a distribution of ζ that leads to decomposition (2.3.1). We can compute the c.d.f. of ε defined by $\varepsilon = \alpha\eta + \zeta$, where η and ζ are independent, and $\eta \sim \mathcal{G}$ has a standard Gumbel distribution: it is $F_\varepsilon(u) = \Pr(\alpha\eta + \zeta \leq u) = \mathbb{E}\left[\mathbb{E}[1\{\eta \leq (u-\zeta)/\alpha\} \,|\zeta]\right] = \mathbb{E}[\exp(-\exp(\zeta/\alpha - u/\alpha))]$; thus $F_\varepsilon(u) = \mathbb{E}[\exp(-\exp(\zeta/\alpha)\exp(-u/\alpha))]$; so denoting $Z = \exp(\zeta/\alpha)$ and $t = \exp(-u/\alpha)$, we have

$$F_\varepsilon(u) = \mathbb{E}\left[\exp\left(-tZ\right)\right], \tag{2.3.3}$$

implying that if Z exists, then its Laplace transform at t should coincide with the c.d.f. of ε at u. Thus for F_ε to be the c.d.f. of a Gumbel distribution, we require that

$$\mathbb{E}\left[\exp\left(-Zt\right)\right] = \exp\left(-t^\alpha\right). \tag{2.3.4}$$

Therefore, the problem of the existence of decomposition (2.3.1) boils down to asking whether there is a positive random variable Z whose Laplace transform is $t \mapsto \exp(-t^\alpha)$. This is a well-known problem, whose answer is yes and follows from theorem A.2.1 in appendix A.2.2, which ensures the existence of the so-called positive stable distribution of parameter α, denoted by $\mathcal{PS}_\alpha$, which has a Laplace transform as in (2.3.4). As a result, the decomposition given in (2.3.1) exists, and we state:

Lemma 2.3.1 (Decomposition lemma). *If $\eta \sim \mathcal{G}$, $Z \sim \mathcal{PS}_\alpha$ for $\alpha \in (0,1]$, and $\eta \perp\!\!\!\perp Z$, then*

$$\alpha(\eta + \log Z) \sim \mathcal{G}.$$

Lemma 2.3.1 suggests that the distributions of random variables of the form $\alpha \log Z$, where $Z \sim \mathcal{PS}_\alpha$, tend to Gumbel distributions as $\alpha \to 0$, in a sense made precise in theorem A.2.2 in appendix A.2.2. In the same appendix, it is shown that these random variables are "α-soft-max stable," and this term is precisely defined.

The nested logit choice model. With these preparations in mind, and in particular thanks to lemma 2.3.1, we are now ready to describe the nested logit model. The nested logit model assumes the following specification:

$$\varepsilon_y = \lambda_{x_y} \log Z_{x_y} + \lambda_{x_y}\eta_y, \tag{2.3.5}$$

where the terms η_y are i.i.d. Gumbel random variables, and the terms Z_x are i.i.d. random variables distributed as $\mathcal{PS}_\lambda$, and independent from η. (Recall that $x_y \in [X]$ is the nest that contains option y.) The following result then follows from expression (2.3.2):

Proposition 2.3.1 (Correlation of the random utilities in the nested logit model). *In the nested logit model, the random utility vector* (ε_y), *which is given by* (2.3.5), *is such that for* $y \neq y'$, *the correlation between* ε_y *and* $\varepsilon_{y'}$ *is* $1-\lambda_x^2$ *if* y *and* y' *are in the same nest* x, *and zero if they are in different nests.*

The decision maker's indirect utility is $u(\varepsilon) = \max_y\{U_y + \varepsilon_y\}$, which is

$$u(\varepsilon) = \max_{y\in[Y]}\{U_y + \lambda_{x_y} \log Z_{x_y} + \lambda_{x_y}\eta_y\} = \max_{x\in[X]}\left\{\lambda_x \log Z_x + \max_{y\in\mathcal{Y}_x}\{U_y + \lambda_x\eta_y\}\right\}.$$

By independence between Z and η, and because of lemma 2.3.1, the value of the inner maximization problem $\max_{y\in\mathcal{Y}_x}\{U_y + \lambda_x\eta_y\}$ equals $\lambda_x \log \sum_{y\in\mathcal{Y}_x} \exp(U_y/\lambda_x) + \lambda_x\hat{\eta}_x$, where $\hat{\eta}_x$ are i.i.d. standard Gumbel independent from the random vector Z. As a result and by lemma 2.3.1 again, the quantities defined by $\hat{\varepsilon}_x := \lambda_x \log Z_x + \lambda_x\hat{\eta}_x$ are i.i.d. standard Gumbel and

$$u(\varepsilon) = \max_{x\in[X]}\left\{\lambda_x \log \sum_{y\in\mathcal{Y}_x} \exp(U_y/\lambda_x) + \hat{\varepsilon}_x\right\}, \tag{2.3.6}$$

which shows that the nests are picked as in a logit model, where the systematic utility associated with nest x is $\lambda_x \log \sum_{y\in\mathcal{Y}_x} \exp(U_y/\lambda_x)$. Taking expectation of (2.3.6) yields an expression of $G(U)$ and related quantities in closed form, namely:

Proposition 2.3.2 (Welfare function and entropy of choice, nested logit model). *The distribution of the random utilities* (ε_y) *given by* (2.3.5) *is a multivariate extreme value distribution, and thus the welfare function* G *and the dependence function* h *are given by*

$$G(U) = \log\Big(\sum_{x\in[X]}\Big(\sum_{y\in\mathcal{Y}_x} \exp(U_y/\lambda_x)\Big)^{\lambda_x}\Big) + \gamma \quad \text{and} \quad h(b) = \sum_{x\in[X]}\Big(\sum_{y\in\mathcal{Y}_x} b_y^{1/\lambda_x}\Big)^{\lambda_x}, \tag{2.3.7}$$

and the corresponding market share map of an alternative y *in nest* x *is*

$$\pi_y(U) = \frac{(\sum_{y\in\mathcal{Y}_x} \exp(U_y/\lambda_x))^{\lambda_x}}{\sum_{x'\in[X]}(\sum_{y\in\mathcal{Y}_{x'}} \exp(U_y/\lambda_{x'}))^{\lambda_{x'}}} \frac{\exp(U_y/\lambda_x)}{\sum_{y'\in\mathcal{Y}_x} \exp(U_{y'}/\lambda_x)}. \tag{2.3.8}$$

The entropy of choice G^* *in the nested logit model is given by*

$$G^*(\pi) = \sum_{x\in[X]} \lambda_x \sum_{y\in\mathcal{Y}_x} \pi_y \log \pi_y + \sum_{x\in[X]} (1-\lambda_x)\Big(\sum_{z\in\mathcal{Y}_x} \pi_z\Big) \log\Big(\sum_{z\in\mathcal{Y}_x} \pi_z\Big) \tag{2.3.9}$$

if $\pi_y \geq 0$ *and* $\sum_{x\in[X]}\sum_{y\in\mathcal{Y}_x}\pi_y = 1$, $G^*(\pi) = +\infty$, *otherwise. Taking the normalization* $G^*(\pi) = 0$, *the inverse market share map for an alternative y in nest x is given by*

$$\pi_y^{-1}(\pi) = \lambda_x \log \pi_y + (1-\lambda_x)\log\Big(\sum_{z\in\mathcal{Y}_x}\pi_z\Big). \tag{2.3.10}$$

The expression of the market share map given in (2.3.8) has a clear behavioral interpretation. Indeed, we see from (2.3.8) that $\pi_y(U)$ is the product of two terms. The first term is the probability of choosing nest x, which is proportional to the quantity $(\sum_{y\in\mathcal{Y}_x}\exp(U_y/\lambda_x))^{\lambda_x}$, while the second term is the conditional probability of picking option y in nest x, which is proportional to $\exp(U_y/\lambda_x)$. We therefore see that we can view the nested logit model as one where the decision-maker picks the nest first, and then chooses the final option among that nest. These formulas are implemented in the coding example in appendix B.2.2.

2.4 The Three Families of Max-Stable Distributions

We saw in proposition 2.1.1 that the Gumbel distribution is the only max-stable distribution whose support is unbounded both above and below. However, there are other max-stable distributions whose support is bounded above or bounded below, and it is easy to construct these distributions from Gumbel random variables. For instance, if the random variable ε has a Gumbel distribution, the distribution of $c + \exp(k\varepsilon)$ for $k > 0$ is max-stable with support bounded below; it is called a *Fréchet distribution.* Likewise, the distribution of $c - \exp(-k\varepsilon)$ for $k > 0$ is also max-stable, this time with support bounded above; it is called a *(reversed) Weibull distribution.* No max-stable distribution can have a support that is bounded at the same time above and below, and it follows from the Fisher–Tippett–Gnedenko theorem, which we are about to state, that the Gumbel, Fréchet, and reversed Weibull distributions exhaust the possibilities of max-stable distributions: they are called type I, type II, and type III, respectively. These three families are the only max-stable distributions. The Fisher–Tippett–Gnedenko theorem actually says more than that: it states that, under certain conditions, the distribution of the maxima of independent and identically distributed random variables can be approximated by one of these three types of distributions.

Theorem 2.4.1 (Fisher–Tippett–Gnedenko). *If* $X_1, X_2, \ldots$ *are i.i.d. real random variables, and* $M_n = \max(X_1, X_2, \ldots, X_n)$ *is their running maximum, and if there exist constants* $a_n > 0$ *and* $b_n \in \mathbb{R}$ *such that* $(M_n - b_n)/a_n$ *converges in distribution as* $n \to \infty$ *to a nondegenerate limiting distribution, then the c.d.f.*

F of that limiting distribution is (up to a location-scale transform) one of the following three types:

- *Type I (Gumbel):* $F(x) = \exp\left(-\exp(-x)\right), \quad x \in \mathbb{R}.$
- *Type II (Fréchet):* $F(x) = \begin{cases} 0 & \text{for } x \le 0 \\ \exp\left(-x^{-\alpha}\right) & \text{for } x > 0 \end{cases}, \quad \alpha > 0.$
- *Type III (Weibull):* $F(x) = \begin{cases} \exp\left(-(-x)^{\alpha}\right) & \text{for } x \le 0 \\ 1 & \text{for } x > 0 \end{cases}, \quad \alpha > 0.$

If the distribution of X has a "light tail," which means that its tail decreases exponentially, then the limiting distribution F of $(M_n - b_n)/a_n$ is the Gumbel distribution; one says that X is in the domain of attraction of the Gumbel class. If X has a heavy tail, which is the case if its distribution is a power law, for example, then it is in the domain of attraction of the Fréchet class. If the support of X is bounded above, then it is in the domain of the Weibull class.

Example 2.2. The following are noteworthy examples of members of domains of attraction of the three extreme value classes:

(i) The exponential distribution of parameter $\lambda > 0$, which has

$$f(x) = \lambda e^{-\lambda x}, \; x \in \mathbb{R}_+,$$

lies in the domain of attraction of the Gumbel distribution.

(ii) The Pareto distribution of parameter $\alpha > 0$, which has

$$f(x) = \frac{\alpha}{x^{\alpha+1}}, \; x \in [1, +\infty),$$

lies in the domain of attraction of the Fréchet distribution.

(iii) The uniform distribution on the unit interval, which has

$$f(x) = 1, \; x \in [0, 1],$$

lies in the domain of attraction of the Weibull class.

In exercise 2.8, the reader is asked to verify these facts. In the coding example of appendix B.2.3, the distribution of M_n is simulated and the convergence is numerically explored.

Generalized extreme value distributions. Note that one can reexpress these three types more compactly, as a family indexed by a single parameter. If we write $F_\gamma(x) = \exp(-(1 + \gamma x)^{-1/\gamma})$, we have that the distribution is

	w.r.t. U	**w.r.t.** E	**w.r.t.** G	**w.r.t.** F	**w.r.t.** W
U	$=U$	$=e^{-E}$	$=e^{-e^{-G}}$	$=e^{-F^{-\alpha}}$	$=\exp(-(-W)^{\alpha})$
E	$=-\ln U$	$=E$	$=e^{-G}$	$=F^{-\alpha}$	$=(-W)^{\alpha}$
G	$=-\ln(-\ln U)$	$=-\ln E$	$=G$	$=\alpha\ln F$	$=-\alpha\ln(-W)$
F	$=(-\ln U)^{-1/\alpha}$	$=E^{-1/\alpha}$	$=e^{G/\alpha}$	$=F$	$=-\frac{1}{W}$
W	$=-(-\ln U)^{1/\alpha}$	$=-E^{1/\alpha}$	$=-e^{-G/\alpha}$	$=-\frac{1}{F}$	$=W$

Figure 2.1: Functional relationships among random variables with distributions of interest: $U\sim\mathcal{U}([0,1])$ is uniform, $E\sim\mathcal{E}$ is exponential with parameter 1, $G\sim\mathcal{G}$ is standard Gumbel, F is Fréchet with parameter $\alpha>0$, and W is Weibull with parameter $\alpha>0$.

type II for $\gamma>0$ (with $\alpha=1/\gamma$), type III for $\gamma<0$ (with $\alpha=-1/\gamma$), and type I for $\gamma=0$; indeed, $-\log(-\log F_\gamma(x))=\log(1+x\gamma)/\gamma$, which tends to x as $\gamma\to 0$. This embedding is what is called the "generalized extreme value distribution" family in mathematical statistics, which can create some confusion given that the term is used to denote multivariate extreme value distributions in economics.

Relations between extreme value distributions. The three variables of the extreme value family, Gumbel, Fréchet, and Weibull, can all be expressed using exponential variables. Recall that E is a (standard) exponential variable, denoted by $E\sim\mathcal{E}$ when its c.d.f. is $F(x)=1-\exp(-x)$. As a result, if $E\sim\mathcal{E}$, then $-\log E\sim\mathcal{G}$ is Gumbel, $E^{-1/\alpha}$ is Fréchet, and $-E^{1/\alpha}$ is Weibull. More generally, it is useful to be able to express quickly one random variable as a function of another one. The table provided in figure 2.1 gives the complete list of such relationships between the uniform, exponential, Gumbel, Fréchet, and Weibull distributions.

Fréchet-based models. Fréchet variables, in particular, are commonly used in some areas of economics such as international trade, as they can describe random elasticities. We can easily convert Gumbel-stability into Fréchet-stability by noting that

$$\max_{y\in[Y]}\left\{U_y+\varepsilon_y\right\}=_D\log\sum_{y\in[Y]}\exp U_y+\tilde{\varepsilon},$$

where the terms ε_y are i.i.d. with a standard Gumbel distribution, which is also the distribution of $\tilde{\varepsilon}$. Letting $\eta_y=\exp\left(\varepsilon_y/\alpha\right)$, $\tilde{\eta}=\exp(\tilde{\varepsilon}/\alpha)$, the entries of η are Fréchet variables of parameter α, and setting $V_y=\exp(U_y/\alpha)$ yields

$$\max_{y\in[Y]}\left\{V_y\eta_y\right\}=_D(\sum_{y\in[Y]}V_y^{\alpha})^{1/\alpha}\tilde{\eta} \tag{2.4.1}$$

which is an important formula, as $(\sum_{y\in[Y]} V_y^\alpha)^{1/\alpha}$ is the constant elasticity of substitution (CES) aggregate of the quantities V^α, a striking property that has made some economists refer to expression (2.4.1) as the "Fréchet magic formula." Problem (5.4) in chapter 5 illustrates this point of view on an important framework in international trade, Eaton–Kortum's model.

2.5 Continuous Logit Model

Models of economics choice often deal with situations where the options are clearly discrete. The binary choice whether to retire or not, the multinomial choice of a transportation mode, or the choice of embarking on a college degree are all decisions agents take from among a finite menu of options, which justifies the term "discrete" choice. However, there are sometimes situations where the underlying choice problem is continuous, as the set of options is continuously infinite. This is arguably the case in the choice of the amount of time worked, for example; or of the age, income, or other characteristic of one's marital partner; etc. While the data always presents itself eventually as a finite number of bits, it matters a great deal from a modeling perspective whether it is sampled from a discrete or a continuous distribution.

If the set of options is continuous, say it is $\mathcal{Y}=\mathbb{R}^d$, we can introduce $U:\mathcal{Y}\rightarrow\mathbb{R}$, a measurable function such that $U(y)$ is the systematic utility associated with option y. For reasons that will be clear soon, we assume that the function $y\mapsto \exp(U(y))$ is summable so that the quantity $\int_{\mathcal{Y}} \exp U(y)dy$ is finite.

If the sample of outcomes is $\{y_i, i\in[I]\}$, one could at first glance assume that the data generating process is $\max_{i\in[I]}\{U(y_i)+\varepsilon_i\}$, where (ε_i) are i.i.d. standard Gumbel variables; but this would lead to an expected ex ante utility equal to $\log\sum_{i\in[I]}\exp U(y_i)$, which would be asymptotically equivalent to $\log\int_{\mathcal{Y}}\exp U(y)dy+\log I$ as $I\rightarrow+\infty$. As a result, to ensure the utility remains finite in the limit, one should replace ε_i by $\varepsilon_i-\log I$, which is not satisfactory as we do not want to make any assumptions on the data generating process that depend on the sample size. The problem is thus to find the right assumptions on the data-generating process that don't require a sample-dependent rescaling. There is a clear continuous generalization of the logit formulas (2.1.4), namely,

$$G(U)=\log\left(\int_{\mathcal{Y}}\exp(U(y))dy\right)+\gamma \quad \text{and} \quad \pi(y)=\frac{\exp(U(y))}{\int_{\mathcal{Y}}\exp(U(y))dy}, \tag{2.5.1}$$

where π should be interpreted as the probability density function with respect to the Lebesgue measure of the chosen option y. While it is easy to extend formulas (2.1.4) to the continuous case, it is much more difficult to give them an interpretation in terms of random utility in the fashion of (2.1.3), such as

$$\max_{y \in \mathcal{Y}} \{U(y) + \varepsilon(y)\}, \tag{2.5.2}$$

which is difficult as it involves building a continuum of independent random variables $\varepsilon(y)$; further, as discussed above, the maximum of independent standard Gumbel random variables grows in the logarithm of the number of terms, and hence the maximum of an infinite number of independent Gumbel random variables is almost surely infinite.

The way out of this conundrum was independently discovered by Dagsvik and Cosslett: it consists of assuming that the decision-maker does not have access to the set of all alternatives $\mathcal{Y}$. Instead, she draws a random subset of $\mathcal{Y}$, an "opportunity set" of options available to her, along with corresponding random utility shocks she associates with each of these opportunities. More precisely, we denote by $\{(y_k, \varepsilon_k), k \in \mathbb{N}\}$ the Poisson point process of options and utilities available to the decision-maker, where y_k is the alternative, and ε_k is the corresponding utility. In problem 2.3, the reader is asked to verify that if this Poisson process has intensity $dy \otimes \exp(-\epsilon) d\epsilon$, then we can define $\varepsilon(y)$ by $\varepsilon(y) = \varepsilon_k$ if $y = y_k$, and by $\varepsilon(y) = -\infty$ otherwise, so that the random choice problem (2.5.2) becomes

$$\max_{k \in \mathbb{N}} \{U(y_k) + \varepsilon_k\}, \tag{2.5.3}$$

and the random choice problem (2.5.2) leads to a probability of choosing option y given by formula (2.5.1).

2.6 Exercises

Exercise 2.1 (Properties of the Gumbel distribution). *Let $(\varepsilon_1, \varepsilon_2)$ be a pair of i.i.d. standard Gumbel random variables.*

(i) *Show that the distribution of $U_1 + \varepsilon_1$ conditional on $U_1 + \varepsilon_1 > U_2 + \varepsilon_2$ coincides with the distribution of $\log\left(\exp U_1 + \exp U_2\right) + \epsilon$, where ϵ is also a standard Gumbel random variable.*

(ii) *Letting $\tilde{y} = 1$ if $U_1 + \varepsilon_1 > U_2 + \varepsilon_2$, and $\tilde{y} = 2$ if $U_1 + \varepsilon_1 \leq U_2 + \varepsilon_2$, deduce that $\tilde{y}$ is independent from $U_{\tilde{y}} + \varepsilon_{\tilde{y}}$.*

(iii) *Show more generally that if $(\varepsilon_1, \dots, \varepsilon_Y)$ are i.i.d. standard Gumbel random variables, and if a random variable $\tilde{y}$ valued in $[Y]$ is a maximizer of $U_y + \varepsilon_y$, then $\tilde{y}$ is independent from $U_{\tilde{y}} + \varepsilon_{\tilde{y}}$.*

Exercise 2.2. *Recall that the dependence function in multivariate extreme value distributions is given by $h(b) = \mathbb{E}[\max_y\{b_y Z_y\}]$, where Z is a Y-dimensional random vector with positive entries. Show that $h(b) = \int_0^\infty 1 - F_Z((\frac{c}{b_y})_y)\,dc$.*

Exercise 2.3 (Characterization of the logit model). *Assume that the distribution $\mathcal{P}$ of $(\varepsilon_y)_{y\in[Y]}$ is such that the market share map π is given by*

$$\pi_y(U) = \exp U_y / (\sum_{z\in[Y]} \exp U_z).$$

Show that the random vector ε can be expressed $\varepsilon = \eta + \zeta \mathbf{1}_Y$, where η is a vector of i.i.d. standard Gumbel random variables, and ζ is any random variable.

Exercise 2.4 (Independence of irrelevant alternatives). *Assume $\mathcal{P}$ satisfies assumption 1.1. Show that if $\pi_y(U)/\pi_{y'}(U)$ only depends on the values of U_y and $U_{y'}$ for any pair of indices y, y' in $[0:Y]$, then $\mathcal{P}$ is the distribution of i.i.d. Gumbel variables.*

Exercise 2.5. *As in example 2.1, set for all $x \in [X]$*

$$U_{x1} = \ln \frac{x^{-\lambda}}{\sum_{x\in[X]} x^{-\lambda}} \text{ and } U_{x2} = \ln \frac{(1+X-x)^{-\lambda}}{\sum_{x\in[X]} x^{-\lambda}},$$

and show that as $X \to +\infty$, the distribution of $(\varepsilon_1, \varepsilon_2)$ obtained by expression (2.2.3) has Pickands representation $h(b) = (b_1^{1/\lambda} + b_2^{1/\lambda})^\lambda$, which means it is the distribution associated with the single-nest binomial nested logit model with parameter λ.

Exercise 2.6 (Nested logit inversion). *Assume that there are three options, one default $y=0$ and two alternatives $y=1$ and $y=2$. We normalize the systematic utility of the default alternative, so $U_0 = 0$. Assume the vector of market share is $\pi = (\pi_1, \pi_2) = (0.3, 0.5)$.*

(i) Compute (U_y) in the standard logit model.

(ii) Same question if we assume that $y=1$ and $y=2$ are in the same nest, and $\lambda = 0.3$.

(iii) Assuming that $y=1$ and $y=2$ are in the same nest, but without fixing the value of λ, what is the set of utilities that U can take?

(iv) *Same question as (ii), but replacing the inverse map associated with the nested logit model by its simulated counterpart, where we shall use formula* (A.2.9) *to simulate the random utilities with 10,000 simulated points.*

Exercise 2.7 (Simulating the nested logit model). *The purpose of this exercise is to simulate the random utilities that appear in the nested logit model. Assume there is one zero option and two alternatives that are part of the same nest with* $\lambda = 0.5$. *Therefore there are two nests,* $x = 0$ *and* $x = 1$, *and* $x_y = 0$ *for* $y = 0$, *as the zero option is alone in its nest, and* $x_y = 1$ *for* $y = 1$ *and* $y = 2$, *which are part of the same nest.*

(i) *Use lemma 2.3.1 to argue that the nested logit model can be formally represented as a mixed logit model.*

(ii) *Use formula* (A.2.9) *to simulate the random utilities, and, in turn, to predict the market shares associated with* $U_1 = 0.3$, $U_2 = -0.1$. *Comment on the precision of the estimator with respect to the number of draws* I.

Exercise 2.8. *Verify the claims in example 2.2, showing that* a_n *and* b_n *in the statement of theorem 2.4.1 are given by:*

(i) *for the exponential random variables with* λ, $a_n = 1/\lambda$ *and* $b_n = \log n/\lambda$;

(ii) *for the Pareto distribution with* $\alpha > 1$ *on* $[1, +\infty)$, $a_n = n^{1/\alpha}$ *and* $b_n = 0$;

(iii) *for the uniform distribution,* $a_n = 1/n$ *and* $b_n = 1$.

Exercise 2.9 (Second derivatives of the logit map). *This exercise requires a good command of the concepts introduced in appendix A.3.1. Let* $U : \mathbb{R}^P \to \mathbb{R}^Y$ *be a smooth function, and define* $\pi : \mathbb{R}^Y \to \mathbb{R}^Y$ *by* $\pi_y(U) = \exp(U_y) / \left(\sum_{z \in [Y]} \exp U_z\right)$, *which is the market share of the logit model. For* $\theta \in \mathbb{R}^P$, *we let* $f(\theta) = \pi(U(\theta))$ *so that* f *is the composition of* π *and* U.

(i) *Show that the first and second derivatives of* f *are expressed by*

$$df = \mathbf{\Delta}_f \mathbf{R}_f \, dU, \tag{2.6.1}$$

$$d^2 f = \mathbf{\Delta}_f \mathbf{R}_f \left((\mathbf{R}_f \, dU) \circledast (\mathbf{R}_f \, dU) + d^2 U \right), \tag{2.6.2}$$

where $\mathbf{\Delta}_f = \mathrm{diag}(f)$ *and* $\mathbf{R}_f = \mathbf{I}_Y - \mathbf{1}_Y f^\top$, *and where* $\circledast$ *is the row-wise Kronecker product defined in appendix A.3.1.*

(ii) *Implement the above formulas in Python, and verify your results using numerical differentiation.*

Now assume that $U : \mathbb{R}^P \to \mathbb{R}^{IY}$ *is a smooth function and that* $\pi : \mathbb{R}^{IY} \to \mathbb{R}^{IY}$ *is defined by* $\pi_{iy}(U) = \exp(U_{iy}) / \left(\sum_{z \in [Y]} \exp U_{iz} \right)$, *so that* $f(\theta) = \pi(U(\theta))$ *remains the composition of* π *and* U.

(i) *Show that the first and second derivatives of* f *are still given by*

$$df = \mathbf{\Delta}_f \mathbf{R}_f dU, \tag{2.6.3}$$

$$d^2 f = \mathbf{\Delta}_f \mathbf{R}_f \left((\mathbf{R}_f dU) \circledast (\mathbf{R}_f dU) + d^2 U \right), \tag{2.6.4}$$

where $\mathbf{\Delta}_f = diag(f)$ *as before, but now* $\mathbf{R}_f = \mathbf{I}_{IY} - (\mathbf{I}_I \otimes \mathbf{1}_{Y \times Y}) \mathbf{\Delta}_f$.

(ii) *Again, implement in Python and verify your results by numerical differentiation.*

2.7 Problems

Problem 2.1 (Ordered logit). *The results established in exercise 2.1 can be used to solve this problem. Let* $(\varepsilon_1, \ldots, \varepsilon_Y)$ *be i.i.d. standard Gumbel random variables with* $Y \geq 3$.

(i) *Show that the probability that the alternatives* 1, 2, *and* 3 *are preferred in that order, namely,* $\mathcal{P}(U_1 + \varepsilon_1 > U_2 + \varepsilon_2 > U_3 + \varepsilon_3)$, *is equal to*

$$\mathcal{P}(\max\{U_2 + \varepsilon_2, U_3 + \varepsilon_3\} \leq U_1 + \varepsilon_1) \, \mathcal{P}(U_2 + \varepsilon_2 > U_3 + \varepsilon_3).$$

(ii) *Deduce that*

$$\mathcal{P}(U_1 + \varepsilon_1 > U_2 + \varepsilon_2 > U_3 + \varepsilon_3) = \frac{e^{U_1}}{e^{U_1} + e^{U_2} + e^{U_3}} \frac{e^{U_2}}{e^{U_2} + e^{U_3}}.$$

(iii) *More generally, show that if we label the options by* $t \in [Y]$ *such that* $U_{y_1} + \varepsilon_{y_1} \geq U_{y_2} + \varepsilon_{y_2} \geq \ldots \geq U_{y_Y} + \varepsilon_{y_Y}$, *the probability of observing a labeling* $(y_t)_{t \in [Y]}$ *is*

$$\prod_{t \in [Y]} \frac{e^{U_{y_t}}}{\sum_{s \in [t:Y]} e^{U_{y_s}}}$$

(iv) *In the remainder of the problem, we assume we observe data under the form* y_{it}, *which means that in observation* $i \in [I]$, *the data are ranked from* y_{i1} *(most preferred) to* y_{iY} *(least preferred). Assuming a linear parametrization of* U, *namely,* $U_y = \sum_k \phi_{yk} \lambda_k$, *describe the maximum* *likelihood estimation procedure to estimate λ. Is the procedure a convex estimation problem?*

(v) *What becomes of the sample log-likelihood above when we replace ε_y by $\sigma\varepsilon_y$, with the same scaling factor $\sigma > 0$ for each of the Gumbel terms? and what is the limit of the maximum likelihood estimation problem for $\sigma \to 0$?*

(vi) *Assume there are $K+1$ basis functions indexed by $k \in [0:K]$, and impose the normalization $\lambda_0 = 1$. Show that the estimation procedure with $\sigma \to 0$ leads to*

$$\max_{\lambda \in \mathbb{R}^K} \left\{ \sum_{it \in [I \times T]} (\phi_{y_{it}0} + \sum_{k \in [K]} \phi_{y_{it}k}\lambda_k) - \sum_{it \in [I \times T]} \max_{s \in [t:Y]} \{\phi_{y_{is}0} + \sum_{k \in [K]} \phi_{y_{is}k}\lambda_k\} \right\}$$

and interpret as a minimax regret procedure.

(vii) *Show that the procedure just described can be recast as a linear programming problem.*

Problem 2.2 (Tiago de Oliveira's formula). *In this problem, one seeks a proof of proposition 2.3.1 without resorting to positive stable distributions, but using a formula due to Tiago de Oliveira: if $k : \mathbb{R} \mapsto \mathbb{R}$ is a function whose limits in both $-\infty$ and $+\infty$ are equal to 1, if $Y = 2$, and if $\mathcal{P}$ is multivariate extreme value with dependence function $h(b_1, b_2) = (b_1 + b_2)k(\log(b_1) - \log(b_2))$, then for $(\varepsilon, \eta) \sim \mathcal{P}$, one has*

$$cov\,(\varepsilon, \eta) = -\int_{-\infty}^{+\infty} \log k\,(w)\,dw. \tag{2.7.1}$$

(i) *Show Hoeffding's formula:*

$$cov\,(\varepsilon, \eta) = \int \left(F\left(x, y\right) - F\left(x\right) F\left(y\right)\right) dxdy.$$

(ii) *Consider the change of variables $w = y - x$ and $v = -\log\left(\exp\left(-x\right) + \exp\left(-y\right)\right)$. Show that*

$$cov\,(\varepsilon, \eta) = \int \exp\left(-\exp\left(-v\right) k\left(w\right)\right) - \exp\left(-\exp\left(-v\right)\right) dvdw.$$

(iii) *Deduce Tiago de Oliveira's formula (2.7.1).*

(iv) *Application: Using the fact that* $\int_0^{+\infty} \log\left(1+e^{-w}\right) dw = \pi^2/12$, *which is* $\eta\,(2)$, *where* η *is Dirichlet's eta function, give a proof of proposition 2.3.1.*

Problem 2.3 (The continuous logit model). *The goal of this problem is to recover formulas* (2.5.1) *when the process* $\varepsilon(y)$ *is the one proposed by Dagsvik and Cosslett as described at the end of section 2.5. As in section 2.5, we consider a continuous choice set, namely,* $\mathcal{Y} = \mathbb{R}^d$. *Let* $\tilde{\mathcal{S}}$ *be the random set formed by the points of a Poisson process on* $\mathcal{Y} \times \mathbb{R}$ *of intensity* $dy \otimes e^{-\epsilon}\,d\epsilon$. *This implies that for a measurable subset* $S \subseteq \mathcal{Y} \times \mathbb{R}$, *the probability that* $\tilde{\mathcal{S}}$ *does not intersect* S *is given by*

$$\Pr(\tilde{\mathcal{S}} \cap S = \emptyset) = \exp\left(-\int_S e^{-\varepsilon}\,dyd\varepsilon\right).$$

(i) *Show that* $\tilde{\mathcal{S}}$ *has almost surely an infinite (though countable, which one is not asked to verify) number of elements. As a result, one shall describe* $\tilde{\mathcal{Y}}$ *by its enumeration—that is, we index it by the set of integers—and we thus write* $\tilde{\mathcal{S}} = \{(y_k, \varepsilon_k), k \in \mathbb{N}\}$. *We let* $\varepsilon(y) = \max_{k\in\mathbb{N}}\{\varepsilon_k : y_k = y\}$, *with the convention that the maximum over an empty set returns* $-\infty$.

(ii) *Show that the values and the maximizers of problems* (2.5.2) *and* (2.5.3) *coincide.*

(iii) *Letting* $\hat{u} := \max_{k\in\mathbb{N}}\left\{U\left(y_k\right) + \varepsilon_k\right\}$ *and denoting* $F_{\hat{u}}(u)$ *its c.d.f., show that*

$$\log F_{\hat{u}}(u) = -\exp\left(-u + \log\int_{\mathcal{Y}} \exp U\left(y\right)\,dy\right),$$

and deduce that $\mathbb{E}[\hat{u}] = \log\int_{\mathcal{Y}} \exp U\left(y\right)\,dy + \gamma$, *which is the first part of claim* (2.5.1).

(iv) *Letting* $\hat{y} \in \arg\max_{y\in\mathcal{Y}}\left\{U(y) + \varepsilon\left(y\right)\right\}$, *show that the density of* $\left(\hat{y}, \hat{u}\right)$ *at* $\left(y, u\right)$ *is* $\exp U\left(y\right) F_{\hat{u}}\left(u\right)\exp(-u)$ *and deduce that* $\hat{y}$ *and* $\hat{u}$ *are independent, and that the density of the random variable* $\hat{y}$ *at* y *is*

$$\frac{e^{U(y)}}{\int_{\mathcal{Y}} e^{U(y)}\,dy},$$

which is the second part of claim (2.5.1).

(v) *Now assume $\mathcal{Y}$ is no longer $\mathbb{R}^d$, but is a finite set instead. Show that for each $y \in \mathcal{Y}$, the set of $k \in \mathbb{N}$ such that $y_k = y$ is non-empty, and show that if $\varepsilon(y)$ is as defined above, the random vector $(\varepsilon(y))_{y\in\mathcal{Y}}$ follows an i.i.d. Gumbel distribution.*

Problem 2.4 (Pickands representation of the nested logit model using Dirichlet distributions). *Recall from appendix A.2.1 that a flat Dirichlet distribution of order J is the uniform distribution over the $(\mathrm{J}-1)$-dimensional subsimplex $\{x \in \mathbb{R}^{\mathrm{J}} : \sum_j x_j = 1\}$, and recall that it follows from proposition A.2.5 in the same appendix that if the J-dimensional random vector $(D_1, \ldots, D_{\mathrm{J}})$ follows a flat Dirichlet distribution of order J, then*

$$\max_{j\in[\mathrm{J}]}\left\{\frac{a_j}{D_j}\right\} =_D \frac{\sum_{j\in[\mathrm{J}]} a_j}{1-U^{1/(\mathrm{J}-1)}},$$

where $U \sim \mathcal{U}([0,1])$. Show that this implies the Pickands representation of the nested logit model

$$h(b) := \left(\sum_{j\in[\mathrm{J}]} b_j^{1/\lambda}\right)^{\lambda} = \mathbb{E}\left[c \max_{j\in[\mathrm{J}]}\left\{\frac{b_j}{D_j^{\lambda}}\right\}\right],$$

where $c = 1/\mathbb{E}[(1-U^{1/(\mathrm{J}-1)})^{-\lambda}]$.

Problem 2.5 (A toy assortment optimization problem). *A monopolistic seller needs to pick a set $\mathcal{B}$ among Y alternatives it offers to its customers. The clients solve a discrete choice problem and maximize $U_y + \varepsilon_y$ among the available options $y \in \mathcal{B} \cup \{0\}$, where 0 is a default option such that $U_0 = 0$. If set $\mathcal{B}$ is offered, then the volume of the sales amounts to*

$$\sum_{y\in\mathcal{B}} \frac{e^{U_y}}{1+\sum_{y\in\mathcal{B}} e^{U_y}}.$$

If the cardinality of $\mathcal{B}$ cannot exceed $K < Y$, what set should the seller pick in order to maximize the sales volume?

2.8 References and Notes

A good general reference for extreme value theory is de Haan and Ferreira (2006), where the Fisher–Tippett–Gnedenko theory is explained. The pioneering use of the Gumbel distribution in discrete choice analysis is widely attributed to Daniel McFadden's seminal work. His groundbreaking paper (McFadden 1974), which appeared in the book *Frontiers in Econometrics*, edited by Paul Zarembka, laid the foundation for what is now known as the logit model in discrete choice analysis by using the Gumbel distribution to model the random utility component in individual choice behavior. However, the logit model predates McFadden and underwent much development with the work of Luce in the 1960s; see a review in Luce (1977). The "blue bus, red bus" paradox is generally credited to Debreu (1960).

A textbook reference for the multivariate extreme value distribution is Joe (1997). Early references are Gumbel (1960), Tiago de Oliveira (1962), and de Haan and Resnick (1977). The dependence function and what is now called Pickands representation were first given in Pickands (1981). In econometrics, McFadden (1977) discusses applications of the multivariate extreme value distribution theory, which he refers to as the "generalized extreme value" distribution theory.

The nested logit model was developed by Ben-Akiva in his 1973 Ph.D. thesis, and an exposition is given in Ben-Akiva and Lerman (1985), where the existence of a decomposition as in (2.3.1) is conjectured. Before that, Tiago de Oliveira (1984) had worked on a class of related extreme value distributions. The decomposition (2.3.1) was further examined by Cardell (1997), who derived an expression for the density of $\log Z$ as an alternating series obtained via inversion of its characteristic function, although without establishing that the resulting candidate density is indeed nonnegative. Lemma 2.3.1, which solved the problem using positive-stable distributions, appeared in Galichon (2022). Many other multivariate extreme value specifications exist; see Bierlaire (2006).

A good reference for extreme value theory discussed in section 2.4 is the book by de Haan and Ferreira (2006). The Fisher–Tippett–Gnedenko is given there as theorem 1.1.3.

Continuous logit models discussed in section 2.5 were pioneered by McFadden in the 1970s. In McFadden (1976), he extended the notion of independence of irrelevant alternatives (IIA) to continuous sets. A few years later, Ben-Akiva and his collaborators worked with continuous logit models in Ben-Akiva and Watanatada (1981) and Ben-Akiva, Litinas, and Tsunekawa (1985) by introducing them as limits of discrete choice models, for the purpose of spatial choice models. Then, in two independent contributions, Cosslett (1988) and Dagsvik (1994) both suggested using the max-stable processes approach to model continuous choice, which is the base of problem 2.3; see also Resnick and Roy (1991).

A reference for the ordered logit model in problem 2.1 is Beggs, Cardell, and Hausman (1981). The reference for problem 2.2 is Tiago de Oliveira (1984), p. 266. A reference for assortment optimization as in problem 2.5 is Talluri and Van Ryzin (2006).

2.9 Summary

SUMMARY OF CHAPTER 2

In this chapter we deal with the logit model and its probabilistic foundation, the theory of extreme values, which studies distributions arising as the limit of the maximum of independent and identically distributed terms.

Section 2.1 introduces the Gumbel distribution, central to the logit model, which arises as a unique max-stable distribution with unbounded support. The logit random utility assumption allows for closed-form solutions for indirect utility, welfare functions, and market share maps, as well as the entropy of choice and inverse market share maps. In section 2.2, the discussion is extended to multivariate extreme value (MEV) distributions, which enable the construction of correlated Gumbel variables. This section highlights how MEV distributions, through a dependence function, allow for modeling choices with correlated shocks, leading to the nested logit model as a prime example. The nested logit model, developed to address the independence of irrelevant alternatives (IIA) property in simple logit models, is discussed in detail in section 2.3. It introduces nests or groups of choices to model decision-making processes more realistically, incorporating correlations within nests to reflect more complex choice behavior.

Section 2.4 provides the fundamental theorem in extreme value theory, theorem 2.4.1, which is due to Fisher, Tippett, and Gnedenko, and which asserts that the maximum of independent and identically distributed random values, after the relevant location-scale transformation, can converge only to three possible distributions, the only one of which to be unbounded above and below is the Gumbel; along with it, the Fréchet is bounded below, and the Weibull is bounded above. There are simple functional relationships between these distributions, which are recalled in the table of figure 2.1 at the end of the section.

Finally, section 2.5 explains the challenges of extending the logit random utility model to the case with a continuum of alternatives; this challenge is due to the difficulty in envisioning a continuously infinite number of i.i.d. random variables. A way out of this difficulty consists in assuming agents draw options along with their corresponding utilities according to a Poisson point process. If the right intensities are well chosen, one is led to formulas for the conditional choice probability that extend the classical (discrete) logit formulas to the continuous case.

CHAPTER 3

Fundamentals of Logistic Regression

In this chapter, we seek to understand the determinants of individual choices by considering a parametric family of systematic utilities. In the case of the logit model, this will lead us to a fundamental topic of this book: logistic regression, which we will explore in depth throughout this chapter, focusing on existence, computation, asymptotic properties, and model selection, as well as various interpretations and links with statistical frameworks. Besides our intrinsic interest in logistic regression, it will serve as an important building block for the equilibrium models introduced later in this book. Many of the ideas presented in this chapter will carry over.

3.1 Parametric Random Utility Models

We consider a parametric version of the random utility model described in chapter 1. In a regression context, the options $y \in [Y]$ are sometimes called *categorical variables*; in this chapter we don't single out any particular default option. An agent $i \in [I]$ must choose from among a set of options $y \in [Y]$. The utility brought to agent i by option y is

$$U_{iy} + \varepsilon_{iy}, \tag{3.1.1}$$

where U_{iy} is the part of the utility that can be predicted based on the observable characteristics of agent i, and $\varepsilon = (\varepsilon_{iy})_y$ is a random draw from probability distribution $\mathcal{P}$ that satisfies assumption 1.1. From section 3.2 on, we shall assume that $\mathcal{P}$ is the distribution of i.i.d. Gumbel shocks; but in the present section, we shall not impose a particular distribution for $\mathcal{P}$ provided that it has a density. Hence, this section is slightly more general than the logistic regression framework, where $\mathcal{P}$ is the distribution of i.i.d. Gumbel shocks.

We assume that U_{iy} is parametrized by a vector $\lambda \in \mathbb{R}^K$ in a linear fashion, so that

$$U_{iy}(\lambda) = \sum_{k\in[K]} \phi_{iyk}\lambda_k = (\Phi\lambda)_{iy}, \tag{3.1.2}$$

where Φ is the $(IY) \times K$ matrix whose rows are indexed by $iy \in [I \times Y]$ and columns by $k \in [K]$.

Example 3.1. In the travel mode choice data from Greene and Hensher taken from Greene (2000), $[Y] = \{1, \ldots, 4\}$ are the travel mode options (air, train, bus, car). Some variables are i-specific, such as the income of individual i or the size of the travelling group; some can be y-specific, such as the in-vehicle time. Others can be i- and y-specific, such as the door-to-door travel time that mode y takes for individual i. In the coding example provided in the notebook Galichon (2025c), we focus on a very simple specification where $K = 3$, and ϕ_{iy1} is the in-vehicle travel time of mode y for individual i, ϕ_{iy2} is the in-vehicle travel time multiplied by the income of individual i, and ϕ_{iy3} is the generalized cost of the transportation mode y for agent i.

Letting G be the welfare function associated with the distribution $\mathcal{P}$, the probability that individual $i \in [I]$ chooses alternative $y \in [Y]$ when the parameter value is λ is given by

$$\pi^\lambda_{iy} := \boldsymbol{\pi}_y\left((\Phi\lambda)_{i.}\right) = \frac{\partial G\left((\Phi\lambda)_{i.}\right)}{\partial U_y}, \tag{3.1.3}$$

while, letting y_i be the actual choice of individual i, one defines the choice dummy $\hat{\pi}$ as an IY-dimensional vector given by

$$\hat{\pi}_{iy} = 1\left\{y = y_i\right\}.$$

The log-likelihood that individual i makes a choice y_i is, as in section 1.5, $\log \pi^\lambda_{iy_i} = \sum_y \hat{\pi}_{iy} \log \pi^\lambda_{iy}$, from which one can deduce an expression for the log-likelihood of the sample as $\ell(\lambda) := \sum_{iy} \hat{\pi}_{iy} \log \pi^\lambda_{iy}$. The maximum likelihood estimator is the parameter vector that maximizes the log-likelihood:

Definition 3.1.1 (Maximum likelihood). *The* maximum likelihood estimator $\hat{\lambda}_{ML}$ *is the solution to*

$$\max_{\lambda\in\mathbb{R}^K} \sum_{iy\in[I\times Y]} \hat{\pi}_{iy} \log \pi^\lambda_{iy}. \tag{3.1.4}$$

Problem (3.1.4) yields first-order optimality conditions

$$\sum_{iyz \in [I \times Y \times Y]} \frac{\hat{\pi}_{iy}}{\pi^{\lambda}_{iy}} \frac{\partial^2 G((\Phi\lambda)_{i.})}{\partial U_y \partial U_z} \phi_{izk} = 0, \quad k \in [K]. \tag{3.1.5}$$

The maximum likelihood is appealing from a statistical standpoint because of its asymptotic statistical efficiency; however, it may not be computationally attractive as $\ell(\lambda)$ may not be a concave function, and thus one may be stuck in a local minimum. A computationally interesting alternative to maximum likelihood estimation consists of matching each predicted kth moment $\sum_{iy} \pi^{\lambda}_{iy} \phi_{iyk} = (\pi^{\lambda\top}\Phi)_k$ with the corresponding observed moment $\sum_{iy} \hat{\pi}_{iy}\phi_{iyk} = (\hat{\pi}^{\top}\Phi)_k$, for $k \in [K]$. We define:

Definition 3.1.2 (Method of moments). *The* method of moments estimator $\hat{\lambda}_{MM}$ *is the parameter vector* $\lambda \in \mathbb{R}^K$ *such that*

$$\sum_{iy \in [I \times Y]} \pi^{\lambda}_{iy} \phi_{iyk} = \sum_{iy \in [I \times Y]} \hat{\pi}_{iy} \phi_{iyk}, \quad k \in [K]. \tag{3.1.6}$$

The computational attractiveness of the method of moments comes from the fact that one can interpret (3.1.6) as optimality conditions associated with a *convex* optimization problem, of which $\hat{\lambda}_{MM}$ is the solution. Indeed, recall that $\pi^{\lambda}_{iy} = \partial G((\Phi\lambda)_{i.}) / \partial U_y$, and therefore $\sum_y \pi^{\lambda}_{iy}\phi_{iyk} = \sum_y (\partial G((\Phi\lambda)_{i.}) / \partial U_y)\phi_{iyk}$ is the derivative of $G((\Phi\lambda)_{i.})$ with respect to λ_k. Therefore, $((\pi^{\lambda} - \hat{\pi})^{\top}\Phi)_k$ arises as the derivative of the convex function $\sum_i G((\Phi\lambda)_{i.}) - \hat{\pi}^{\top}\Phi\lambda$ with respect to λ_k. Hence:

Theorem 3.1.1. *Under assumption 1.1:*

(i) The method of moments estimator $\hat{\lambda}_{MM}$ *is the solution to*

$$\max_{\lambda \in \mathbb{R}^K} \left\{ \hat{\pi}^{\top}\Phi\lambda - \sum_{i \in [I]} G((\Phi\lambda)_{i.}) \right\}. \tag{3.1.7}$$

(ii) The value of problem (3.1.7) *is equal to the value of its dual*

$$\min_{\pi_{iy} \geq 0} \sum_{i \in [I]} G^*(\pi_{i.}) \tag{3.1.8}$$

$$s.t.\ ((\pi - \hat{\pi})^{\top}\Phi)_k = 0\ [\lambda_k], \quad k \in [K],$$

and $\hat{\lambda}_{MM}$ *is also the vector of Lagrange multipliers associated with the constraints in that problem.*

Note that the dual formulation (3.1.8) has a crisp interpretation as the problem of minimizing the generalized entropy of choice subject to moment matching conditions.

Sketch of proof of theorem 3.1.1. We have argued above that system (3.1.6) arose as the optimality conditions for problem (3.1.7). To see that (3.1.8) is the dual problem to (3.1.7), we use proposition 1.3.2 to rewrite the primal as

$$\max_{\lambda_k} \min_{\pi_{iy} \geq 0} \left\{ (\hat{\pi} - \pi)^\top \Phi\lambda + \sum_{i \in [I]} G^* (\pi_{i.}) \right\}$$

$$= \min_{\pi_{iy} \geq 0} \max_{\lambda_k} \left\{ \sum_{i \in [I]} G^* (\pi_{i.}) + (\hat{\pi} - \pi)^\top \Phi\lambda \right\},$$

where equality follows from strong duality that is implied by Slater's theorem, recalled in appendix A.1.6, and the right-hand side is exactly the dual problem (3.1.8). The optimal solution (λ_k) to the primal then arises as the Lagrange multipliers of the constraints in the dual problem. ∎

3.2 Multinomial Logistic Regression

Our discussion in section 3.1 was fully general, in the sense that it allowed for any distribution $\mathcal{P}$ of the Y-dimensional random vector ε. From now on, we shall assume that the heterogeneity is logistic, that is, $\mathcal{P}$ is the distribution of a Y-dimensional vector of i.i.d., centered Gumbel random variables. In that case, we get the celebrated *conditional logit model* of McFadden, and equation (3.1.3) becomes

$$\pi^\lambda_{iy} = \frac{\exp(\Phi\lambda)_{iy}}{\sum_{z \in [Y]} \exp(\Phi\lambda)_{iz}}, \tag{3.2.1}$$

and therefore $\log\left(\pi^\lambda_{iy}\right) = (\Phi\lambda)_{iy} - \log\left(\sum_z \exp(\Phi\lambda)_{iz}\right) = (\Phi\lambda)_{iy} - G\left((\Phi\lambda)_{i.}\right)$, and thus we see that the log-likelihood of the sample is

$$\ell(\lambda) = \hat{\pi}^\top \Phi\lambda - \sum_{i \in [I]} \log\left(\sum_{z \in [Y]} \exp(\Phi\lambda)_{iz} \right), \tag{3.2.2}$$

which is the objective of both problem (3.1.4) and problem (3.1.7), as these objectives coincide in the logit case, and in that case only. Therefore, one may state:

Proposition 3.2.1. *In the logit case, the maximum likelihood estimator coincides with the method of moments estimator, and consists of the following optimization problem, called* multinomial logistic regression:

$$\max_{\lambda \in \mathbb{R}^K} \left\{ \hat{\pi}^\top \Phi\lambda - \sum_{i\in[I]} \log\left(\sum_{y\in[Y]} \exp(\Phi\lambda)_{iy} \right) \right\}. \tag{3.2.3}$$

The solution $\hat{\lambda}$ *to the primal problem* (3.2.3), *provided it exists, is such that equalities* (3.1.6) *hold. It is called the* multinomial logistic regression estimator,[1] *or just* logistic estimator.

Note that the claim makes no promise as to the existence and the uniqueness of $\hat{\lambda}$, and we will return to this question in sections 3.5 and 3.6 below. Until then, we shall assume without further verbal precaution that this estimator exists and is unique when making statements about it.

We next investigate the asymptotic distribution of $\hat{\lambda}$ as $I \to +\infty$. This follows from textbook results on maximum likelihood inference, of which we recall the intuition. Define a $K \times K$ matrix $\hat{B}$ by

$$\hat{B}_{kl} = \frac{1}{I} \sum_{iy\in[I\times Y]} \pi^{\lambda}_{iy}\phi_{iyk}\phi_{iyl} - \frac{1}{I} \sum_{iyz\in[I\times Y\times Y]} \pi^{\lambda}_{iy}\pi^{\lambda}_{iz}\phi_{iyk}\phi_{izl},$$

so that $\hat{B} = I^{-1}\sum_i \mathbb{V}(\tilde{\phi}_i|i)$, where $\tilde{\phi}_i$ is the K-dimensional random vector of regressors $(\phi_{iyk})_k$ for a fixed observation i, when y is distributed according to the true choice distribution π^{λ}_{iy}. The matrix $\hat{B}$ is therefore the average over the observations $i \in [I]$ of the conditional variance-covariance matrix $\mathbb{V}(\tilde{\phi}_i|i)$ of $\tilde{\phi}_i$. Define its population analog, the matrix $B = \mathbb{E}[\mathbb{V}(\tilde{\phi})]$ as the expectation of the aforementioned variance-covariance matrices. The matrix B is called the *Fisher information matrix*, and it is the limit of $\hat{B}$ when $I \to +\infty$ of $\hat{B}$. As I gets large, the maximum likelihood estimator $\hat{\lambda}$ gets close to the true parameter vector λ, and an expansion of (3.1.6) yields

$$\sum_{l\in[K]} B_{kl}(\hat{\lambda}_l - \lambda_l) \approx \frac{1}{I} \sum_{iy\in[I\times Y]} \phi_{iyk}\left(\hat{\pi}_{iy} - \pi_{iy}\right), \quad \forall k \in [K],$$

or, in vector form,

$$B(\hat{\lambda}_l - \lambda_l) \approx \Phi^\top \left(\frac{\hat{\pi} - \pi}{I}\right);$$

[1]Our definition of "multinomial logistic regression" is slightly more general than a more common acceptance which sets $U_{iy} = \sum_{l\in[L]} \Psi_{il}\theta_{ly}$, so that the utility associated with each option is independently parametrized. This is recast as a special case of our formulation in exercise 3.2.

but as $cov\left(\hat{\pi}_{iy}, \hat{\pi}_{iz}\right) = \pi^{\lambda}_{iy} 1\left\{y = z\right\} - \pi^{\lambda}_{iy}\pi^{\lambda}_{iz}$, and as the average of the vector $\hat{\pi}_{iz}$ over the observations is asymptotically Gaussian, the asymptotic distribution of the right-hand side is $\mathcal{N}(0, B/I)$, and thus the asymptotic distribution of $\sqrt{I}(\hat{\lambda}_l - \lambda_l)$ is $\mathcal{N}\left(0, B^{-1}\right)$. These heuristic considerations can be made rigorous, and we can state:

Theorem 3.2.1. *As $I \to +\infty$, one has:*

(i) the convergence $\hat{\lambda} \to \lambda$ holds almost surely; and
(ii) the convergence

$$I^{1/2}(\hat{\lambda} - \lambda) \Rightarrow \mathcal{N}\left(0, B^{-1}\right)$$

holds in distribution.

The computation of the estimator $\hat{B}$ of the Fisher information matrix is coded in appendix B.3.2.

3.3 Computation with Gradient Descent

Recall that the estimation of logistic regression, stated in problem (3.2.3), consists of maximizing the log-likelihood $\ell(\lambda)$ provided in expression (3.2.2). In section 3.4, we shall see how to compute this problem as a generalized linear model using a Python library such as `scikit-learn`, but this library will be used as a black box, and it is worth examining how this computation is performed by the gradient descent algorithm. Recall that the predicted assignment π^{λ}_{iy} was given by (3.1.3), and we have $\partial \ell(\lambda) / \partial \lambda_k = \sum_{iy} (\hat{\pi}_{iy} - \pi^{\lambda}_{iy}) \phi_{iyk}$ and $\partial^2 \ell(\lambda) / \partial \lambda_k \partial \lambda_{k'} = -\sum_{iy} \pi^{\lambda}_{iy} \phi_{iyk} \phi_{iyk'} + \sum_{iyy'} \pi^{\lambda}_{iy} \pi^{\lambda}_{iy'} \phi_{iyk} \phi_{iy'k'}$. In vector notations, defining $\mathbf{\Delta}_{\pi} := diag\left(\pi^{\lambda}\right)$ as the $IY \times IY$ matrix whose diagonal terms are the entries of π^{λ}_{iy}, one has

$$\begin{cases} \nabla \ell(\lambda) = \Phi^{\top}\left(\hat{\pi} - \pi^{\lambda}\right) \\ D^2 \ell(\lambda) = -\Phi^{\top}\left(\mathbf{\Delta}_{\pi} - \mathbf{\Delta}_{\pi}\left(\mathbf{I}_I \otimes \mathbf{1}_{Y \times Y}\right) \mathbf{\Delta}_{\pi}\right) \Phi. \end{cases} \tag{3.3.1}$$

Gradient descent is one of the most important numerical methods for solving optimization problems such as (3.2.3). If we have determined an approximation λ^t of the maximizer $\hat{\lambda}$ of (3.2.3) at step t, gradient descent consists of replacing $\ell(\lambda)$ by the following ansatz

$$\ell(\lambda^t) + \nabla \ell(\lambda^t)^{\top}(\lambda - \lambda^t) - \alpha \frac{|\lambda - \lambda^t|^2}{2}, \tag{3.3.2}$$

and computing λ^{t+1} as the maximizer of that ansatz yields the *gradient descent step*

$$\lambda^{t+1} = \lambda^t + \alpha \nabla \ell(\lambda^t) = \lambda^t + \alpha \Phi^\top (\hat{\pi} - \pi^{\lambda^t}), \tag{3.3.3}$$

where $\alpha > 0$ is called the *learning parameter.*

The objective function ℓ is concave and smooth, so its Hessian $D^2\ell(\lambda)$ has only nonpositive eigenvalues. As recalled in definition A.1.4 in appendix A.1.2, one says that a concave function is β-smooth if at all points, all the eigenvalues of the Hessian are greater than or equal to $-\beta$. As a result, if ℓ is $1/\alpha$-smooth, then all the eigenvalues of $\mathbf{I}_K + \alpha D^2\ell(\lambda)$ lie in the interval $[0, 1]$, and thus the map T defined by $T(\lambda) = \lambda + \alpha \nabla \ell(\lambda)$, whose Jacobian is given by this matrix, is 1-Lipschitz. The gradient descent step can be interpreted as $\lambda^{t+1} = T(\lambda^t)$. T is not a strict contraction mapping as the Lipschitz constant is not strictly less than 1, but the convergence of gradient descent nevertheless holds, as implied by theorem A.1.1. (Alternatively, one could also choose a slightly smaller α to ensure that T is a strict contraction mapping and therefore that convergence occurs.) Thus the gradient descent algorithm, as defined by (3.3.3), converges when ℓ is $1/\alpha$-smooth.

To find a suitable α, we use $u^\top D^2\ell(\lambda) u \geq -u^\top \Phi^\top diag(\pi^\lambda) \Phi u \geq -u^\top \Phi^\top \Phi u$; thus, letting $\|\Phi\| = \sup\{|\Phi u| : |u| = 1\}$ be the operator norm of Φ, which is the largest singular value of Φ, one has $u^\top D^2\ell(\lambda) u \geq -\|\Phi\|^2 |u|^2$, which shows that ℓ is β-smooth with $\beta = \|\Phi\|^2$. As a result, we get:

Proposition 3.3.1. *Assume problem* (3.2.3) *has a solution, and let* $\alpha = 1/\|\Phi\|^2$. *Then the gradient descent scheme* (3.3.3) *for the multinomial logistic regression converges to a maximizer* $\hat{\lambda}$ *of problem* (3.2.3).

The computation of logistic regression by gradient descent is implemented in appendix B.3.1.

3.4 Logistic Regression as a Generalized Linear Model

In this section, we argue that the multinomial logistic regression just discussed can be recast as a generalized linear model (GLM) with a logarithmic link function—or, in other words, as a Poisson regression *with individual fixed effects.* Basic facts about the Poisson regression are recalled in appendix A.2.8. As seen there, the Poisson regression of dependent variable $\hat{\pi}$ on design matrix R corresponds to the following optimization problem:

$$\max_{\theta \in \mathbb{R}^P} \left\{ \hat{\pi}^\top R\theta - \sum_{\omega \in [\Omega]} \exp{(R\theta)_\omega} \right\}, \tag{3.4.1}$$

where the sample has Ω observations, and for observation $\omega \in [\Omega]$, $\hat{\pi}_\omega$ is a draw of the dependent variable, while the vector $\left(R_{\omega p}\right)_p$ is the corresponding draw of the vector of explanatory variables. Now, we can map to the Poisson regression setting described in appendix A.2.8 as follows:

- the observations are $\omega = iy$; hence the set of observations is the product of the set of individuals and the set of options, so that $\Omega = IY$;
- the dependent variables $\hat{\pi}_\omega$ become $\hat{\pi}_{iy} = 1\{y_i = y\}$, so the dependent variable is a dummy for "i chooses option y in observation iy";
- the parameter vector $\theta \in \mathbb{R}^P$ is decomposed as $\theta = (\lambda^\top, u^\top)^\top$, where $\lambda \in \mathbb{R}^K$ is the parameter of systematic utilities and $u \in \mathbb{R}^I$ is the vector of individual fixed effects, so that $P = K + I$;
- the regressors R include the individual-option specific moments ϕ_{iyk} and the categorical dummies, so that $(R\theta)_{iy} = \sum_k \phi_{iyk}\lambda_k - u_i = \Phi\lambda - (\mathbf{I}_I \otimes \mathbf{1}_Y)u$, where we recall that $\otimes$ denotes the Kronecker product (see appendix A.3.1), and thus

$$R = \begin{pmatrix} \Phi & -\mathbf{I}_I \otimes \mathbf{1}_Y \end{pmatrix}. \tag{3.4.2}$$

One can then rewrite problem (3.4.1), which becomes

$$\max_{\lambda \in \mathbb{R}^K, u \in \mathbb{R}^I} \left\{ \hat{\pi}^\top \Phi\lambda - \sum_{i \in [I]} u_i - \sum_{iy \in [I \times Y]} \exp\left((\Phi\lambda)_{iy} - u_i \right) \right\}. \tag{3.4.3}$$

Let's hold the value of $\lambda \in \mathbb{R}^K$ fixed and consider the partial maximization with respect to u only. The optimality condition with respect to u_i yields $1 = \sum_y \exp((\Phi\lambda)_{iy} - u_i)$, which in turn implies

$$u_i = \log\left(\sum_{y \in [Y]} \exp(\Phi\lambda)_{iy} \right). \tag{3.4.4}$$

The notation u_i for the i-fixed effect was intentionally chosen: indeed, u_i is the expected indirect utility of individual i in the logit model. After substitution of the optimal value of u_i given by (3.4.4) in problem (3.4.3), we obtain (up to an irrelevant constant)

$$\max_{\lambda \in \mathbb{R}^K} \left\{ \hat{\pi}^\top \Phi\lambda - \sum_{i \in [I]} \log\left(\sum_{y \in [Y]} \exp(\Phi\lambda)_{iy} \right) \right\}, \tag{3.4.5}$$

which is exactly the multinomial regression. We have therefore:

Theorem 3.4.1. *The multinomial logistic regression can be estimated as a Poisson regression with an individual fixed effect.*

The Python code displayed in appendix B.3.2 shows how to compute a multinomial logistic regression as a Poisson regression using the Python library `scikit-learn`, providing the regressors from expression (3.4.2). Beware, however, that the expression of the standard errors returned by the library will be incorrect, as they assume that the observations are independent; yet in the equivalence that we saw, the iy are treated as observations, although iy and iy' are not independent. As a result, the equivalence can be used for estimation, but should not be used for inference.

Thanks to the dual expression (A.2.51) for the Poisson regression in appendix A.2.8, we can get the dual formulation to expression (3.2.3) for the multinomial logistic regression in proposition 3.2.1 as:

$$\begin{aligned} \min_{\pi_{iy}\geq 0} & \sum_{iy\in[I\times Y]} \pi_{iy}\log \pi_{iy} \\ s.t. & \sum_{y\in[Y]} \pi_{iy}=1 \quad [u_i], \\ & \sum_{iy\in[I\times Y]} \pi_{iy}\phi_{iyk}=\sum_{iy\in[I\times Y]} \hat{\pi}_{iy}\phi_{iyk} \quad [\lambda_k], \end{aligned} \tag{3.4.6}$$

where the Lagrange multipliers, indicated in brackets, are the solutions to the primal problem (3.2.3). We had seen a first interpretation of the fixed effect u_i as the expected indirect utility of individual i. Formula (3.4.6) shows a nice complementary interpretation, this time as the Lagrange multiplier of the scarcity constraint $\sum_y \pi_{iy}=1$, which imposes one choice per individual.

3.5 Identification

Logistic regression is identified if problem (3.2.3) does not have multiple solutions. If the problem has multiple solutions, one says that identification "fails." Clearly, the condition that Φ should be of full rank is necessary for identification. As we shall assume $K\leq IY$, this means that the rank of Φ is K. It is also clear that identification fails if an individual fixed effect is included among the regressors: indeed, both specifications—namely, the systematic utilities $U_{iy}=(\Phi\lambda)_{iy}$ and $U_{iy}=a_i+(\Phi\lambda)_{iy}$—yield the same choice probabilities, and therefore the individual fixed effects (a_i) are not identified.

As we shall see, the two occurrences of an identification failure we just described—rank deficiency and individual fixed effects—are the only possible ones, which we now prove formally. Indeed, if identification fails, then there are two values $\lambda\neq\lambda'$ such that $\ell(\lambda)=\ell(\lambda')$, which implies that $\ell(t\lambda+(1-t)\lambda')$ is constant for $t\in[0,1]$. As a result, $(\lambda-\lambda')^\top D^2\ell(\lambda)(\lambda-\lambda')=0$, and thus, setting $\varphi = \Phi(\lambda - \lambda')$, we have $\varphi \neq 0$ because Φ is of full rank, and we have

$$\sum_{iy \in [I \times Y]} \pi^{\lambda}_{iy} \varphi_{iy} \varphi_{iy} - \sum_{iyy' \in [I \times Y \times Y]} \pi^{\lambda}_{iy} \pi^{\lambda}_{iy'} \varphi_{iy} \varphi_{iy'} = 0, \tag{3.5.1}$$

which implies that for each i, the variance of φ_{iy} is zero, and hence that $\varphi_{iy} = a_i$ for some vector a of size I. Thus, if identification fails and Φ is of rank K, then $\operatorname{Im} \Phi$ contains a vector of the form $a \otimes \mathbf{1}_Y$ for a a nonzero vector of size I. Conversely, if the rank of Φ is K and $\operatorname{Im} \Phi$ contains a vector of the form $a \otimes \mathbf{1}_Y$ for $a \neq 0$, then letting δ be the vector of size K such that $a \otimes \mathbf{1}_Y = \Phi \delta$, we have $\ell(\lambda + \delta) = \ell(\lambda)$. This yields:

Theorem 3.5.1. *The logistic regressor $\hat{\lambda}$ is identified if and only if* rank $(\Phi) = K$, *and* $\operatorname{Im} \Phi$ *does not contain any vector of the form $a \otimes \mathbf{1}_Y$ for $a \in \mathbb{R}^I \backslash \{0\}$.*

One way to impose that $\operatorname{Im} \Phi$ does not contain any vector of the form $a \otimes \mathbf{1}_Y$ for $a \in \mathbb{R}^I \backslash \{0\}$ is to require that $\sum_y \phi_{iyk} = 0$ for all $i \in [I]$ and $k \in [K]$. Indeed, if that is the case, $a_i = \sum_k \phi_{iyk} \lambda_k$ for some $\lambda \in \mathbb{R}^K$ implies $a_i Y = \sum_y a_i = \sum_k (\sum_y \phi_{iyk}) \lambda_k = 0$, thus $a = 0$.

3.6 Existence and the Zero-Cell Problem

At this stage, it is important to note that the existence of a logistic regression estimator is not guaranteed due to the *zero-cell problem.* Take, for example, the binomial case, when there are two observations $I = 2$ and $Y = 2$, so that there are two options $y = 1$ and $y = 2$, and the regressors consist only of a dummy of $y = 2$, so that Φ_{iy} is simply $\mathbf{1}\{y = 2\}$. In this case, $\hat{\pi}_2 = \sum_{i \in [2]} \hat{\pi}_{i2}$, and the logistic regressor (3.2.3) becomes

$$\max_{\lambda \in \mathbb{R}} \left\{ \hat{\pi}_2 \lambda - I \log \left(1 + \exp \lambda\right) \right\}, \tag{3.6.1}$$

and we see that if $\hat{\pi}_2 = 0$, meaning that the second option is not observed, there is no solution to problem (3.6.1) and the logistic regressor does not exist. This comes from the fact that, although concave, the objective function is decreasing for λ large positive, but not increasing for λ large negative; one says that it is not *coercive*; see appendix A.1.1. Here, the optimal λ is $-\infty$, which leads to predicting no occurrence of $y = 2$. Similarly, if $\hat{\pi}_2 = 2$, option $y = 2$ occurs twice, and thus option $y = 1$ does not occur in the data, and in this case, the objective function is increasing for λ large negative but not decreasing for λ large positive, and the objective function is again not coercive. On the contrary, if $\hat{\pi}_2 = 1$, the zero-cell problem disappears as both options are observed. In that case, the concave objective function is increasing when λ is large negative, and decreasing when it is large positive; as a result it is coercive, and problem (3.6.1) has a solution.

The intuition behind this simple example is formalized in appendix A.2.8, which provides a theory of when the objective function in problem (3.2.3) is coercive. Theorems A.2.6 and A.2.7 in that appendix imply the following result:

Theorem 3.6.1. *The following conditions are equivalent:*

(i) the logistic estimator (3.2.3) *exists,*
(ii) there exists $\bar{\pi} \in \mathbb{R}^{I \times Y}$ *such that* $\bar{\pi}_{iy} > 0$ *for all* $iy \in [I \times Y]$ *and*

$$\sum_{y \in [Y]} \bar{\pi}_{iy} = \sum_{y \in [Y]} \hat{\pi}_{iy} \ \forall i \in [I] \quad \textit{and}$$

$$\sum_{iy \in [I \times Y]} \bar{\pi}_{iy} \phi_{iyk} = \sum_{iy \in [I \times Y]} \hat{\pi}_{iy} \phi_{iyk} \ \forall k \in [K],$$

(iii) the value V of the linear programming problem

$$V = \max_{t \geq 0, \bar{\pi}_{iy} \geq 0} t \tag{3.6.2}$$

$$\text{s.t. } \bar{\pi}_{iy} \geq t,$$

$$\sum_{y \in [Y]} \bar{\pi}_{iy} = \sum_{y \in [Y]} \hat{\pi}_{iy} \ \forall i \in [I],$$

$$\sum_{iy \in [I \times Y]} \bar{\pi}_{iy} \phi_{iyk} = \sum_{iy \in [I \times Y]} \hat{\pi}_{iy} \phi_{iyk} \ \forall k \in [K]$$

is finite and strictly greater than zero.

Appendix B.3.3 implements the linear programming problem (3.6.2). We see from the theorem that adding new regressors—by stacking them onto the existing ones—amounts to adding constraints to the program (3.6.2). As a result, the value V of the program decreases. Hence the less parsimonious the model, the more likely it is that the logistic estimator does not exist.

3.7 Shrinkage and Regularization

A common problem with every regression setting (linear or nonlinear) is *overfitting*, which occurs when the dimension of the parameter space K is large with respect to the sample size I. In this case, it will be difficult for the model to match the large number of moments to be equated, and the fitted parameter value may be unrealistically large. The problem may be further increased by the zero-cell problem discussed in the previous section, where we saw that the value of the parameters may be infinite when the zero-cell problem is acute, due to the ensuing lack of the coercivity of the objective function.

For these reasons, one may seek to impose restrictions on the parameter space, in particular, to ensure the fitted parameter vector remains bounded. *Shrinkage estimators* look among those parameter vectors $\lambda \in \mathbb{R}^K$ that are "localized," that is, not "too large" in the sense that they maximize the same objective function as (3.2.3) but on a restricted set of parameters:

$$\max_{\lambda \in \mathbb{R}^K} \left\{ \hat{\pi}^\top \Phi\lambda - \sum_{i \in [I]} \log \left(\sum_{y \in [Y]} \exp(\Phi\lambda)_{iy} \right) \right\} \tag{3.7.1}$$
$$s.t. \; \varphi(\lambda) \leq t,$$

where φ is a convex function, often a norm, called a *regularizing function.* Common choices are:

- the squared L2 norm (a.k.a. Euclidean norm) $|.|_2$, which is such that $\varphi(\lambda) = |\lambda|_2^2 = \sum_k \lambda_k^2$—this is called the *ridge regularization*, or *Tikhonov regularization*;
- the L1 norm $|.|_1$ such that $|\lambda|_1 = \sum_k |\lambda_k|$, which is called the *least absolute shrinkage and selection operator* (LASSO) *regularization*;
- a combination of the previous two: the *elastic net regularization* is such that $\varphi(\lambda) = \kappa|\lambda|_2^2 + (1-\kappa)|\lambda|_1$. Clearly, the ridge regularization is recovered by taking $\kappa = 1$ and the LASSO regularization by taking $\kappa = 0$.

Letting $\gamma \geq 0$ be the Lagrange multiplier of the constraint, problem (3.7.1) can be equivalently reformulated as

$$\max_{\lambda \in \mathbb{R}^K} \left\{ \hat{\pi}^\top \Phi\lambda - \sum_{i \in [I]} \log \left(\sum_{y \in [Y]} \exp(\Phi\lambda)_{iy} \right) - \gamma\varphi(\lambda) \right\}, \tag{3.7.2}$$

where γ is the smallest scalar so that the solution λ to this problem satisfies the shrinkage constraint $\varphi(\lambda) \leq t$. Formulation (3.7.2) explains the denomination "regularizing function" for the term $\varphi(\lambda)$: it is added to the objective function, which is noisy, as it is plagued by sample uncertainty, and this addition thus makes the objective function more "regular." We shall impose that φ be coercive (see definition A.1.2), which is the case for the three regularization schemes discussed above, so that unlike the original logistic regression problem (3.2.3), its regularized version (3.7.2) always has solutions.

The optimality conditions in problem (3.7.2) are

$$\Phi^\top(\hat{\pi} - \pi^\lambda) \in \gamma\partial\varphi(\lambda), \tag{3.7.3}$$

where π^λ_{iy} is expressed by (3.2.1), and $\partial\varphi(\lambda)$ is the subdifferential of the regularizer at λ; in the case when the regularizer is differentiable, this boils down to simply $\Phi^\top(\hat{\pi} - \pi^\lambda) = \gamma\nabla\varphi(\lambda)$. In the case of the ridge regularization, this is $\sum_{iy}\phi_{iyk}(\hat{\pi}_{iy} - \pi^\lambda_{iy}) = 2\gamma\lambda_k$.

As we see from equation (3.7.3), introducing a regularization term for shrinkage purposes deteriorates the fit between the predicted moments $\sum_{iy}\pi^\lambda_{iy}\phi_{iyk}$ and the observed ones $\sum_{iy}\hat{\pi}_{iy}\phi_{iyk}$, and the discrepancy increases with γ. With the ridge regularization, we have just seen that this discrepancy is $2\gamma\lambda$, which is in theory not bounded. Things are different, however, with the LASSO regularization, as $\max_{\lambda\in\mathbb{R}^K}\{\ell(\lambda) - \gamma\,|\lambda|_1\}$ has optimality conditions

$$\begin{cases} \lambda_k > 0 \Longrightarrow \frac{\partial\ell(\lambda)}{\partial\lambda_k} = \gamma, \\ \lambda_k < 0 \Longrightarrow \frac{\partial\ell(\lambda)}{\partial\lambda_k} = -\gamma, \\ \lambda_k = 0 \Longrightarrow \frac{\partial\ell(\lambda)}{\partial\lambda_k} \in [-\gamma, \gamma], \end{cases} \tag{3.7.4}$$

which implies that if λ is optimal for (3.7.2) under LASSO, the difference between any predicted moments, and the corresponding observed one does not exceed γ in absolute value.

The L1 regularization leads to a regularization function φ that is nonsmooth, in which case the standard gradient descent algorithm described in section 3.3 fails. Indeed, one can easily verify that the gradient descent method fails to optimize even the real function of the real variable $\lambda \mapsto |\lambda|$, no matter how small the learning parameter. As a result, we need to use another algorithm. One that is particularly well-suited to the L1 regularization is the proximal gradient algorithm, which we now describe.

Conditions (3.7.4) show that the LASSO regularization induces sparsity: indeed, $\partial\ell(\lambda)/\partial\lambda_k \in (-\gamma, \gamma)$ implies $\lambda_k = 0$. As discussed in appendix A.1.3, the proximal gradient algorithm, which converges to the LASSO-regularized logistic estimator, consists of replacing in the standard gradient descent method the explicit scheme by a semi-implicit scheme, where

$$\text{explicit scheme: } \lambda^{t+1} = \lambda^t + \alpha\nabla\ell(\lambda^t) - \alpha\gamma\nabla\varphi(\lambda^t), \tag{3.7.5}$$

$$\text{semi-implicit scheme: } \lambda^{t+1} = \lambda^t + \alpha\nabla\ell(\lambda^t) - \alpha\gamma\nabla\varphi(\lambda^{t+1}). \tag{3.7.6}$$

Scheme (3.7.5) is called "explicit" because it allows for the direct computation of λ^{t+1} as a function of λ^t, because the right-hand side of (3.7.5) is a function of λ^t. If we replaced λ^t by λ^{t+1} in the right-hand side of (3.7.5), we would obtain a fully implicit scheme, which would simply be the optimality conditions in problem (3.7.2). Hence the implicit scheme is not solvable unless the optimization problem is solvable in one single step. Scheme (3.7.6), however, replaces λ^t by λ^{t+1}, but only in $\nabla\varphi$, which is why it is called semi-implicit.

In fact, the explicit scheme is actually not properly defined due to the fact that φ is not differentiable, and thus expression (3.7.5) refers to a quantity $\nabla\varphi$ that does not exist. The interest of the semi-implicit scheme is that one can rewrite it without making any reference to $\nabla\varphi$, reformulating it instead as an optimization problem. To see this, define

$$\tilde{\lambda}^{t+1} = \lambda^t + \alpha\nabla\ell(\lambda^t),$$

and notice that the semi-implicit scheme arises as the solution to $\tilde{\lambda}^{t+1} + \alpha\gamma\nabla\varphi\left(\lambda^{t+1}\right) - \lambda^{t+1} = 0$, which may be seen as first-order conditions of a convex optimization problem,

$$\lambda^{t+1} = \arg\min_{\lambda}\left\{\frac{1}{2}\left|\tilde{\lambda}^{t+1} - \lambda\right|^2 + \alpha\gamma\varphi\left(\lambda\right)\right\},$$

and when $\varphi(\lambda) = \sum_k |\lambda_k|$, one can see (see appendix A.1.3) that λ_k^{t+1} is obtained from $\tilde{\lambda}_k^{t+1}$ by the *solf-thresholding operator*, defined as

$$\lambda_k^{t+1} = \begin{cases} 0 & \text{if } \tilde{\lambda}_k^{t+1} \in [-\alpha\gamma, \alpha\gamma], \\ \tilde{\lambda}_k^{t+1} - \alpha\gamma & \text{if } \tilde{\lambda}_k^{t+1} \geq \alpha\gamma, \\ \tilde{\lambda}_k^{t+1} + \alpha\gamma & \text{if } \tilde{\lambda}_k^{t+1} \leq -\alpha\gamma, \end{cases} \tag{3.7.7}$$

which is another illustration of the sparsity induced by the LASSO penalization, as the algorithm shrinks the kth dimension of the parameter toward zero, and replaces it with zero if it is within $\alpha\gamma$ of zero. The proximal gradient descent algorithm for the L1 regularization of the logistic regression is displayed in appendix A.1.3. Its convergence is guaranteed by the following result:

Proposition 3.7.1. *The proximal gradient descent algorithm defined by the semi-implicit scheme (3.7.6) converges to the solution to the regularized optimization problem of logistic regression (3.7.2).*

While the parameter vector λ is determined as the solution of the optimization problem (3.7.2), that problem depends on the scalar parameter $\gamma \geq 0$ and it remains to discuss how that parameter should be picked. The key notion here is *out-of-sample validity*, and the method of *cross-validation*. The idea is to split the sample set into two samples, the *training sample* and the *validation sample*. For each value of $\gamma \geq 0$, one computes $\lambda(\gamma)$ as the solution to problem (3.7.2). One assesses the generalization ability of each choice of γ by evaluating the quality of the prediction made on the validation set based on the parameter $\lambda(\gamma)$ that has been trained on the training sample, but not on the validation sample. This is done by computing a measure of discrepancy between the predictor on the validation sample and the observations on the validation sample. The parameter γ is then chosen in order to minimize this discrepancy, that is, maximize the out-of-sample prediction.

3.8 Minimax Regret and Small-Noise Limit

In this section, we introduce another paradigm, called *minimax regret estimation*, to carry out the parametric estimation of the systematic utilities, and we discuss its connection with logistic regression.

Consider a framework in which, in contrast to (3.1.1), there is no unobserved heterogeneity; hence, if in observation i the decision-maker chooses option y, it gets only the systematic part of the utility, and no random utility term. Using the same linear parametrization of the systematic utilities as before, the utility that $i \in [I]$ associates to option $y \in [Y]$ is $u_{iy} = U_{iy}(\lambda) = \sum_k \Phi_{iyk}\lambda_k$, where the systematic part $U_{iy}(\lambda)$ is just as in formula (3.1.2), but the random utility term ε_{iy} is set to zero. For reasons that will become clear shortly, one shall freeze the value of one of the entries of λ. We do this by adding a regressor $\Phi_0 = (\phi_{iy0})_{iy}$ as an additional column at the left of the $IY \times K$ matrix Φ. We use index $k=0$ to refer to this column and we set the corresponding entry of λ to 1, so $\lambda_0 = 1$ and

$$u_{iy} = (\Phi_0 + \Phi\lambda)_{iy} = \phi_{iy0} + \sum_{k\in[K]} \phi_{iyk}\lambda_k. \tag{3.8.1}$$

The additional regressor (ϕ_{iy0}) will be called a "reference regressor"; it could, for instance, be the dollar cost associated by individual i to alternative y; such a choice would impose the utility to be measured in dollars.

Recall that we observe the choice actually made by the agent in observation i, which is y_i. The *regret* with respect to option y in observation i is therefore equal to the utility that the agent would have obtained if y had been chosen, minus the utility that the agent has actually obtained by choosing y_i, that is,

$$\rho_{iy}^{\lambda} := u_{iy} - u_{iy_i} = (\Phi_0 + \Phi\lambda)_{iy} - (\Phi_0 + \Phi\lambda)_{iy_i}, \tag{3.8.2}$$

and the *maximum regret* in observation i is $\max_y \rho_{iy}^{\lambda}$, a nonnegative quantity. The total maximum regret $\sum_i \max_y \rho_{iy}^{\lambda}$ is the sum over observations of the previous quantity. It is a function of the parameter because it compares the predictions of the model (which depend on the parameter value) with the data (which depend on the observations y_i).

The *minimax regret framework* is an estimation method for the parameter vector λ. It is based on the assumption that agents make choices that maximize their utility, which would lead to $\max_y \rho^{\lambda}_{iy} = 0$ for each observation. As a result, a minimax regret estimator minimizes over λ the maximum regret, and therefore is a solution to

$$\min_{\lambda \in \mathbb{R}^K} \sum_{i \in [I]} \max_{y \in [Y]} \rho^{\lambda}_{iy}, \tag{3.8.3}$$

whose solution is generally not unique, which is why we wrote "a" and not "the" maximum regret estimator. Also note that an inspection of problem (3.8.3) reveals why we fixed the value of the parameter associated with the additional column of regressors ϕ_{iy0} associated with $k=0$ by setting $\lambda_0 = 1$. If we had not done that, the function to minimize, the total regret $\sum_i \max_y \rho^{\lambda}_{iy}$, would have been homogeneous of degree 1, and setting $\lambda_k = 0$ for $k \in [0{:}K]$ would have provided a trivial yet uninformative solution to problem (3.8.3).

Recalling the definition of $\hat{\pi}$ as the vector of choice dummy data $\hat{\pi}_{iy} = 1\{y_i = y\}$, we can rewrite problem (3.8.3) as

$$\min_{\lambda \in \mathbb{R}^K} \sum_{i \in [I]} \max_{y \in [Y]} \left\{ \phi_{iy0} + \sum_{k \in [K]} \phi_{iyk} \lambda_k \right\} - \sum_{iy \in [I \times Y]} \hat{\pi}_{iy} \Big(\phi_{iy0} + \sum_{k \in [K]} \phi_{iyk} \lambda_k \Big) \tag{3.8.4}$$

and we note that if we replace the maximum by the soft maximum (as defined in expression (A.2.10) in appendix A.2.2), problem (3.8.4) becomes a logistic regression—a connection to which we shall return below. For now, let us argue that problem (3.8.3) has a linear programming formulation, which can be obtained by introducing the auxiliary variable $u_i = \max_y \{\phi_{iy0} + \sum_k \phi_{iyk} \lambda_k\}$, which allows one to rewrite the problem as

$$\min_{u \in \mathbb{R}^I, \lambda \in \mathbb{R}^K} \sum_{i \in [I]} u_i - \sum_{iy \in [I \times Y]} \hat{\pi}_{iy} \Big(\phi_{iy0} + \sum_{k \in [K]} \phi_{iyk} \lambda_k \Big) \tag{3.8.5}$$

$$s.t.\ u_i \geq \phi_{iy0} + \sum_{k \in [K]} \phi_{iyk} \lambda_k, \ \forall iy \in [I \times Y],$$

which is dual to a "primal formulation" given by

$$\max_{\pi_{iy} \geq 0} \sum_{iy \in [I \times Y]} \pi_{iy} \phi_{iy0} \tag{3.8.6}$$

$$s.t.\ \sum_{y \in [Y]} \pi_{iy} = 1,$$

$$\sum_{iy \in [I \times Y]} \pi_{iy} \phi_{iyk} = \sum_{iy \in [I \times Y]} \hat{\pi}_{iy} \phi_{iyk} \quad [\lambda_k].$$

The code for the minimax regret estimation procedure is provided in appendix B.3.5.

Small-noise limit. We return to the random utility model as specified in expression (3.1.1), but with a twist: we impose a scaling coefficient σ in front of the random part. As a result, the utility of an individual i choosing alternative y becomes

$$U_{iy} + \sigma \varepsilon_{iy}, \tag{3.8.7}$$

and we are interested in the "small-noise limit," that is, the limit in which σ is small, and eventually tends to zero. Obviously, $\sigma = 0$ is the case described in the previous paragraph when we discussed the minimax regret estimation framework.

If we adopted the parametrization given by expression (3.1.2) in section 3.2, we would get an identification issue, as the model with random utility $\sum_k \phi_{iyk}\lambda_k + \sigma\varepsilon_{iy}$ is observationally equivalent with a model where the random utility is $\sum_k \phi_{iyk}(\lambda_k/\sigma) + \varepsilon_{iy}$. Hence in order to guarantee identification, we shall include a reference regressor for which we will fix the value of the corresponding parameter to 1, and impose, as in specification (3.8.1), that $U = \Phi_0 + \Phi\lambda$. The utility of an individual i choosing alternative y is then

$$\phi_{iy0} + \sum_{k\in[K]} \phi_{iyk}\lambda_k + \sigma\varepsilon_{iy}, \quad (\varepsilon_y) \sim \text{i.i.d. Gumbel}, \tag{3.8.8}$$

from which, following the argument in sections 3.2 and 3.4, we see that the maximum likelihood estimator is given by

$$\max_{\lambda\in\mathbb{R}^K} \left\{ \hat{\pi}^\top (\Phi_0 + \Phi\lambda) - \sigma \sum_{i\in[I]} \log \left(\sum_{y\in[Y]} \exp \frac{(\Phi_0 + \Phi\lambda)_{iy}}{\sigma} \right) \right\}. \tag{3.8.9}$$

As $\sigma \to 0$, the soft maximum in expression (3.8.9) tends to the true maximum, which provides intuition for the fact that $\hat{\lambda}^\sigma$, the maximizer of the regularized problem (3.8.9), converges to a maximizer of the unregularized problem (3.8.3), allowing us to state the following theorem, whose formal proof relies on the gamma convergence of the soft max toward the max:

Theorem 3.8.1. *As $\sigma \to 0$, the maximum likelihood estimator $\hat{\lambda}^\sigma$ in model* (3.8.8) *converges to a solution of* (3.8.3), *which is a minimax regret estimator.*

Computation and the log-sum-exp trick. When computing the logistic regression in the small-noise limit, one faces a well-known numerical

Table 3.1: Low-noise logistic regression. Based on the data simulated using the code of appendix B.3.5 and B.3.6 with various values of σ.

σ	Objective	λ_1	λ_2	λ_3
1	851.223	−0.00639669	−0.105702	0.0236205
0.1	501.289	0.0527662	−0.06673	0.0498178
0.01	496.256	0.0556363	−0.0625979	0.0545422
0.001	496.194	0.0574922	−0.0629233	0.0544917
0.0001	496.192	0.0575033	−0.0622268	0.0549091
1e-05	496.192	0.0574758	−0.0621303	0.0549299
1e-06	496.192	0.0574731	−0.0621207	0.054932
0	496.192	0.0574728	−0.0621196	0.0549322

challenge. As σ is small, the terms in ϕ/σ that appear in the argument of the exponential may become very large positive numbers, provided that the numerator is positive. In this case, the exponentials will "blow up" and will return infinity, and the calculation of the gradient of the objective function will involve infinity at the numerator and at the denominator, and therefore will be indefinite. Fortunately, a simple way out exists, called the "log-sum-exp trick" and based on the idea that $\log(\exp(a-t)+\exp(b-t)) = \log(\exp(a)+\exp(b)) - t$. Applying this idea with $t=\max(a,b)$ ensures that the arguments of the exponentials are always less than or equal to zero, preventing a blowup.

Introducing $m_i = \max_y\{(\Phi_0 + \Phi\lambda)_{iy}\}$ and computing the objective function in (3.8.9) using its equivalent but more numerically stable expression

$$\hat{\pi}^\top(\Phi_0 + \Phi\lambda) - \sum_{i\in[I]} m_i - \sigma \sum_{i\in[I]} \log\left(\sum_{y\in[Y]} \exp\frac{(\Phi_0 + \Phi\lambda)_{iy} - m_i}{\sigma}\right),$$

and returning its gradient still as $\Phi^\top\left(\hat{\pi} - \pi^\lambda\right)$, but where π^λ is numerically computed using its equivalent yet numerically preferable expression,

$$\pi^\lambda_{iy} = \frac{\exp\frac{(\Phi_0+\Phi\lambda)_{iy}-m_i}{\sigma}}{\sum_{z\in[Y]}\exp\frac{(\Phi_0+\Phi\lambda)_{iz}-m_i}{\sigma}}.$$

Demo code is provided in appendix B.3.6, where one can verify numerically the convergence of λ^σ toward a minimax regret estimator computed for the same data as the one generated in appendix B.3.5 in the previous paragraph. Table 3.1 reports results for various small values of σ, including $\sigma = 0$.

3.9 Exercises

Exercise 3.1. *Show that problem* (3.1.7) *is a maximum likelihood estimator if and only if the random vector of utility shocks* (ε_y) *has an i.i.d. Gumbel distribution.*

Exercise 3.2. *The multinomial logistic regression is sometimes presented under a more restrictive form, which assumes that the utility provided by option y to the decision maker i is*

$$U_{iy} = \sum_{l \in [L]} \Psi_{il} \theta_{ly}$$

or, in matrix form, $U = vec_R(\Psi\Theta)$, where U is a column vector of size IY, Ψ is an $I \times L$ matrix of full rank, and Θ denotes the $L \times Y$ matrix of term θ_{ly}. In this specification, the utility associated with each option y is controlled by a different set of parameters, which corresponds to the column y of matrix θ.

(i) *Letting $\lambda = vec_R(\Theta)$, which is a column vector of size LY, show that $U = \Phi\lambda$, where $\Phi = \Psi \otimes \mathbf{I}_Y$.*

(ii) *If R is the $(IY) \times (LY + I)$ matrix given by $R = \begin{pmatrix} \Phi & -\mathbf{I}_I \otimes \mathbf{1}_Y \end{pmatrix}$, show that the rank deficiency of R is L, or, in other words, that the rank of R is $LY + I - L$.*

(iii) *The code in appendix B.3.7 implements the more restrictive version of the multinomial logistic regression based on the* `LogisticRegression` *class of the* `sklearn.linear_model` *module of the* `scikit-learn` *Python library. Reproduce its output using the Poisson regression, that is, using the* `PoissonRegressor` *class.*

Exercise 3.3. *Discuss the existence and identification of the multinomial logistic regressor in the case when there is a single constant regressor term, that is, $K = 1$ and $\phi_{iy1} = 1$.*

Exercise 3.4. *Write the logistic regression of $\hat{\mu}_{txy}$ on ϕ_{txyk} plus y-fixed effects, i.e.,*

$$\pi_{txy}^{(\lambda,\nu)} = \frac{\exp\left(\sum_k \phi_{txyk}\lambda_k - \nu_y\right)}{\sum_y \exp\left(\sum_k \phi_{txyk}\lambda_k - \nu_y\right)}$$

as a Poisson regression. Using the `scikit-learn` *Python library, compute λ and ν numerically based on the code given in appendix B.3.8.*

Exercise 3.5. *Verify the vectorized formula given in* (3.3.1) *for the Hessian* $D^2\ell(\lambda)$ *of the objective function of the logistic regression from section 3.3.*

Exercise 3.6. *Express the linear programs* (3.8.6) *and* (3.8.5) *using a vectorized fashion.*

Exercise 3.7. *Consider the minimax regret estimation procedure discussed in section 3.8.*

(i) *Is the solution* λ *to problem* (3.8.3) *unique? If yes, give a proof; if no, provide an example.*

(ii) *Assume that for each* $\sigma > 0$, *the vector* $\hat{\lambda}^\sigma$ *defined in section 3.8 exists and is unique, and define*

$$\pi^\sigma_{iy} = \frac{\exp(\Phi\hat{\lambda})_{iy}}{\sum_{z\in[Y]}\exp(\Phi\hat{\lambda})_{iz}}$$

to be the predictor of the choice of y by i. Show that the limit π^0_{iy} *of* π^σ_{iy} *as* $\sigma \to 0$ *is unique and show that it is the solution to an optimization problem.*

Exercise 3.8. *Assume that* $I = 2$, $Y = 2$, *and* $K = 2$, *and let*

$$\phi_{iy1} = \begin{pmatrix} 0 & 2 & 0 & 3 \end{pmatrix} \text{ and } \phi_{iy2} = \begin{pmatrix} 4 & 2 & -3 & 0 \end{pmatrix}.$$

Does identification hold? Assume $\hat{\pi}_{iy} = \begin{pmatrix} 0 & 1 & 1 & 0 \end{pmatrix}$. *Does a logistic estimator exist?*

Exercise 3.9. *Assume that we have a panel where* y_{it} *is the dummy for individual* $i \in [I]$ *choosing option* $y \in [Y]$ *at time* $t \in [T]$. *Assume that we have* K *time-independent regressors, so that the regressor matrix is still* $\Phi_{iy,k}$, *but we would like to include time fixed effects. Reformulate this model as a Poisson regression using vectorization and Kronecker products for the new regressor matrix.*

3.10 Problems

Problem 3.1 (Choice of travel mode). *We use the dataset of Greene and Hensher taken from Greene (2000) where 210 individuals reported their choice of* *a travel mode between various cities, and measures of costs (in terms of time and money) associated with each travel alternative. The variables are* `wait`*, the waiting time;* `vcost`*, the in-vehicle cost;* `travel`*, the in-vehicle travel time;* `gcost`*, the generalized cost defined as the sum of* `vcost` *and the value associated with time savings;* `income`*, the income of the household, in thousands of dollars; and* `size`*, the size of the travel group. We assume that we observe individual characteristics that are relevant for individual choices, that is, $U_{iy} = \sum_k \Phi_{iyk}\beta_k$, or, in matrix form, $U = \Phi\beta$, where $\beta \in \mathbb{R}^p$ is a parameter, and Φ is an $IY \times p$ matrix. The random utility associated by individual i to alternative y is therefore $u_{iy} = U_{iy} + \varepsilon_{iy} = \sum_k \Phi_{iyk}\beta_k + \varepsilon_{iy}$, where the entries ε_{iy} are i.i.d. Gumbel. We denote by $\hat{\pi}_{iy}$ the indicator that i chooses alternative y. You should run the code in appendix B.3.9 that loads the dataset, and answer the following questions:*

(i) *With Φ given above, compute the estimator of β obtained by logistic regression by running an optimization procedure.*

(ii) *Same question, but this time using a generalized linear model (GLM) formulation and the* `scikit-learn` *Python library.*

(iii) *Report the confidence interval associated with your estimator.*

(iv) *We consider a slight modification from the model considered above. Assume now $u_{iy} = U_{iy} + T\varepsilon_{iy} = \sum_k \Phi_{iyk}\beta_k + T\varepsilon_{iy}$, where T is an additional parameter, but where we normalize to 1 the last entry of β, that is, $\beta_K = 1$. For various values of T between 0 and 1, compute the estimator of β and report β/T. Why does it fail for small values of T? Report the smallest value of T for which your procedure works.*

(v) *Suggest a fix for the problem with small (but positive) values of T.*

(vi) *Show that the limiting problem for $T \to 0$ is a linear programming problem. Write it and code the resulting estimator.*

Problem 3.2 (Comparing descent algorithms). *On the dataset simulated in appendix B.3.1, use the method* `minimize` *available from the library* `scipy.optimize` *(in the* `scipy` *Python library) in order to compute the logistic regressor $\hat{\lambda}$ using the following methods: (i) Nelder–Mead, (ii) Powell, (iii) conjugate gradient, (iv) BFGS, and (v) Newton-CG. Compare the various numerical performances of these methods.*

3.11 References and Notes

There are many book references for the multinomial logistic regression in econometrics, including the excellent textbooks by Greene (2000), Wooldridge (2010), and Cameron and Trivedi (2005).

A reference for the method of moments approach described in section 3.1 is Gourieroux, Monfort, and Trognon (1984). Theorem 3.1.1 is a generalization of the maximum entropy principle (MaxEnt) developed by the physicist E. T. Jaynes in the 1960s; see Jaynes (1957). The foundations of Poisson regression—and, more broadly, of generalized linear models—are covered in the monographs by Agresti (2012) and McCullagh and Nelder (1989). For econometric applications, see Winkelmann (2008) and Cameron and Trivedi (2013). The analysis presented in section 3.2 was pioneered by McFadden's 1974 study of the conditional logit model.

Section 3.3 contains basic notions of convex optimization by gradient descent. Textbook references abound; we recommend the classic texts by Luenberger and Ye (2015) and Boyd and Vandenberghe (2004), and the more recent one by Carlier (2022).

References for the link between the Poisson regression and the multinomial logistic regression made in section 3.4 include the papers by Palmgren (1981), Baker (1994), Lang (1996), and Guimarães (2004). The connection is popular among the machine learning community, where it is known as the "Poisson trick."

Early references for section 3.6 are the work of Albert Verbeek (1989, 1992); more recently, and in connection with the gravity equation in international trade, see the paper by Santos Silva and Tenreyro (2011), as well as the one by Zylkin (2019).

The minimax loss criterion was pioneered by Savage (1951). Regret analysis is often used in machine learning, in particular in the context of online learning; see Bubeck and Cesa-Bianchi (2012). A recent application to economics is the paper by Kitagawa, Lee, and Qiu (2022). I am not aware of existing references for the minimax estimation procedure for discrete choice models described in section 3.8, nor for its connection to maximum likelihood estimation in the logit model. This does not necessarily imply that the idea is original, as it is a natural and intuitive one.

Shrinkage estimators discussed in section 3.7 are an old idea in statistics; an early reference is the paper by James and Stein (1961), who introduced shrinkage estimation. The interest of the L1 penalization (LASSO) in relation with sparsity was pioneered in Tibshirani (1996). A well-known textbook reference is Hastie, Tibshirani, and Friedman (2009). A textbook reference on proximal gradient descent algorithms is Bauschke and Combettes (2017).

3.12 Summary

SUMMARY OF CHAPTER 3

In this chapter, we consider parametric random utility models and their estimation. In particular, we focus on logistic regression, its estimation, inference, and computation.

Section 3.1 considers a random utility model where the utility that an agent derives from choosing a particular option is linearly parametrized by observable characteristics and a vector of parameters. This formulation allows for the systematic part of the utility to be modeled and estimated. In this section, one does not restrict the distribution of ε_i, the vector of utility shocks drawn by individual $i \in [I]$. It shows that the utilities with a linear parametrization can be estimated either by maximum likelihood estimation (3.1.4) based on the observed conditional choice probabilities, or by a method of moments (3.1.6) to match the predicted and observed moments. While maximum likelihood inference is statistically efficient, the method of moments is computationally attractive as it relies on a convex optimization problem, whose dual formulation, spelled out in theorem 3.1.1, has a nice interpretation as the maximization of a generalized entropy under moment matching constraints.

In section 3.2, we specify the distribution of ε_i to be the Gumbel distribution, leading to the conditional logit model. This model forms the basis for the logistic regression. In that case, proposition 3.2.1 shows that the maximum likelihood estimation procedure coincides with the method of moments estimation procedure, and we call that method the multinomial logistic regression. The asymptotic distribution for the multinomial logistic regression is given in theorem 3.2.1.

Computational methods for estimating logistic regression parameters are discussed in section 3.3, where the use of gradient descent techniques for maximizing the likelihood function is highlighted.

In section 3.4, we explore the connection between the multinomial logistic and generalized linear models, specifically the Poisson regression. Theorem 3.4.1 shows that the logistic regression can be estimated as a Poisson regression with individual fixed effects. We can use that equivalence to estimate the logistic regression using generalized linear model libraries.

Section 3.5 discusses conditions necessary for the unique identification of the logistic regression model's parameters. Identification issues can arise when the model is not specified correctly or when certain data configurations occur.

We next provide in section 3.6 necessary and sufficient conditions for the existence of a solution to the multinomial regression. Theorem 3.6.1

provides a criterion based on linear programming, which we implement in section B.3.3 of the appendix.

To address overfitting and improve model generalization, section 3.7 discusses incorporating regularization techniques into the estimation process, such as LASSO (L1 regularization) and ridge regression (L2 regularization). We motivate and discuss the regularization of the multinomial logistic regression, in particular in the LASSO case.

Lastly, in section 3.8, we look at the limit of the logistic regression procedure in the zero-noise limit, that is, when the scale of the heterogeneity in front of the Gumbel error term tends to zero. We show in theorem 3.8.1 that the logistic estimator, which is a maximum likelihood estimator, tends in the zero-noise limit toward a minimax regret estimator, which can be computed as a linear programming problem.

CHAPTER 4

Characteristics-Based Models

As we saw in chapter 2, models based on multivariate extreme value distributions like the nested logit model are a first way out of the limitations of the logit model: going back to the discussion at the start of section 2.2 on the demand for apartments, one may assume that a first nest contains smaller apartments, and another one contains larger apartments, and that individuals then refine their choices within the nest. However, the distinction between "small" and "large" apartments will be arbitrary; also this approach prevents us from realistically modeling trade-offs between size and other amenities. Instead, it may make more sense to assume that while apartments differ by their size, agents differ by their *valuation* of size, in which case the random utility shock associated with an apartment depends on the interaction between the size of the good and valuation of size by the decision-maker. A simple way of capturing this interaction would be to write that the random utility shock equals the valuation of a unit of size multiplied by the size of the good. Goods will usually have multidimensional characteristics, whose interaction with the consumer's own features will generate utility. This is the idea of the *characteristics approach*, pioneered by Lancaster in the 1960s, which we are about to describe. One of the benefits of the approach is that it allows us to predict demand for new goods, which the standard logit model would not permit. In Lancaster's forceful words, quoted from his 1966 *Journal of Political Economy* paper: "Perhaps the most important aspects of consumer behavior . . . are those of consumer reactions to new commodities and to quality variations. Traditional theory has nothing to say on these. In the case of new commodities, the theory is particularly helpless. We have to expand from a commodity space of dimension n to one of dimension $n+1$, replacing the old utility function by a completely new one, and even a complete map of the consumer's preferences among the n goods provides absolutely no information about the new preference map. A theory which can make no use of so much information is a remarkably empty one."

4.1 The Pure Characteristics Model

The Lancasterian framework, which we call the "pure characteristics model" following the terminology of Berry and Pakes, assumes that the vector of characteristics of option y, ξ_y, is an M-dimensional vector whose entries are ξ_{ym}, and η_i, also an M-dimensional vector, is the vector of valuations of the respective characteristics by agent i, such that η_{im} is i's valuation of a unit of characteristic m. We assume that the M-dimensional random vector η_i, which is sometimes called a "utility shifter," has a distribution $\mathbf{P}_\eta$ that can in general depend on the observable characteristics of i. An agent $i \in [I]$ draws a utility shock ε_{iy} associated with option y that is therefore given by the sum of the valuations of the various characteristics of option y,

$$\varepsilon_{iy} = \sum_{m \in [M]} \eta_{im} \xi_{ym}.$$

Hence, if η stands for the M-dimensional random vector of distribution $\mathbf{P}_\eta$, and if $\varepsilon \sim \mathcal{P}$ denotes the Y-dimensional random vector of utility shocks associated with each option, then one has

$$\varepsilon = \boldsymbol{\xi} \eta, \tag{4.1.1}$$

where $\boldsymbol{\xi}$ is the $Y \times M$ matrix of term ξ_{ym}. When the distribution of η is Gaussian, this model is the well-known *probit model.*

Example 4.1 (Probit model). When $\mathbf{P}_\eta$ is the $\mathcal{N}(0, \Sigma)$ distribution, then the pure characteristics model is called a probit model; in this case, $\varepsilon \sim \mathcal{N}(0, \boldsymbol{\xi} \Sigma \boldsymbol{\xi}^\top)$.

It will also be interesting to consider the case where the distribution of η is uniform over the hypercube. We call this model the "uniform cubic model," but this is not a particularly standard terminology.

Example 4.2 (Uniform cubic model). When $\mathbf{P}_\eta = \mathcal{U}([0,1]^M)$, namely, when the components η_m of the random vector $\eta \sim \mathbf{P}_\eta$ are independent and all uniformly distributed over the unit interval, the model is called the *uniform cubic model.*

Note the distribution ε will not necessarily have full support, so assumption 1.2 does not necessarily hold. In particular, in the probit model, ε will not have full support unless $M \geq Y$ and $\boldsymbol{\xi}$ is of full rank. For the uniform cubic coefficient model, assumption 1.2 never holds, as the support of ε is compact.

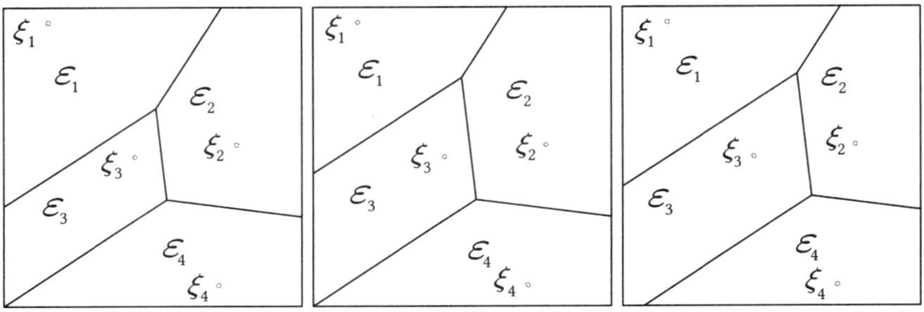

Figure 4.1: Laguerre diagrams, showing the cells $\mathcal{E}_y$ when $Y=4$, $M=2$, and the support of $\mathbf{P}_\eta$ is $[0,1]^2$. The circles are the locations of the ξ_y. From the left to the center, U_1 has decreased, which is why the frontier between $\mathcal{E}_1$ and both $\mathcal{E}_2$ and $\mathcal{E}_3$ has moved closer to ξ_1. On the right, the utilities U have been adjusted so that $U=\boldsymbol{\pi}^{-1}(\mathbf{1}_Y/Y)$, which means that all cells $\mathcal{E}_y$ have the same area.

Recall that for $y\in[Y]$, ξ_y is the M-dimensional vector whose mth entry is ξ_{ym}. Given U a vector of systematic utilities, we let $\mathcal{E}_y\subset\mathbb{R}^M$ be the set of vectors of utility-shifters $\eta\in\mathbb{R}^M$ that lead the consumer to prefer y weakly over the other options. We have

$$\mathcal{E}_y=\left\{\eta\in\mathbb{R}^M: U_y+\xi_y^\top\eta\geq U_{y'}+\xi_{y'}^\top\eta,\ \forall y'\in[Y]\right\}. \tag{4.1.2}$$

The division of $\mathbb{R}^M$ into the cells $\mathcal{E}_y$ is called a *Laguerre diagram*, or *power diagram*, and each $\mathcal{E}_y$ is called a *Laguerre cell.* Various Laguerre diagrams corresponding to the same $\boldsymbol{\xi}$ and $\mathbf{P}_\eta$, but different values of the systematic utilities U_y, are displayed in figure 4.1, which was generated using the code in appendix B.4.1.

Several remarks are in order:

- If we replace the weak inequality by a strict inequality in definition (4.1.2), we get the set of vectors of utility-shifters η that lead to preferring y strictly, which is simply the interior of $\mathcal{E}_y$. If $\mathbf{P}_\eta$ has a density, then the probability of being indifferent is zero, and one should not be concerned about the distinction between "strict" versus "weak" preference.
- It is easy to see that $\mathcal{E}_y$ is the intersection over all alternatives $y'\in[Y]$ of the set of η that prefer option y to alternative y'; this set, which is defined by $(\xi_{y'}-\xi_y)^\top\eta\leq U_y-U_{y'}$, is a half-space, and as a result $\mathcal{E}_y$, which is the intersection of a finite number of half-spaces, is a convex polyhedron.

- If one decreases the value of U_y for some $y \in [Y]$ holding the other entries of U fixed, then the set of η that induces indifference between alternatives y and y', which is a hyperplane orthogonal to $\xi_y - \xi_{y'}$, will undergo parallel displacement to move closer to y; therefore $\mathcal{E}_y$ will shrink, and the adjacent cells will increase. This is a situation we observe in figure 4.1, where the value of U_1 is lower on the left pane than on the center pane, while the other entries of U remain constant.

By substitution of the expression of ε given by (4.1.1) into the welfare function G defined in (1.1.2), one has the following expression:

$$G(U) = \mathbb{E}\Big[\max_{y \in [Y]} \{U_y + (\boldsymbol{\xi}\eta)_y\}\Big] = \sum_{y \in [Y]} \boldsymbol{\pi}_y(U) \left(U_y + \xi_y^\top \bar{\eta}^y\right), \qquad (4.1.3)$$

where we define $\bar{\eta}^y := \mathbb{E}_{\mathbf{P}_\eta}[\eta \mid \eta \in \mathcal{E}_y]$ to be the barycenter of $\mathcal{E}_y$ with respect to the distribution $\mathbf{P}_\eta$. Therefore, only two quantities are needed in order to compute G: first, the probability weight $\boldsymbol{\pi}_y(U)$ that $\mathbf{P}_\eta$ assigns to each Laguerre cell; and second, the barycenter $\bar{\eta}^y$ of each Laguerre cell with respect to $\mathbf{P}_\eta$. These quantities are more or less easy to compute numerically depending on the nature of the distribution $\mathbf{P}_\eta$. In general we need to resort to simulation; in some cases, such as in the probit model, some techniques such as importance sampling are available to improve on naive simulations, as seen in example 4.1, and in some situations, such as for the uniform cubic model of example 4.2, there are exact methods available.

Example 4.1 (Probit model, continued). In the case of the probit model, the *Geweke–Hajivassiliou–Keane simulator*, based on importance sampling ideas described in appendix A.2.3, consists of the following steps.

First, note that computing the market shares in the probit model amounts to computing the c.d.f. of a multinomial Gaussian variable. Indeed, letting $\pi_y = \boldsymbol{\pi}_y(U)$, we note that π_y is the probability that $\varepsilon_{y'} - \varepsilon_y \leq U_y - U_{y'}$ for all $y' \in [Y] \backslash \{y\}$, and therefore, defining $\epsilon_{y'} = \varepsilon_{y'} - \varepsilon_y$ and $z_{y'} = U_y - U_{y'}$ for $y' \in [Y] \backslash \{y\}$, π_y is equal to $F_\epsilon(z)$, the c.d.f. of the $(Y-1)$-dimensional random vector ϵ evaluated at $z \in \mathbb{R}^{Y-1}$.

To ease exposition, let us momentarily restrict ourselves to the case $Y = 3$, so that $\epsilon = (\epsilon_1, \epsilon_2)$ is a bivariate Gaussian vector with variances σ_1^2 and σ_2^2, and correlation ρ. In that case, we can use the i.i.d. factor representation of ϵ using $(g_1, g_2) \sim \mathcal{N}(0, \mathbf{I}_2)$:

$$\begin{cases} \epsilon_1 = \sigma_1 g_1, \\ \epsilon_2 = \sigma_2 \rho g_1 + \sigma_2 \sqrt{1-\rho^2}\, g_2. \end{cases}$$

Then, if we define $p = \Pr(\epsilon_1 \le z_1) = \Phi(z_1/\sigma_1)$, where $\Phi(\cdot)$ is the c.d.f. of the standard normal variable and $\varphi(x) = \Phi\big((z_2 - \rho\sigma_2\Phi^{-1}(x))/(\sqrt{1-\rho^2}\,\sigma_2)\big)$, the reader can verify that

$$F_\epsilon(z) = \mathbb{E}[\Pi_1] \text{ where } \Pi_1 = 1\left\{X \le p\right\}\varphi(X) \text{ and } X \sim \mathcal{U}([0,1]). \tag{4.1.4}$$

But $\mathbb{E}[\Pi_1] = p\mathbb{E}\left[\varphi(X)1\left\{X \le p\right\}/p\right] = p\mathbb{E}[\varphi(X) \mid X \le p] = p\mathbb{E}\left[\varphi\left(pX\right)\right]$, and thus

$$F_\epsilon(z) = \mathbb{E}[\Pi_2] \text{ where } \Pi_2 = p\varphi\left(pX\right) \text{ and } X \sim \mathcal{U}([0,1]). \tag{4.1.5}$$

However, $\operatorname{Var}(\Pi_1) = p\mathbb{E}\left[\varphi^2\left(pX\right)\right] - \mathbb{E}[\Pi_1]^2$ and $\operatorname{Var}(\Pi_2) = p^2\mathbb{E}\left[\varphi^2\left(pX\right)\right] - \mathbb{E}[\Pi_1]^2$, as the reader shall verify, and it follows that

$$F_\epsilon(z) = \mathbb{E}[\Pi_1] = \mathbb{E}[\Pi_2] \text{ and } \operatorname{Var}(\Pi_1) > \operatorname{Var}(\Pi_2). \tag{4.1.6}$$

Using a random number generator, which is a sequence (x_i) of I numbers independently sampled from the uniform distribution over the unit interval, one can then approximate the value of $F_\epsilon(z)$ by two different ways of simulation. The accept-reject simulator of $\Pr(\epsilon_1 \le z_1, \epsilon_2 \le z_2)$, based on simulating (4.1.4), is

$$\frac{1}{I}\sum_{i\in[I]} 1\left\{x_i \le \Phi\left(\frac{z_1}{\sigma_1}\right)\right\}\Phi\left(\frac{z_2 - \rho\sigma_2\Phi^{-1}(x_i)}{\sqrt{1-\rho^2}\,\sigma_2}\right), \tag{4.1.7}$$

while the GHK estimator, based on simulating formula (4.1.5), is

$$\frac{1}{I}\sum_{i\in[I]} \Phi\left(\frac{z_1}{\sigma_1}\right)\Phi\left(\frac{z_2 - \rho\sigma_2\Phi^{-1}\left(x_i\Phi\left(\frac{z_1}{\sigma_1}\right)\right)}{\sqrt{1-\rho^2}\,\sigma_2}\right), \tag{4.1.8}$$

and inequality (4.1.6) ensures that the GHK simulator (4.1.8) converges faster than the accept-reject simulator (4.1.7).

In the general case ($Y \ge 3$), $F_\epsilon(z)$ is computed by simulation using a combination of lemma A.2.3 and the expression of the Rosenblatt quantile of the multivariate normal distribution given in example A.2. As a result, setting $J = Y - 1$, the GHK estimator of $F_\epsilon(z)$ is given by

$$\frac{1}{I}\sum_{i\in[I]}\prod_{j\in[J]}\Phi\left(\frac{z_j - L_{j1}\Phi^{-1}(\hat{x}_{i1}) - \cdots - L_{j,j-1}\Phi^{-1}(\hat{x}_{i(j-1)})}{L_{jj}}\right),$$

where the lower triangular matrix L is the Cholesky factor of Σ_ϵ, the variance-covariance matrix of ϵ, and where the expression of $\hat{x}_{ij}$ is recursively given by

$$\begin{cases} \hat{x}_{i1} = x_{i1}\Phi\left(\frac{z_1}{L_{11}}\right), \; i \in [I], \\ \hat{x}_{ij} = x_{ij}\Phi\left(\frac{z_j - L_{j1}\Phi^{-1}(\hat{x}_{i1}) - \cdots - L_{j,j-1}\Phi^{-1}(\hat{x}_{i(j-1)})}{L_{jj}}\right), \; i \in [I], \; j \in [2:J-1] \end{cases}$$

where, in turn, the x_{ij} terms are i.i.d. draws from the uniform distribution over $[0, 1]$. Demo code is given in appendix B.4.2, where the accept-reject simulator is compared with the GHK simulator. In appendix B.4.3, the GHK simulator is used to compute the market share map in the probit model.

As indicated, dedicated methods can be used in the case of the uniform cubic model. These originate from the field of computational geometry, which provides algorithms to compute the volume and barycenter of any *polytope*—that is, a bounded polyhedron. Since the underlying measure is the Lebesgue measure, these algorithms apply directly to the distribution of η in the uniform cubic model, which is the uniform measure over $[0, 1]^M$.

Example 4.2 (Uniform cubic model, continued). In the demo code provided in appendix B.4.1, we use the library `sdot` by Hugo Leclerc and Quentin Mérigot. Computational geometry libraries allow the computation of Laguerre (power) diagrams. In their simplest form, these algorithms typically involve three steps: *vertex enumeration*, *triangulation*, and *aggregation*.

In the **vertex enumeration** step, given the Y vectors of M-dimensional characteristics ξ_y and the systematic utilities U_y, one computes the extreme points of the Laguerre cells $\mathcal{E}_y$. The Laguerre cells as defined in (4.1.2) are in a so-called *half-space representation*, or H-representation, as they are defined as the intersection of the half-spaces representing the vector of utility-shifters η that prefer y over y'. By enumerating the extreme points of the Laguerre cells, we obtain another representation called the *vertex representation* or V-representation, as the Laguerre cell is also the convex hull of its extreme points (or vertices). A Python library that performs vertex enumeration is, for example, the `pypoman` library by Stéphane Caron. The precise description of these algorithms is beyond our scope here, but, broadly speaking, they go from a vertex to an adjacent one using the pivot method from the simplex algorithm.

In the **triangulation** step, the Laguerre cells (based on their V-representation) are subdivided into elementary triangles (in dimension 2) or more generally into elementary simplices (in higher dimension), which do not intersect, except on the boundary. Thus, the Laguerre cells are now represented as the union of nonoverlapping triangles. There are simple formulas such as (A.2.1) that allow for the computation of the Lebesgue measure of these triangles, as well as their barycenters.

Lastly, in the **aggregation** step, the measures of the elementary triangles are summed up and their barycenters are averaged, which gives the measure and the barycenter of each Laguerre cell.

One can obtain the expression of the entropy of choice $G^*(\pi)$ directly from its definition (1.3.3) by

$$G^*(\pi) = \max_{U \in \mathbb{R}^Y} \left\{ \sum_{y \in [Y]} \pi_y U_y - \mathbb{E}\left[\max_{y \in [Y]} \left\{ U_y + (\boldsymbol{\xi}\eta)_y \right\} \right] \right\}, \tag{4.1.9}$$

and it follows from theorem 1.3.3 that if $\pi_y \geq 0$ is a vector of market shares (so that $\sum_y \pi_y = 1$), the value of the entropy of choice $G^*(\pi)$ is given (up to a minus sign) by the following optimal transport problem:

$$-G^*(\pi) = \max_{\lambda \in \mathcal{M}(\mathbf{P}_\eta, \mathcal{P}_{\xi,\pi})} \mathbb{E}_\lambda \left[\xi^\top \eta \right], \tag{4.1.10}$$

where $\mathbf{P}_\eta$ is the distribution of the M-dimensional random vector η, $\mathcal{P}_{\xi,\pi}$ is the distribution of the M-dimensional random vector $\xi_{\tilde{y}}$ when $\tilde{y} \sim \pi$, λ is the joint distribution of (η, ξ), and, we recall, $\mathcal{M}(\mathbf{P}_\eta, \mathcal{P}_{\xi,\pi})$ is defined as the set of distributions that are such that $\eta \sim \mathbf{P}_\eta$ and $\xi \sim \mathcal{P}_{\xi,\pi}$. More precisely, this is called a *semi-discrete* optimal transport problem, as the distribution $\mathbf{P}_\eta$ is continuous while the distribution $\mathcal{P}_{\xi,\pi}$ is discrete.

Computation by simulation. Following the approach sketched in section 1.4, we can draw I samples of $\mathbf{P}_\eta$, and if we denote by η_{im} the mth dimension of the utility-shifter in draw i, we recall that the random utility is given by $\varepsilon_{iy} = \sum_m \eta_{im} \xi_{ym}$. The sample analog of the entropy of choice is given by the linear programming problem

$$\begin{aligned} -G_I^*(\pi) = \max_{\lambda_{iy} \geq 0} & \sum_{iy \in [I \times Y]} \lambda_{iy} \varepsilon_{iy} \\ s.t. & \sum_{i \in [I]} \lambda_{iy} = \pi_y, \; y \in [Y], \\ & \sum_{y \in [Y]} \lambda_{iy} = \frac{1}{I}, \; i \in [I], \end{aligned} \tag{4.1.11}$$

whose dual expression is

$$\begin{aligned} & \min_{u_i, U_y} \frac{1}{I} \sum_{i \in [I]} u_i - \sum_{y \in [Y]} \pi_y U_y \\ & s.t. \; u_i - U_y \geq \varepsilon_{iy}, \; iy \in [I \times Y], \end{aligned} \tag{4.1.12}$$

allowing one to recover U_y as a solution to the linear program (4.1.12).

Continuous extension of the pure characteristics model. An advantage of the characteristics approach is that it allows for a straightforward extension to the case where the set of options is continuous, much more so than in the logit case, for which the subtleties associated with building a continuous logit model were discussed in section 2.5. Assume that the set $\mathcal{Y}$ of options y is infinite and continuous, and large enough so that the set $\{\xi_y, y \in \mathcal{Y}\}$ coincides with $\mathbb{R}^M$. In that case one may redefine $U(\xi)$ as the systematic utility U_y associated with the option y such that $\xi = \xi_y$, and the indirect utility of a consumer with heterogeneity η is

$$u(\eta) = \max_{\xi \in \mathbb{R}^M} \{U(\xi) + \xi^\top \eta\}, \tag{4.1.13}$$

which is well defined, and does not require any renormalization. The function u so defined is convex, and the envelope theorem implies that the quality vector ξ chosen by the consumer with heterogeneity η is $\xi = \nabla u(\eta)$.

4.2 Random Coefficient Logit Specification

In the random coefficient logit model, also called the mixed logit model, developed by Berry, Levinsohn, and Pakes in the 1990s, the random utility term ε_y is a weighted sum of a pure characteristic term and a logit term. Hence a random coefficient logit utility vector $\varepsilon \sim \mathcal{P}$ has by definition the following expression:

$$\varepsilon = \boldsymbol{\xi} \eta + \sigma \epsilon,$$

where $\eta \sim \mathbf{P}_\eta$ is an M-dimensional vector of utility-shifters, and $\epsilon \sim \mathcal{G}$ is a random vector of i.i.d. Gumbel terms, independent of the term η. As a result, the welfare function G has expression $G(U) = \mathbb{E}[\max_y \{U_y + (\boldsymbol{\xi}\eta)_y + \sigma \epsilon_y\}]$, and by the law of iterated expectations, one can compute that expectation by taking expectations on ϵ first and then on η, which yields

$$G(U) = \mathbb{E}\left[\sigma \log \sum_{y \in [Y]} \exp\left(\frac{U_y + (\boldsymbol{\xi}\eta)_y}{\sigma}\right)\right]. \tag{4.2.1}$$

The following result establishes a precise connection between the random coefficient logit model and the framework of entropic optimal transport described in appendix A.1.8:

Proposition 4.2.1. *The value of the entropy of choice associated with the random coefficient logit model coincides (up to a sign inversion) with the value of the entropic regularized optimal transport problem,*

$$-G^*(\pi) = \max_{\lambda \in \mathcal{M}(\mathbf{P}_\eta, \mathcal{P}_{\xi,\pi})} \left\{\mathbb{E}_\lambda\left[\eta^\top \xi\right] - \sigma \mathbb{E}_\lambda\left[\log \lambda(\eta, \xi)\right]\right\}. \tag{4.2.2}$$

As in the pure characteristics model, if (η_{im}) are I realizations of M-dimensional vectors sampled from the distribution $\mathbf{P}_\eta$, and if one lets $\nu_{iy} = \sum_m \eta_{im}\xi_{ym}$, the sample analogs of the welfare function and the market share map are respectively given by

$$\begin{cases} G_{I,\sigma}(U) = \frac{\sigma}{I} \sum_{i\in[I]} \log \sum_{y\in[Y]} \exp((U_y + \nu_{iy})/\sigma), \\ (\boldsymbol{\pi}_{I,\sigma})_y(U) = \frac{1}{I} \sum_{i\in[I]} \frac{\exp((U_y+\nu_{iy})/\sigma)}{\sum_{y'\in[Y]} \exp((U_{y'}+\nu_{iy'})/\sigma)}, \end{cases} \tag{4.2.3}$$

and the entropy of choice in the random coefficient logit model is given as a solution to the sample analog of the regularized optimal transport problem (4.2.2), whose primal formulation is

$$\begin{aligned} -G^*_{I,\sigma}(\pi) = \max_{\lambda_{iy}\geq 0} & \left\{ \sum_{iy\in[I\times Y]} \lambda_{iy}\nu_{iy} - \sigma \sum_{iy\in[I\times Y]} \lambda_{iy} \log \lambda_{iy} \right\} \\ s.t. \quad & \sum_{i\in[I]} \lambda_{iy} = \pi_y, \; y \in [Y], \\ & \sum_{y\in[Y]} \lambda_{iy} = \frac{1}{I}, \; i \in [I], \end{aligned} \tag{4.2.4}$$

which has an equivalent dual expression equal to

$$-G^*_{I,\sigma}(\pi) = \min_{u_i,U_y} \left\{ \frac{1}{I} \sum_{i\in[I]} u_i - \sum_{y\in[Y]} \pi_y U_y + \sigma \sum_{iy\in[I\times Y]} \exp\left(\frac{\nu_{iy} - u_i + U_y}{\sigma}\right) \right\}, \tag{4.2.5}$$

allowing one to recover U_y as a solution to the problem. We can use the iterated proportional fitting procedure, also called Sinkhorn's algorithm and described in appendix A.1.8, in order to compute U. For this, we come up with an initial guess of U^0_y and we iterate

$$\begin{cases} u_i^{t+1} = \sigma \log\left(I \sum_{y\in[Y]} \exp((\nu_{iy} + U^t_y)/\sigma)\right), \\ U^{t+1}_y = -\sigma \log\left(\frac{1}{\pi_y} \sum_{i\in[I]} \exp((\nu_{iy} - u_i^{t+1})/\sigma)\right) \end{cases} \tag{4.2.6}$$

over $t \geq 0$. Alternatively, we can combine both steps, and run what Berry, Levinsohn, and Pakes call the *contraction mapping algorithm*, which consists of iterating

$$U^{t+1}_y = -\sigma \log\left(\frac{1}{I\pi_y} \sum_{i\in[I]} \frac{\exp(\nu_{iy}/\sigma)}{\sum_{y'\in[Y]} \exp((\nu_{iy'} + U^t_{y'})/\sigma)} \right). \tag{4.2.7}$$

Theorem A.1.9 in appendix A.1.8 implies that the solution $(u^\star_i, U^\star_y)$ to (4.2.5) is unique up to a constant, and that the iteration given in (4.2.7) is a contraction for the Hilbert metric, that is, $d_H(\exp(U^t_y/\sigma), \exp(U^\star_y/\sigma)) \leq K^t d_H(\exp(U^0_y/\sigma), \exp(U^\star_y/\sigma))$, where the expression of K is given in appendix A.1.8. If we pick $y^* \in [Y]$ and take the normalization $U_{y^*} = 0$, and if we define $\tilde{U}^t_y = U^t_y - U^t_{y^*}$, we have

$$|\tilde{U}^t_y - U^\star_{y^*}| \leq \sigma\, d_H(\exp(U^t_y/\sigma), \exp(U^\star_y/\sigma)),$$

and therefore $|\tilde{U}^t_y - U_{y^*}| \leq K^t C$ for $C = \sigma\, d_H(\exp(U^0_y/\sigma), \exp(U^\star_y/\sigma))$. As a result, $\tilde{U}^t_y = U^t_y - U^t_{y^*} \to U^\star_y$ as $t \to +\infty$, and we can summarize the above discussion into:

Theorem 4.2.1. *The solution* $(u^\star_i, U^\star_y)$ *to* (4.2.5) *exists and is unique up to a constant. Further, as* $t \to +\infty$, $\tilde{U}^t_y = U^t_y - U^t_{y^\star}$ *converges to* $U^\star_y$ *at rate* K^t.

Code implementing the IPFP is provided in appendix B.5.2.

Remark 4.2.1 (Use of the Gauss–Hermite quadrature). The previous discussion did not make specific assumptions on $\mathbf{P}_\eta$, the distribution of the vector η of utility-shifters. We simply assumed that this distribution could be sampled conveniently, which led to the simulated version (4.2.4). However, when $\mathbf{P}_\eta$ is a Gaussian distribution, we can improve the accuracy of the simulation by resorting to the Gauss–Hermite quadrature, as described in appendix A.2.4. Assume for simplicity that $M = 1$ and $\mathbf{P}_\eta$ is the $\mathcal{N}(0,1)$ distribution. In that case, we can use the quadrature to find a better approximation of $G(U)$ in (4.2.1) than the one provided by naive Monte Carlo; indeed, we can use formula (A.2.19) with $N = I$ and $f(\eta) = \sigma \log \sum_y \exp((U_y + \xi_y \eta)/\sigma)$, which means that we shall take as sample points the η_i, $i \in [I]$, which are the roots of H_I, the Hermite polynomial of degree I, and weights w_i given by

$$w_i = \frac{2^{I-1} I!}{I^2 \left(H_{I-1}(\eta_i)\right)^2}, \tag{4.2.8}$$

in which case the expression of G is modified from (4.2.3) into

$$\begin{cases} G_{I,\sigma}(U) = \sigma \sum_{i \in [I]} w_i \log \sum_{y \in [Y]} \exp((U_y + \xi_y \eta_i)/\sigma), \\ (\boldsymbol{\pi}_{I,\sigma})_y(U) = \sum_{i \in [I]} w_i \frac{\exp((U_y + \xi_y \eta_i)/\sigma)}{\sum_{y' \in [Y]} \exp((U_{y'} + \xi_{y'} \eta_i)/\sigma)}, \end{cases} \tag{4.2.9}$$

and the entropy of choice is (better) approximated by

$$-G^*_{I,\sigma}(\pi) = \max_{\lambda_{iy} \geq 0} \left\{ \sum_{iy \in [I \times Y]} \lambda_{iy} \xi_y \eta_i - \sigma \sum_{iy \in [I \times Y]} \lambda_{iy} \log \lambda_{iy} \right\} \tag{4.2.10}$$

$$\text{s.t.} \sum_{i\in[I]} \lambda_{iy} = \pi_y, \ y \in [Y],$$

$$\sum_{y\in[Y]} \lambda_{iy} = w_i, \ i \in [I].$$

4.3 Demand Estimation with Endogenous Characteristics

In this section we focus on the case when some product characteristics are observed by agents, but not by the analyst. A particular concern is endogeneity, which happens when this unobserved component is correlated with another characteristic included in the regressors. Endogeneity may lead to biased parametric estimates, in a way that we now discuss. This section will introduce step-by-step, and on a small example, the main ideas of the method of structural estimation of Berry, Levinsohn, and Pakes, which will be described in full in the next section.

Assume for the sake of the discussion that there is only one unit good, which leaves the consumer with two options, not purchasing (the default option) or purchasing (the alternative). The good is present in T markets, and the price of the good in market $t \in [T]$ is p_t (we drop the subscript y as there is only one good). A consumer i in market t associates a random utility $\lambda_1 + \lambda_2 p_t + \epsilon_{ti1}$ with getting the good, and a reservation utility ϵ_{ti0} with not getting it, where ϵ_{ti0} and ϵ_{ti1} are independent and have a standard Gumbel distribution. The logit model predicts that the market share of the good in market t is

$$\pi_t^\lambda = \frac{\exp(\lambda_1 + \lambda_2 p_t)}{1 + \exp(\lambda_1 + \lambda_2 p_t)},$$

and if the observed market share of the (unique) good in market t is $\hat{\pi}_t$, then we can follow the approach suggested in chapter 3 and estimate λ_1 and λ_2 by logistic regression,

$$\max_{(\lambda_1,\lambda_2)\in\mathbb{R}^2} \sum_{t\in[T]} \hat{\pi}_t(\lambda_1 + \lambda_2 p_t) - \sum_{t\in[T]} \log(1 + \exp(\lambda_1 + \lambda_2 p_t)),$$

which, as explained in chapter 3, provides the maximum likelihood estimator $\hat{\lambda}$.

Unobservable quality characteristics. In the model presented above, goods share the same observable characteristics across markets. However, we would like to allow for variation in their quality. To that end, assume that there is a market-specific quality characteristic κ_t that uniformly affects each consumer's valuation for the good in market t, but that it is not observed by the econometrician, so that the utility associated with purchasing the good becomes $\lambda_1 + \kappa_t + \lambda_2 p_t + \epsilon_{ti1}$, while the utility of not possessing the good is still ϵ_{ti0}. If we are unwilling to assume a specific distribution for κ, but if we assume that κ_t is centered and orthogonal with p_t, we get two moment conditions, $\mathbb{E}[\kappa_t] = 0$ and $\mathbb{E}[\kappa_t p_t] = 0$. If we define the systematic utility associated with the good $U_t = \lambda_1 + \lambda_2 p_t + \kappa_t$, so that we have $\kappa_t = U_t - \lambda_1 - \lambda_2 p_t$, we get from the logit structure that U_t is obtained from the market shares by $\hat{U}_t = \ln(\hat{\pi}_t/(1-\hat{\pi}_t))$. In that case, the two moment conditions $\mathbb{E}[U_t - \lambda_1 - \lambda_2 p_t] = 0$ and $\mathbb{E}[(U_t - \lambda_1 - \lambda_2 p_t)p_t] = 0$ have empirical analogs

$$\sum_{t\in[T]} (\hat{U}_t - \lambda_1 - \lambda_2 p_t) = 0 \quad \text{and} \quad \sum_{t\in[T]} (\hat{U}_t - \lambda_1 - \lambda_2 p_t) p_t = 0, \tag{4.3.1}$$

whose solution is $\hat{\lambda}^{OLS}$, the ordinary least squares regressor of $\hat{U}_t = \ln(\hat{\pi}_t/(1-\hat{\pi}_t))$ on the constant 1 and the price p_t.

Dealing with endogeneity. Our approach has rested so far on the assumption that the market-specific quality characteristic κ_t is exogenous, that is, uncorrelated with the price p_t, which is somewhat unrealistic. It is reasonable to assume that the price of the good in those markets where the quality is higher will experience a higher demand, and that resellers (especially if they have some market power) have the ability to impose higher prices on them. In that case, the approach suggested by Berry in his 1994 paper consists of using a valid instrument ζ_t, which is a centered random variable that is correlated with p_t, but uncorrelated with κ_t, so that $\mathbb{E}[\zeta_t] = 0$, $\mathbb{E}[\kappa_t \zeta_t] = 0$, and $cov(p_t, \zeta_t) \neq 0$. For instance, cost shifters, such as labor costs or prices of raw materials, are valid instruments as they are correlated with the price, but arguably independent from the market-specific quality.

In that case, the moment conditions (4.3.1) are replaced by

$$\sum_{t\in[T]} (\hat{U}_t - \lambda_1 - \lambda_2 p_t) = 0 \quad \text{and} \quad \sum_{t\in[T]} (\hat{U}_t - \lambda_1 - \lambda_2 p_t) \zeta_t = 0, \tag{4.3.2}$$

whose solution (as recalled in appendix A.2.7) is now the *instrumental variable regressor* $\hat{\lambda}^{IV}$ of $\hat{U}_t = \ln(\hat{\pi}_t/(1-\hat{\pi}_t))$ on two endogenous regressors, the constant and the price p_t, using as instruments the constant and ζ_t. Hence the value of $\hat{\lambda}_2^{IV}$ is given by

$$\hat{\lambda}_2^{IV} = \frac{cov_T(\zeta_t, \hat{U}_t)}{cov_T(\zeta_t, p_t)},$$

where cov_T denotes the sample covariance, from which $\hat{\lambda}_1^{IV}$ can easily be deduced. The code demo in appendix B.4.4 implements this example. One sees in that example that $\hat{\lambda}_2^{OLS}$ is positive: due to the correlation between the unobserved quality variable κ_t and the price p_t, if one does not use instruments, one is led to mistakenly conclude that the higher the price, the higher the consumer utility, which is obviously nonsensical. The instrumental variable regression produces the right conclusion, and leads to an unbiased estimate of λ.

IV-GMM for demand estimation. In typical applications, there will be many instruments, more than the dimension of the parameter, so we will need to select a combination of instruments. Let us continue our intuitive presentation based on the same example, but assume that we have a third instrument ζ_t' in addition to 1 and ζ_t. As before, we have two parameters, λ_1 and λ_2, so the model is overidentified in the sense that the two parameters are not enough to satisfy the three moment restrictions indicating the orthogonality of the three instruments:

$$\sum_{t\in[T]}(\hat{U}_t-\lambda_1-\lambda_2 p_t)=\sum_{t\in[T]}(\hat{U}_t-\lambda_1-\lambda_2 p_t)\zeta_t=\sum_{t\in[T]}(\hat{U}_t-\lambda_1-\lambda_2 p_t)\zeta_t'=0. \tag{4.3.3}$$

To solve this problem efficiently, Berry, Levinsohn, and Pakes have suggested using Hansen's generalized method of moments (GMM), which in this context is called IV-GMM, and recalled in appendix A.2.7. For this, we pick a 3×3 symmetric positive definite matrix W, and define our estimator $\hat{\lambda}$ as the value of (λ_1,λ_2) minimizing the W-norm of the vectors of the three terms that we were hoping to nullify in expression (4.3.3); that is, we solve

$$\min_{(\lambda_1,\lambda_2)\in\mathbb{R}^2}\left\|\sum_{t\in[T]}(\hat{U}_t-\lambda_1-\lambda_2 p_t)(1,\zeta_t,\zeta_t')^\top\right\|_W^2, \tag{4.3.4}$$

where $\|.\|_W$ is the quadratic norm associated with W, defined as $\|a\|_W^2=a^\top Wa$. Letting $\boldsymbol{\zeta}=(\mathbf{1}_T,\zeta,\zeta')$ be the $T\times 3$ matrix of instruments stacked column-wise, $\boldsymbol{\Phi}=(\mathbf{1}_T,p)$ the $T\times 2$ matrix of the two regressors (a constant and prices) stacked column-wise, and $\Pi=\boldsymbol{\zeta}W\boldsymbol{\zeta}^\top$, the problem reformulates as

$$\min_{\lambda\in\mathbb{R}^2}(\hat{U}-\boldsymbol{\Phi}\lambda)^\top\Pi(\hat{U}-\boldsymbol{\Phi}\lambda), \tag{4.3.5}$$

whose solution is straightforward and yields the IV-GMM estimator:

$$\hat{\lambda}=\left(\boldsymbol{\Phi}^\top\Pi\boldsymbol{\Phi}\right)^{-1}\left(\boldsymbol{\Phi}^\top\Pi\hat{U}\right). \tag{4.3.6}$$

While any choice of a positive symmetric definite W leads to a consistent estimate of λ, one can show (as recalled in appendix sections A.2.6 and A.2.7) that the efficient estimator—in the sense that it minimizes the asymptotic variance of the estimator—is obtained by taking W as the inverse of the variance-covariance matrix of the vector $\kappa(1,\zeta,\zeta')^\top$. Hence

$$W=\mathbb{E}[\kappa^2(1,\zeta,\zeta')^\top(1,\zeta,\zeta')]^{-1}. \tag{4.3.7}$$

If κ is independent from ζ and ζ' (the homoskedastic case), one has $W=\mathbb{E}[\kappa^2]^{-1}\tilde{W}$, where the matrix $\tilde{W}$ is defined as $\tilde{W}=\mathbb{E}[(1,\zeta,\zeta')^\top(1,\zeta,\zeta')]^{-1}$. In that case, the scaling by the positive number $\mathbb{E}[\kappa^2]$ does not matter and we can use the matrix $\tilde{W}$ in place of the matrix of weights W. For practical purposes, we shall use the sample analog of $\tilde{W}$, namely $(\boldsymbol{\zeta}^\top\boldsymbol{\zeta}/T)^{-1}$.

If the conditional variance of κ given the value of the instruments is not constant (heteroskedastic case), then the previous simplification of expression (4.3.7) does not work. That expression requires the computation of the covariance of powers of ζ and ζ' with κ^2. But as $\kappa_t=U_t-\lambda_1-\lambda_2 p_t$, knowing κ requires knowing the coefficients λ_1 and λ_2, so computing W is not directly feasible. As a way out of this problem, one uses a two-step procedure: in the first step, one runs the estimation problem (4.3.4) using the weights matrix $\tilde{W}$ as one would in the homoskedastic case. This provides a first-stage estimate $(\hat{\lambda}_1,\hat{\lambda}_2)$, which we can then in turn use in the second stage to form the residuals $\hat{\kappa}_t=\hat{U}_t-\hat{\lambda}_1-\hat{\lambda}_2 p_t$, and use $\hat{\kappa}$ to form the sample analog of expression (4.3.7), which will then be used for the efficient GMM estimation procedure. A code demo is provided in appendix B.4.5.

Adding a random coefficient term to the utility shocks. One may want to allow for dependence between the random utility and the observable characteristics of the good. For instance, if we know the distribution of consumer incomes in a particular market, we may draw I sample points of income y_i from that distribution, let $\eta_i=-1/y_i$ be minus the inverse of the income, and assume that the utility associated with purchasing the good in market t is

$$U_t+\tau\eta_i p_t+\epsilon_{ti1},$$

where, as before, $U_t=\lambda_1+\lambda_2 p_t+\kappa_t$. In other words, we are adding a term $\tau\eta_i p_t=-\tau p_t/y_i$ that captures the interaction between the income of consumer i and the price of the good p_t, where τ is a scalar parameter to be estimated along with λ_1 and λ_2. We expect $\tau>0$, as utility should be decreasing with respect to the price; also, because the income appears in the denominator, the price sensitivity is decreasing with respect to the income. The utility associated with not purchasing the good is still ϵ_{ti0}.

The approach suggested in the previous paragraph is unchanged for the most part; only the market share inversion is modified, as we are moving from a logit model to a random coefficient logit model. Thus, for each market $t \in [T]$, we solve for the systematic utility U_t in each of the T scalar equations

$$\frac{1}{I} \sum_{i \in [I]} \frac{1}{1 + \exp(U_t + \tau \eta_i p_t)} = 1 - \hat{\pi}_t. \tag{4.3.8}$$

However, one should note that the result depends on the scalar τ that enters in the specification of the random coefficient logit term. We therefore call $\hat{U}_t^\tau$ the resulting solution of equation (4.3.8).

Once $\hat{U}_t^\tau$ has been solved for, one uses the IV-GMM approach described in the previous paragraph. We now have three parameters $(\lambda_1, \lambda_2, \tau)$, and we assume, for the sake of discussion, that we have four instruments: the constant 1, and three variables ζ_t, ζ_t', and ζ_t''. Thus, we still have more instruments than parameters and we are not in the just-identified case. Problem (4.3.4) now becomes the minimization of

$$\left\| \sum_{t \in [T]} (\hat{U}_t^\tau - \lambda_1 - \lambda_2 p_t)(1, \zeta_t, \zeta_t', \zeta_t'')^\top \right\|_W^2 = (\hat{U}^\tau - \mathbf{\Phi}\lambda)^\top \Pi (\hat{U}^\tau - \mathbf{\Phi}\lambda) \tag{4.3.9}$$

with respect to $(\lambda_1, \lambda_2, \tau) \in \mathbb{R}^3$, where now W is a 4×4 symmetric positive definite matrix, and $\boldsymbol{\zeta} = (\mathbf{1}_T, \zeta, \zeta', \zeta'')$ is the $T \times 4$ matrix of instruments stacked column-wise, while we still have $\Pi = \boldsymbol{\zeta} W \boldsymbol{\zeta}^\top$ and $\mathbf{\Phi} = (\mathbf{1}_T, p)$, the $T \times 2$ matrix of regressors.

We note that the dependence of the predicted error $\hat{U}_t^\tau - \lambda_1 - \lambda_2 p_t$ is linear with respect to parameters λ_1 and λ_2, but nonlinear with respect to τ. We therefore minimize with respect to τ the value of the minimum with respect to λ_1 and λ_2, as the problem admits a closed-form solution, and we thus compute the minimization of expression (4.3.9) as $\min_\tau F(\tau)$, where

$$F(\tau) = \min_{\lambda \in \mathbb{R}^2} \ (\hat{U}^\tau - \mathbf{\Phi}\lambda)^\top \Pi (\hat{U}^\tau - \mathbf{\Phi}\lambda), \tag{4.3.10}$$

and a quick calculation recovers the result of formula (A.2.36) in appendix A.2.7: the value of the inner minimization problem is given by

$$F(\tau) = (\hat{U}^\tau)^\top (\Pi - \Pi\mathbf{\Phi}(\mathbf{\Phi}^\top \Pi \mathbf{\Phi})^{-1} \mathbf{\Phi}^\top \Pi) \hat{U}^\tau.$$

It remains to solve the minimization problem over the nonlinear parameter τ, namely, minimize F. This is done numerically by a gradient descent algorithm, and it is therefore desirable to obtain the derivative of F in closed form. We have $F'(\tau) = 2(\hat{U}^\tau)^\top (\Pi - \Pi\mathbf{\Phi}(\mathbf{\Phi}^\top \Pi \mathbf{\Phi})^{-1} \mathbf{\Phi}^\top \Pi) d\hat{U}^\tau / d\tau$, and the derivative of $\hat{U}_t^\tau$ with respect to τ can be obtained by implicit differentiation in equation (4.3.8), from

$$\sum_{i\in[I]} \frac{\exp(\hat{U}_t^\tau + \tau\eta_i p_t)}{(1+\exp(\hat{U}_t^\tau + \tau\eta_i p_t))^2}\left(\frac{d\hat{U}_t^\tau}{d\tau} + \eta_i p_t\right) = 0.$$

Once the optimal value of $\hat{\tau}$ has been determined, one deduces the values of $\hat{\lambda}_1$ and $\hat{\lambda}_2$ using the analog of formula (4.3.6). As in the previous paragraph, one may run a second stage to be robust to heteroskedasticity. A code demo is provided in appendix B.4.6.

4.4 A Cornerstone in Structural Estimation: BLP's Method

In this section we build on the intuition developed in the previous section and we present the estimation method of Berry, Levinsohn, and Pakes (BLP) in its full generality. The BLP method has undoubtedly become an industry standard in empirical industrial organization and structural econometrics, and has been extensively used to model demand for differentiated products such as cars, computers, foods, health services, housing goods, etc. In its core form, it focuses on the demand side by combining the use of random coefficient logit unobservable heterogeneity, and dealing with potential endogeneity issues through the use of an IV-GMM method. Endogeneity, here, comes from the presence of unobservable characteristics of goods (quality or reputation, for instance) that cannot be measured by the econometrician, and that may be correlated with some observable product characteristics (such as prices). The method rests on an equilibrium model that incorporates the supply side under oligopolistic competition (Bertrand), while also addressing potential endogeneity. A distinctive feature of our presentation is its emphasis on mathematical elegance, made possible by leveraging the tensor calculus developed in appendix A.3.1. The reader should ensure familiarity with the material presented there, particularly the rules of matrix differentiation.

Example 4.3. In their seminal 1995 paper, Berry, Levinsohn, and Pakes took on the central question of how to estimate consumer demand and measure market power in the U.S. automobile market, a prime example of a market for differentiated products. The paper introduced most of the tools we are discussing, including the random coefficients logit model, and the treatment of endogeneity, as well as the use of cost shifters, in order to identify how consumer preferences vary across individuals and how firms set prices in response to this heterogeneity. The notebook Galichon (2025d) revisits the original study based on the original data, and illustrates the steps described in this section and in the next one.

Demand-side specification. We first focus on the demand side. In market $t \in [T]$, we model agent i's utility u_{tiy} associated with alternative $y \in [Y_t]$ as the sum of three terms: one that is systematic in the sense that it does not vary across agents; another that results from the interactions between the agents' taste variations and the products' observed characteristics; and one that is purely idiosyncratic and Gumbel-distributed. We thus specify u_{tiy} as

$$\begin{cases} u_{tiy} = U_{ty} + \nu^{\tau}_{tiy} + \epsilon_{tiy}, \\ u_{ti0} = \epsilon_{ti0}, \end{cases} \tag{4.4.1}$$

where we assume that:

- the first component of u_{tiy} is U_{ty}, a term that is specific to alternatives and markets, but does not vary across agents. This term may come from observed product characteristics, denoted by ϕ_{tyk}, such as capacity, amenities, size, price, etc., but it may also come from some unobserved residual variations in quality denoted by κ_{ty} that are neither directly observed, nor captured in the observed product characteristics. We thus postulate that

$$U_{ty} = \sum_{k \in [K]} \phi_{tyk} \lambda_k + \kappa_{ty}, \tag{4.4.2}$$

where $\lambda \in \mathbb{R}^K$ is a parameter vector to be estimated, and the distribution of the term κ_{ty} is left unspecified. We will not assume that κ is uncorrelated from the observable characteristics, as some of these, such as the price of the product, or the delay in obtaining the product, may be affected by the level of unobservable quality. A number D of "demand instruments" ζ_{tyd} for $d \in [D]$ are available and allow us to write the exclusion restrictions that are needed for identification. These instruments are correlated with the regressors ϕ_{tyk}, but uncorrelated with κ_{ty}, so that

$$\mathbb{E}[\kappa_{ty} \zeta_{tyd}] = 0, \tag{4.4.3}$$

which provides a set of moments conditions that we will use to estimate the model parameters using IV-GMM estimation;

- the second component of u_{tiy}, denoted by ν^{τ}_{tiy}, captures various interactions between agent-specific unobserved tastes and observed characteristics. We assume

$$\nu^{\tau}_{tiy} = \sum_{m\in[M]} \eta_{im}\xi_{tym}\tau_m = \sum_{m\in[M]} \nu_{tiy,m}\tau_m, \tag{4.4.4}$$

where $\nu_{tiy,m} := \eta_{im}\xi_{tym}$, the M-dimensional vector $(\xi_{tym})_m$ represents various characteristics of alternative y in market t, and $(\eta_{im})_m$ is a vector of the same dimension that captures agent i's valuation of the characteristic.[1] The term η_{im} measures the valuation by consumer i of the mth characteristic, and the vector η is distributed according to $\mathbf{P}_\eta$. The parameter vector $\tau \in \mathbb{R}^M$ controls the relative importance of the valuation of the mth dimension in the overall preference;
- finally, the third term, ϵ_{tiy}, is an idiosyncratic preference shock associated to option y by agent i in market t. It is assumed that for each i and t, the (Y_t+1)-dimensional vector $(\epsilon_{tiy})_y$ has an i.i.d. Gumbel distribution.

Supply-side specification. We now move on to analyzing the supply side. We shall assume that sellers are engaged in imperfect competition, specifically Bertrand price competition. Assume that each market t is structured into firms, each of which controls a disjoint subset of options for which it is free to set prices. Let $\mathbf{H}$ be the $TY \times TY$ matrix of term $H_{ty,t'y'} = 1$ if y and y' are controlled by the same firm and if $t = t'$, and $H_{ty,t'y'} = 0$ otherwise. Letting c_{ty} be the unit cost to produce good y in market t, the markup on good y is $p_{ty} - c_{ty}$ and the profit of a firm producing y in market t is thus

$$\sum_{y'\in[Y]} H_{ty,ty'}\pi_{ty'}(p_{ty'} - c_{ty'}), \tag{4.4.5}$$

where the market share π_{ty} depends on the entries of the price vector p through the systematic utility $U_{ty} = \sum_{k\in[K]} \phi_{tyk}\lambda_k + \kappa_{ty}$, but also through the random coefficient part $\nu^{\tau}_{tiy} = \sum_{m\in[M]} \nu_{tiym}\tau_m$. Indeed, we recall that

$$\pi_{ty} = \frac{1}{I}\sum_{i\in[I]} \frac{\exp\left(\sum_{k\in[K]} \phi_{tyk}\lambda_k + \kappa_{ty} + \sum_{m\in[M]} \nu_{tiym}\tau_m\right)}{1 + \sum_{y'\in[Y]} \exp\left(\sum_{k\in[K]} \phi_{ty'k}\lambda_k + \kappa_{ty'} + \sum_{m\in[M]} \nu_{tiy'm}\tau_m\right)}, \tag{4.4.6}$$

[1] We may take $M = K$ and $\xi_{tym} = \phi_{tym}$, so that the mth consumer characteristic η_{im} interacts with the corresponding product characteristic ξ_{tym}. However, the formulation is general enough to allow for more complex interactions; see exercise 4.4.

or, in vectorized terms,

$$\pi = \frac{1}{I}\mathbf{\Sigma}_I f \text{ with } f = \exp\left(V - \mathbf{\Sigma}_Y^\top u\right), \tag{4.4.7}$$

where we have defined

$$V = \mathbf{\Sigma}_I^\top \mathbf{\Phi}\lambda + \mathbf{\Sigma}_I^\top \kappa + \boldsymbol{\nu}\tau \text{ and } u = \log\left(\mathbf{1}_{TI} + \mathbf{\Sigma}_Y \exp V\right). \tag{4.4.8}$$

As both the regressors of the systematic utility, ϕ, and the components of the random coefficients, ν, may depend on prices p, it follows that the choice probabilities π also depend on p through these channels. The producer of y in market t sets the prices $p_{ty'}$ of the goods y' it controls in order to maximize profit. These goods are the y' such that $H_{ty,ty'} = 1$. The profit of the producer of y in market t is $\sum_{y'} H_{ty,ty'}\pi_{ty'}(p_{ty'} - c_{ty'})$, and by the first-order conditions for profit maximization under Bertrand competition, one obtains:

$$\sum_{y' \in [Y]} H_{ty,ty'} \frac{\partial \pi_{ty'}}{\partial p_{ty}}(p_{ty'} - c_{ty'}) + \pi_{ty} = 0. \tag{4.4.9}$$

Letting $\mathbf{D}$ be the square matrix of size TY with entries $D_{ty,t'y'} = \partial \pi_{t'y'}/\partial p_{ty}$ if y and y' are controlled by the same firm and if $t = t'$, and with entries $D_{ty,t'y'} = 0$ otherwise, expression 4.4.9 can be recast in vectorized terms as

$$\mathbf{D}(c - p) = \pi, \text{ where } \mathbf{D} = \mathbf{H} \odot (\frac{\partial \pi}{\partial p})^\top, \tag{4.4.10}$$

where $\odot$ is the Hadamard (element-wise) product, as recalled in appendix A.3.1. By matrix inversion, equation (4.4.10) yields the following expression for the cost vector:

$$c = p + \mathbf{D}^{-1}\pi. \tag{4.4.11}$$

Having specified both the demand and supply sides, we are now ready to describe the method of Berry, Levinsohn, and Pakes, which we divide into four steps.

(a) Recovering consumer utilities via demand inversion. If we fix $\tau \in \mathbb{R}^M$ for now (its determination will be discussed in step (c) below), we can simulate ν_{tiy}^τ from (4.4.4) using $(\eta_{im})_m \sim \mathbf{P}_\eta$. In that case, the discrete choice model obtained from system (4.4.1) is a random coefficient logit model, as described in section 4.2 above. The probability that agent i in market t opts for alternative y is therefore $\pi_{tiy} = \exp(U_{ty} + \nu_{tiy}^\tau - u_{ti})$, where u_{ti} is the expected ex-interim utility of agent i, computed conditionally on the random coefficient ν_{tiy}^τ and by integrating over the logit shock, which is given, as usual, by $u_{ti} = \log(1 + \sum_y \exp(U_{ty} + \nu^{\tau}_{tiy}))$, and the market share map is given by $\pi_{ty}(U) = I^{-1} \sum_i \exp(U_{ty} + \nu^{\tau}_{tiy} - u_{ti})$. We are interested in the inverse market share map, which is done by solving for U and u in the system

$$\begin{cases} \hat{\pi}_{ty} = \sum_{i \in [I]} \frac{1}{I} \exp(U_{ty} + \nu^{\tau}_{tiy} - u_{ti}), \\ \frac{1}{I} = \frac{1}{I} \exp(-u_{ti}) + \sum_{y \in [Y]} \frac{1}{I} \exp(U_{ty} + \nu^{\tau}_{tiy} - u_{ti}). \end{cases} \tag{4.4.12}$$

We denote by $(\hat{U}^{\tau}, \hat{u}^{\tau})$ the solution to that system, where the superscript τ indicates that the market share inversion is done conditional on τ. In vectorized notation, the probability that consumer i in market t chooses alternative y is given by

$$f = \exp(\mathbf{\Sigma}_I^{\top} U + \boldsymbol{\nu}\tau - \mathbf{\Sigma}_Y^{\top} u), \tag{4.4.13}$$

where we define $\mathbf{\Sigma}_I := \mathbf{I}_T \otimes \mathbf{1}_I^{\top} \otimes \mathbf{I}_Y$ and $\mathbf{\Sigma}_Y := \mathbf{I}_T \otimes \mathbf{I}_I \otimes \mathbf{1}_Y^{\top}$, and where we denote by $\boldsymbol{\nu}$ the matrix of size $TIY \times M$ whose term is $\nu_{tiy,m} = \eta_{im}\xi_{tym}$. We recall that $\otimes$ is the Kronecker product; that $(\hat{\pi}_{ty}) \in \mathbb{R}^{TY}$ is the vector of market shares; that $(U_{ty}) \in \mathbb{R}^{TY}$ is the vector of systematic utilities; that $(\tau_m) \in \mathbb{R}^M$ is the parameter of the random coefficient; and that $(u_{ti}) \in \mathbb{R}^{TI}$ is the individual welfare. The system (4.4.12) can be reexpressed using vectorized notation as

$$\begin{cases} \hat{\pi} = \frac{1}{I} \mathbf{\Sigma}_I \exp(\mathbf{\Sigma}_I^{\top} U + \boldsymbol{\nu}\tau - \mathbf{\Sigma}_Y^{\top} u), \\ \mathbf{1}_{TI} = \exp(-u) + \mathbf{\Sigma}_Y \exp(\mathbf{\Sigma}_I^{\top} U + \boldsymbol{\nu}\tau - \mathbf{\Sigma}_Y^{\top} u). \end{cases} \tag{4.4.14}$$

The first equation of system (4.4.14) ensures that the conditional choice probabilities f_{tiy} of individual i, defined by (4.4.13), aggregate to the observed market shares of each product y in each market t. The second equation implies that these are indeed conditional probabilities. As discussed in section 4.2, the iterated fitting procedure from optimal transport (seen in appendix A.1.8), which consists of iteratively solving for U in the first set of equations in (4.4.14) and for u in the second set of equations in the same system, and repeating until convergence, is an efficient and practical algorithm for solving for (U, u), which is recalled in this section in algorithm 4.4.1.[2]

(b) Recovering sellers' costs from profit maximization. Next, we would like to estimate the vector of costs, c_{ty}, using expression (4.4.11), which follows from the assumption of profit maximization. This expression involves the market share vector π and the price vector p, both of which are directly observable. However, it also depends on the matrix $\mathbf{D}$, which contains the

[2] In exercise 4.5, the reader is asked to verify that the IPFP algorithm is equivalent to an alternative formulation of the "contraction mapping algorithm" proposed in the original BLP paper.

Algorithm 4.4.1 IPFP algorithm for BLP market share inversion

Input : $\boldsymbol{\nu}, u_{ti}^{0}$
Parameter: ϵ (tolerance)
while $\max_{ty \in [T \times Y]} \| U_{ty}^{s} - U_{ty}^{s-1} \| \geq \epsilon$ **do**
  $U^{s+1} = -\log(\boldsymbol{\Delta}_{\hat{\pi}}^{-1} I^{-1} \boldsymbol{\Sigma}_I \exp(\boldsymbol{\nu}\tau - \boldsymbol{\Sigma}_Y^\top u^s))$
  $u^{s+1} = \log(\mathbf{1}_{TI} + \boldsymbol{\Sigma}_Y \exp(\boldsymbol{\Sigma}_I^\top U^{s+1} + \boldsymbol{\nu}\tau))$
  $s = s+1$
end

derivatives $\partial\pi/\partial p$ of market shares with respect to prices. These derivatives are not directly observable from the data. However, exercise 4.6 asks the reader to verify, recalling that $\boldsymbol{\Delta}_f$ denotes the diagonal matrix whose diagonal is the vector f, that the differential of π with respect to p is given by

$$\frac{\partial \pi}{\partial p} = \frac{1}{I} \boldsymbol{\Sigma}_I \boldsymbol{\Delta}_f \mathbf{R}_f \frac{\partial V}{\partial p}, \quad \text{where } \mathbf{R}_f = \mathbf{I}_{TIY} - (\mathbf{I}_{TI} \otimes \mathbf{1}_{Y \times Y})\boldsymbol{\Delta}_f, \tag{4.4.15}$$

where we recall that $V = \boldsymbol{\Sigma}_I^\top \boldsymbol{\Phi}\lambda + \boldsymbol{\Sigma}_I^\top \kappa + \boldsymbol{\nu}\tau$ is the vector of individual utilities. The differential of V with respect to the price vector p (which enters through both $\boldsymbol{\Phi}$ and $\boldsymbol{\nu}$) is obtained by applying formula A.3.14 in appendix A.3.1, yielding

$$\frac{\partial V}{\partial p} = \boldsymbol{\Sigma}_I^\top \frac{\partial \boldsymbol{\Phi}}{\partial p} (\mathbf{I}_{TY} \otimes \lambda) + \frac{\partial \boldsymbol{\nu}}{\partial p} (\mathbf{I}_{TY} \otimes \tau). \tag{4.4.16}$$

Typically, however, only a few entries of $\boldsymbol{\Phi}$ are endogenous (i.e., depend on prices), while others are exogenous (in the sense that they do not depend on prices). For that reason, we write

$$\boldsymbol{\Phi} = \left(\boldsymbol{\Phi}^e, \boldsymbol{\Phi}^x\right), \tag{4.4.17}$$

where $\boldsymbol{\Phi}^e$ is a $TY \times K^e$ matrix and $\boldsymbol{\Phi}^x$ is a $TY \times K^x$ matrix, with $K^e + K^x = K$. We assume that $\partial \boldsymbol{\Phi}^x / \partial p = 0$, and we write the parameter vector as $\lambda = (\alpha^\top, \beta^\top)^\top$, where α is a vector of size K^e and β is a vector of size K^x. We have

$$\boldsymbol{\Phi}\lambda = \boldsymbol{\Phi}^e \alpha + \boldsymbol{\Phi}^x \beta, \tag{4.4.18}$$

and expression (4.4.16) simplifies into

$$\frac{\partial V}{\partial p} = \boldsymbol{\Sigma}_I^\top \frac{\partial \boldsymbol{\Phi}^e}{\partial p} (\mathbf{I}_{TY} \otimes \alpha) + \frac{\partial \boldsymbol{\nu}}{\partial p} (\mathbf{I}_{TY} \otimes \tau). \tag{4.4.19}$$

While $f=\exp(\mathbf{\Sigma}_I^\top U+\boldsymbol{\nu}\tau-\mathbf{\Sigma}_Y^\top u)$ is not directly observable, it can be estimated by substituting $\hat{U}^\tau$ and $\hat{u}^\tau$, computed in step (a), for U and u. Let $\hat{f}$ denote the resulting estimator of f. We can then form $\mathbf{R}_{\hat{f}}$ and $\mathbf{\Delta}_{\hat{f}}$, and obtain an estimator $\hat{\mathbf{D}}$ of the matrix $\mathbf{D}$ by combining expressions (4.4.15) and (4.4.10). As a result, the cost vector c is estimated by

$$\hat{c}^{\tau,\alpha}=p+\widehat{\mathbf{D}}^{-1}\hat{\pi},\ \text{where}\ \widehat{\mathbf{D}}=\mathbf{H}\odot\left(\frac{1}{I}\mathbf{\Sigma}_I\mathbf{\Delta}_{\hat{f}}\mathbf{R}_{\hat{f}}\frac{\partial V}{\partial p}\right)^\top. \tag{4.4.20}$$

The superscript (τ,α) in the notation $\hat{c}^{\tau,\alpha}$ indicates that the estimator $\widehat{\mathbf{D}}$ depends on τ, because the market share inversion is performed conditionally on τ, and also on α, since the price elasticities depend on it, as seen in formulas (4.4.15) and (4.4.19).

(c) IV-GMM estimation. The central step of the method consists of running an IV-GMM estimation procedure: regressing $\hat{U}^\tau$ on $\mathbf{\Phi}$ and $\hat{c}^{\tau,\alpha}$ on $\mathbf{\Psi}$, in order to estimate the demand parameters $\lambda=(\alpha^\top,\beta^\top)^\top$, the supply parameters γ, and the demand parameters τ. The overall objective is to minimize

$$\left\|\begin{pmatrix}\boldsymbol{\zeta}^\top(\hat{U}^\tau-\mathbf{\Phi}\lambda)\\ \boldsymbol{\chi}^\top(\hat{c}^{\tau,\alpha}-\mathbf{\Psi}\gamma)\end{pmatrix}\right\|_W^2=\left\|\begin{pmatrix}\boldsymbol{\zeta}^\top(\hat{U}^\tau-\mathbf{\Phi}^e\alpha-\mathbf{\Phi}^x\beta)\\ \boldsymbol{\chi}^\top(\hat{c}^{\tau,\alpha}-\mathbf{\Psi}\gamma)\end{pmatrix}\right\|_W^2 \tag{4.4.21}$$

over $(\tau,\lambda,\gamma)\in\mathbb{R}^{M+K+L}$, where $\lambda=(\alpha^\top,\beta^\top)^\top$. We recall that $\|a\|_W^2=a^\top Wa$, and W is a $(D+S)\times(D+S)$ matrix that is symmetric positive definite.

We observe that the quantity appearing inside the norm in expression (4.4.21) is nonlinear in $\theta=(\tau^\top,\alpha^\top)^\top$ and linear in β and γ. Thus, if we fix θ—a parameter vector of size $N=M+K^e$—we may run a linear IV-GMM procedure with dependent variable $\mathbf{Y}^\theta$, defined as the concatenation of $\hat{U}^\tau-\mathbf{\Phi}^e\alpha$ and $\hat{c}^{\tau,\lambda}$, on regressors $\mathbf{X}$, given by the aggregation of $\mathbf{\Phi}^x$ and $\mathbf{\Psi}$, and instruments $\mathbf{Z}$, given by the aggregation of $\boldsymbol{\zeta}$ and $\boldsymbol{\chi}$. We define

$$Y^\theta=\begin{pmatrix}\hat{U}^\tau-\mathbf{\Phi}^e\alpha\\ \hat{c}^{\tau,\alpha}\end{pmatrix},\ X=\begin{pmatrix}\mathbf{\Phi}^x & \mathbf{0}_{TY,L}\\ \mathbf{0}_{TY,K^x} & \mathbf{\Psi}\end{pmatrix},\ \text{and}\ Z=\begin{pmatrix}\boldsymbol{\zeta} & \mathbf{0}_{TY,S}\\ \mathbf{0}_{TY,D} & \boldsymbol{\chi}\end{pmatrix}, \tag{4.4.22}$$

which are matrices of respective sizes $2TY\times 1$, $2TY\times(K^x+L)$, and $2TY\times(D+S)$.

With these notations, the expression of the IV-GMM objective displayed in (4.4.21) is $\|\mathbf{Z}^\top(\mathbf{Y}^\theta-\mathbf{X}(\beta^\top,\gamma^\top)^\top)\|_W^2$, to be minimized over $\theta=(\tau,\alpha)$, β, and γ. If we minimize over β and γ in an inner step, and over θ in an outer step, the inner step boils down to linear IV-GMM. In that case, we can rewrite the optimization problem as the minimization of $F(\theta)$ over $\theta \in \mathbb{R}^N$, where

$$F(\theta) = \min_{(\beta,\gamma)\in\mathbb{R}^{K^x+L}} \{\|\mathbf{Z}^\top(\mathbf{Y}^\theta - \mathbf{X}(\beta^\top,\gamma^\top)^\top\|_W^2\}/(2TY)^2, \tag{4.4.23}$$

so that $F(\theta)$ is the value of the objective function of the linear part of the IV-GMM procedure. If we denote $\Pi = \mathbf{Z}W\mathbf{Z}^\top$ and $\Upsilon = \Pi - \Pi^\top\mathbf{X}\left(\mathbf{X}^\top\Pi X\right)^{-1}X^\top\Pi$, we have an explicit expression for the value of the minimizers β and γ given θ. Denoting by $\beta^\star(\theta)$ and $\gamma^\star(\theta)$ these minimizers, we have the following expressions:

$$\begin{pmatrix}\beta^\star(\theta)\\ \gamma^\star(\theta)\end{pmatrix} = \left(\mathbf{X}^\top\Pi\mathbf{X}\right)^{-1}\mathbf{X}^\top\Pi\mathbf{Y}^\theta \text{ and } F(\theta) = \left(\mathbf{Y}^\theta\right)^\top\Upsilon\mathbf{Y}^\theta/(2TY)^2. \tag{4.4.24}$$

The function $F(\theta)$ is minimized numerically using gradient descent. It is therefore useful to obtain a formula for the differential of F, and we have

$$\frac{\partial F(\theta)}{\partial\theta} = \frac{1}{2}\frac{1}{(TY)^2}\left(\mathbf{Y}^\theta\right)^\top\Upsilon\frac{\partial\mathbf{Y}^\theta}{\partial\theta} \text{ and } \frac{\partial\mathbf{Y}^\theta}{\partial\theta} = \begin{pmatrix}\frac{\partial\hat{U}^\tau}{\partial\tau} & -\mathbf{\Phi}^e\\ \frac{\partial\hat{c}^{\tau,\alpha}}{\partial\tau} & \frac{\partial\hat{c}^{\tau,\alpha}}{\partial\alpha}\end{pmatrix}. \tag{4.4.25}$$

The quantities that appear in the expression of $\partial Y^\theta/\partial\theta$ are computed in exercise 4.7, where formula 4.5.1 is derived for $\partial\hat{U}/\partial\tau$, and in exercise 4.8, where formula (4.5.2) is obtained for $\partial\hat{c}^\theta/\partial\theta$. These steps are summarized in table 4.5 and implemented in Python in appendix B.4.7.

(d) Heteroskedasticity-robust two-step approach. It remains to explain how we choose the weight matrix W. This is done following the textbook approach to GMM, as recalled in appendix A.2.6. In the first step, we set $W = (\mathbf{Z}^\top\mathbf{Z})^{-1}$ and perform the minimization of $F(\theta)$ as described in paragraph (c) above. This yields a first-step estimator $\hat{\theta}$. We then proceed with a heteroskedasticity-robust second step, by repeating the procedure in paragraph (c), but this time using the updated weight matrix

$$W = \left(\mathbf{Z}^\top \operatorname{diag}\left(\left(\mathbf{Y}^{\hat{\theta}} - \mathbf{X}\begin{pmatrix}\hat{\beta}\\ \hat{\gamma}\end{pmatrix}\right)^2\right)\mathbf{Z}\right)^{-1}. \tag{4.4.26}$$

4.5 Exercises

Exercise 4.1. *Consider the pure characteristics model of section 4.1 when there is only one characteristic, that is, $K = 1$. Give simplified expressions for the usual objects (welfare function, market share map, entropy of choice, inverse market share).*

Table 4.1: BLP's method: Dimensions.

Variable	Description
T	Number of markets.
Y	Number of products.
I	Number of simulation points.
M	Number of dimensions of idiosyncratic utility.
K^e	Number of endogenous demand-relevant systematic characteristics.
K^x	Number of exogenous demand-relevant systematic characteristics.
K	Number of demand systematic characteristics, $K = K^e + K^x$.
L	Number of supply-relevant characteristics.
D	Number of demand-side instruments.
S	Number of supply-side instruments.
N	Number of nonlinear parameters, $N = M + K^e$.

Table 4.2: BLP's method: Required data.

Object	Size	Description
$\hat{\pi}$	$(TY, 1)$	$\hat{\pi}_{ty}$ is the market share of product y in market t.
p	$(TY, 1)$	p_{ty} is the price of product y in market t.
$\mathbf{\Phi}^e$	(TY, K^e)	ϕ^e_{ty,k^e} is the k^eth observable demand-relevant endogenous (i.e., price-dependent) characteristic of y in market t entering the systematic utility U.
$\mathbf{\Phi}^{e\prime}$	(TY, K^e)	$\phi^{e\prime}_{ty,k^e} = \partial\phi^e_{ty,k^e}/\partial p_{ty}$ is the price derivative of the previous quantity.
$\mathbf{\Phi}^{\mathbf{x}}$	(TY, K^x)	ϕ^x_{ty,k^x} is the k^xth observable demand-relevant exogenous (i.e., price-independent) characteristic of y in market t entering in the systematic utility U.
η	$(IM, 1)$	η_{im} is the mth entry of the random characteristics of individual i (across markets) that interacts with ξ below to enter the idiosyncratic utility $\boldsymbol{\nu}\tau$.
ξ	$(TYM, 1)$	ξ_{tym} is the mth entry of the characteristics of product y in market t that interacts with η above to enter the idiosyncratic utility $\boldsymbol{\nu}\tau$.

(*continued*)

Table 4.2: (*continued*)

Object	Size	Description
ξ'	$(TYM, 1)$	$\xi'_{tym} = \partial \xi_{tym} / \partial p_y$ is the price derivative of the previous quantity.
ζ	(TY, D)	$\zeta_{ty,d}$ is the dth instrument for the demand for product y in market t.
$\mathbf{\Psi}$	(TY, L)	$\Psi_{ty,l}$ is the lth supply-relevant characteristic of product y in market t.
$\mathbf{H}$	(TY, TY)	$H_{ty,t'y'}$ is 1 if y and y' are controlled by the same firm and $t = t'$, and 0 otherwise.
χ	(TY, S)	$\chi_{ty,s}$ is the sth instrument for the supply of product y in market t.

Table 4.3: BLP's method: Parameters to be estimated.

Object	Size	Description
α	$(K^e, 1)$	α parametrizes the endogenous part of the systematic consumer utility $\mathbf{\Phi^e}\alpha$.
β	$(K^x, 1)$	β parametrizes the exogenous part of the systematic consumer utility $\mathbf{\Phi^x}\beta$.
λ	$(K, 1)$	$\lambda = (\alpha^\top, \beta^\top)^\top$ parametrizes the overall systematic utility $U = \mathbf{\Phi}\lambda$; see below.
γ	$(L, 1)$	γ_l parametrizes the seller cost $c = \mathbf{\Psi}\gamma + \sigma$, where σ is a noise term.
τ	$(M, 1)$	τ_m parametrizes the idiosyncratic utility of the consumer, $\boldsymbol{\nu}\tau$.
θ	$(N, 1)$	$\theta = (\tau^\top, \alpha^\top)^\top$ is a nonlinear GMM parameter $(N = M + K^e)$.

Exercise 4.2 (GHK simulator for a probit model). *Assume the characteristics space is two-dimensional* ($K = 2$)*, and there are four alternatives, with characteristics* $\xi_1 = (0.2, 0.2)$, $\xi_2 = (0.8, 0.7)$, $\xi_3 = (0.1, 0.8)$, *and* $\xi_4 = (0.4, 0.6)$. *Assume that the vector of utility shifter* η *is distributed as* $\mathcal{N}(0, \mathbf{I}_2)$. *When* $U = (0.0, -0.1, 0.3, -0.2)$, *compute the market share map respectively using accept-reject and GHK. Report the relative precision of the estimators for various sample sizes.*

Table 4.4: BLP's method: Construction of matrices.

Object	Formula	Size	Comment
$\mathbf{\Sigma}_I$	$\mathbf{\Sigma}_I = \mathbf{I}_T \otimes \mathbf{1}_I^\top \otimes \mathbf{I}_Y$	(TY, TIY)	Summation over $i \in [I]$.
$\mathbf{\Sigma}_Y$	$\mathbf{\Sigma}_Y = \mathbf{I}_T \otimes \mathbf{I}_I \otimes \mathbf{1}_Y^\top$	(TI, TIY)	Summation over $y \in [Y]$.
$\mathbf{\Phi}$	$\mathbf{\Phi} = (\Phi^e, \Phi^x)$	(TY, K)	Demand-relevant characteristics of y (systematic part). $K = K^e + K^x$.
$\frac{\partial \mathbf{\Phi^e}}{\partial p}$	$\frac{\partial \mathbf{\Phi^e}}{\partial p} = \mathbf{I}_{TY} \circledast \Phi^{e\prime}$	$(TY, TY\,K^e)$	Price derivative of Φ^e.
$\boldsymbol{\nu}$	$\nu_{tiy,m} = \eta_{im}\xi_{tym}$	(TIY, M)	$\boldsymbol{\nu}_{tiy,m}$ is the mth component of the random utility associated by i to y.
$\boldsymbol{\nu}'$	$\nu'_{tiy,m} = \eta_{im}\frac{\partial \xi_{tym}}{\partial p_y}$	(TIY, M)	$\nu'_{tiy,m} = \partial \nu_{tiy,m} / \partial p_y$ is the price derivative of the previous quantity.
$\frac{\partial \boldsymbol{\nu}}{\partial p}$	$\frac{\partial \boldsymbol{\nu}}{\partial p} = (\mathbf{I}_T \otimes \mathbf{1}_I \otimes \mathbf{I}_Y) \circledast \boldsymbol{\nu}'$	(TIY, TYM)	Price derivative of $\boldsymbol{\nu}$.

Table 4.5: BLP's method: Main procedure.

Step	**Given $\theta = (\tau, \alpha)$ where $\tau \in \mathbb{R}^M$ and $\alpha \in \mathbb{R}^{K^e}$, and W is a $(D+S) \times (D+S)$ matrix, do ...**
1.	Following algorithm 4.4.1, iterate $$\begin{cases} U^{s+1} = -\log(\mathbf{\Delta}_{\hat{\pi}}^{-1} I^{-1} \mathbf{\Sigma}_I \exp(\boldsymbol{\nu}\tau - \mathbf{\Sigma}_Y^\top u^s)) \\ u^{s+1} = \log(\mathbf{1}_{TI} + \mathbf{\Sigma}_Y \exp(\mathbf{\Sigma}_I^\top U^{s+1} + \boldsymbol{\nu}\tau)) \end{cases}$$ over $s \in \mathbb{N}$ until $\max_{ty \in [T \times Y]} \Vert U_{ty}^s - U_{ty}^{s-1} \Vert$ falls below tolerance level.
2.	Set $\hat{f} = \exp(\mathbf{\Sigma}_I^\top \hat{U}^\tau + \boldsymbol{\nu}\tau - \mathbf{\Sigma}_Y^\top \hat{u}^\tau)$.
3.	Set $\mathbf{R}_{\hat{f}} = \mathbf{I}_{TIY} - (\mathbf{I}_{TI} \otimes \mathbf{1}_{Y \times Y}) \mathbf{\Delta}_{\hat{f}}$.
4.	Letting $V = \mathbf{\Sigma}_I^\top (\mathbf{\Phi}\lambda + \kappa) + \boldsymbol{\nu}\tau$, set $$\frac{\partial V}{\partial p} = \mathbf{\Sigma}_I^\top \frac{\partial \Phi^e}{\partial p} (\mathbf{I}_{TY} \otimes \alpha) + \frac{\partial \boldsymbol{\nu}}{\partial p} (\mathbf{I}_{TY} \otimes \tau).$$

(*continued*)

Table 4.5: *(continued)*

Step	Do ...
5.	Set $\widehat{\mathbf{D}} = \mathbf{H} \odot \left(\frac{1}{I} \boldsymbol{\Sigma}_I \boldsymbol{\Delta}_{\hat{f}} \mathbf{R}_{\hat{f}} \frac{\partial V}{\partial p} \right)^{\top}.$
6.	Set $\hat{c}^{\tau,\alpha} = p + \widehat{\mathbf{D}}^{-1} \hat{\pi}.$
7.	Set $\mathbf{Y}^{\theta} = \begin{pmatrix} \hat{U}^{\tau} - \boldsymbol{\Phi}^{e} \alpha \\ \hat{c}^{\tau,\alpha} \end{pmatrix}, \ \mathbf{X} = \begin{pmatrix} \boldsymbol{\Phi}^{x} & \mathbf{0}_{TY,L} \\ \mathbf{0}_{TY,K^{x}} & \boldsymbol{\Psi} \end{pmatrix}, \ \mathbf{Z} = \begin{pmatrix} \boldsymbol{\zeta} & \mathbf{0}_{TY,S} \\ \mathbf{0}_{TY,D} & \boldsymbol{\chi} \end{pmatrix}.$
8.	Set $\Pi = \mathbf{Z} W \mathbf{Z}^{\top}$ and $\Upsilon = \Pi - \Pi^{\top} \mathbf{X} \left(\mathbf{X}^{\top} \Pi \mathbf{X} \right)^{-1} \mathbf{X}^{\top} \Pi.$
9.	Set $F(\theta) = \left(\mathbf{Y}^{\theta} \right)^{\top} \Upsilon \mathbf{Y}^{\theta} / (2TY)^{2}.$
10.	Set $\frac{\partial \hat{U}^{\tau}}{\partial \tau} = \left(-I \boldsymbol{\Delta}_{\hat{\pi}} + \boldsymbol{\Sigma}_I \boldsymbol{\Delta}_{\hat{f}} \boldsymbol{\Sigma}_Y^{\top} \boldsymbol{\Sigma}_Y \boldsymbol{\Delta}_{\hat{f}} \boldsymbol{\Sigma}_I^{\top} \right)^{-1} \boldsymbol{\Sigma}_I \boldsymbol{\Delta}_{\hat{f}} \mathbf{R}_{\hat{f}} \boldsymbol{\nu}.$
11.	Letting $\hat{V} = \boldsymbol{\Sigma}_I^{\top} U^{\tau} + \boldsymbol{\nu} \tau$, set $\frac{\partial \hat{V}}{\partial \theta} = \left(\frac{\partial \hat{V}}{\partial \tau}, \ \mathbf{0}_{TIY,K^{e}} \right)$ where $\frac{\partial \hat{V}}{\partial \tau} = \boldsymbol{\Sigma}_I^{\top} \frac{\partial \hat{U}^{\tau}}{\partial \tau} + \boldsymbol{\nu}.$
12.	Recall that $V = \boldsymbol{\Sigma}_I^{\top} (\boldsymbol{\Phi} \lambda + \kappa) + \boldsymbol{\nu} \tau$ and set $\left(\frac{\partial^2 V}{\partial \tau \partial p} \right)_{tiy,mt'y'} = \frac{\partial \boldsymbol{\nu}_{tiy,m}}{\partial p_{t'y'}}$ and $\left(\frac{\partial^2 V}{\partial \alpha \partial p} \right)_{tiy,k^{e}t'y'} = \left(\boldsymbol{\Sigma}_I^{\top} \frac{\partial \boldsymbol{\Phi}^{e}}{\partial p} \right)_{tiy,t'y'k^{e}}.$
13.	Set $\frac{\partial^2 V}{\partial \theta \partial p} = \left(\frac{\partial^2 V}{\partial \tau \partial p}, \ \frac{\partial^2 V}{\partial \alpha \partial p} \right).$
14.	Set $\frac{\partial \widehat{\mathbf{D}}^{\top}}{\partial \theta} = \left(\mathbf{1}_N^{\top} \otimes \mathbf{H} \right) \odot \left(\frac{1}{I} \boldsymbol{\Sigma}_I \boldsymbol{\Delta}_{\hat{f}} \mathbf{R}_{\hat{f}} \left(\left(\mathbf{R}_{\hat{f}} \frac{\partial \hat{V}}{\partial \theta} \circledast \mathbf{R}_{\hat{f}} \frac{\partial V}{\partial p} \right) + \frac{\partial^2 V}{\partial \theta \partial p} \right) \right).$

Table 4.5: *(continued)*

Step	**Do** ...
15.	Set $$\left(\frac{\partial \widehat{\mathbf{D}}}{\partial \theta}\right)_{ty,nt'y'} = \left(\frac{\partial \widehat{\mathbf{D}}}{\partial \theta}^{\top}\right)_{t'y',nty}.$$
16.	Set $$\frac{\partial \hat{c}^{\theta}}{\partial \theta} = -\widehat{\mathbf{D}}^{-1} \frac{\partial \widehat{\mathbf{D}}}{\partial \theta} \left(\mathbf{I}_N \otimes \left(\hat{c}^{\theta} - p\right)\right).$$
17.	Set $$\frac{\partial F(\theta)}{\partial \theta} = \frac{1}{2} \frac{1}{(TY)^2} \left(Y^{\theta_1}\right)^{\top} \Upsilon \begin{pmatrix} \frac{\partial \hat{U}^{\tau}}{\partial \tau} & -\mathbf{\Phi}^{e} \\ \frac{\partial \hat{c}^{\tau,\alpha}}{\partial \tau} & \frac{\partial \hat{c}^{\tau,\alpha}}{\partial \alpha} \end{pmatrix}.$$

Exercise 4.3 (Modeling error). *In the setting of exercise 4.2, compute the vector of market shares associated with* $U = (0.0, -0.1, 0.3, -0.2)$ *in the probit model using the GHK simulator. Next, apply the inverse market share map* of the logit model *with the normalization* $U_1 = 0$. *Do you recover the original* U? *Comment on the result.*

Exercise 4.4. *In most cases, the expression of the random coefficient term given in expression* (4.4.4) *will boil down to* $\nu^{\tau}_{tiy} = \sum_{k\in[K]} \tilde{\eta}_{ik}\phi_{tyk}\tau_k$, *where* $\tilde{\eta}_{ik}$ *are multidimensional idiosyncratic consumer characteristics, and* ϕ_{tyk} *are product-specific characteristics, which is consistent with expression* (4.4.4) *with* $M = K$, $\eta = \tilde{\eta}$, *and* $\xi_{tym} = \phi_{tym}$. *However, in some applications, it is desirable to allow for interactions between all pairs of consumer and product characteristics, that is, between any* $\tilde{\eta}_{ik}$ *and any* ϕ_{tyk}. *Assume that the second component of* u_{tiy} *in expression* (4.4.1) *can be written* $\nu^{\tau}_{tiy} = \sum_{kl\in[K\times K]} \tilde{\eta}_{il}\phi_{tyk}\tau_{lk}$. *Show that it can be recast as expression* (4.4.4) *with* $M = K^2$, $\eta = (\mathbf{I}_{IK} \otimes \mathbf{1}_K)\tilde{\eta}$, *and* $\xi = (\mathbf{I}_{TY} \otimes \mathbf{1}_K \otimes \mathbf{I}_K)\phi$, *and where the index* $m \in [M]$ *corresponds to the double index* lk *indexing the parameter* τ_{lk}.

Exercise 4.5 (BLP's contraction mapping). *Recall the expression of the market share map associated with the random coefficient logit model:*

$$\boldsymbol{\pi}_y(U) = \frac{1}{I} \sum_{i\in[I]} \frac{\exp(U_{ty} + \nu^{\tau}_{tiy})}{\sum_{y'\in[Y]} \exp(U_{ty'} + \nu^{\tau}_{tiy'})}.$$

Show that the IPFP algorithm 4.4.1 can equivalently be rewritten as

$$U_y^{t+1} = U_y^t - \ln \frac{\boldsymbol{\pi}_y(U^t)}{\hat{\pi}_y},$$

which is classically called "BLP's contraction mapping algorithm" in the economic literature. Comment on the resemblance with a gradient descent procedure.

Exercise 4.6 (Price elasticity of demand). *Using the result of exercise 2.9, show formula 4.4.15 which expresses the price elasticity of demand, and from which expression* (4.4.20), *and step 5 in table 4.5, can be deduced.*

Exercise 4.7 (Sensitivity of demand inversion). *In this exercise, we compute the differential $\partial \hat{U}/\partial \tau$ of $\hat{U}^\tau$ with respect to τ, as defined in section 4.4, establishing the formula in step 10 of table 4.5, that is,*

$$\frac{\partial \hat{U}^\tau}{\partial \tau} = \left(-I\boldsymbol{\Delta}_{\hat{\pi}} + \boldsymbol{\Sigma}_I \boldsymbol{\Delta}_{\hat{f}} \boldsymbol{\Sigma}_Y^\top \boldsymbol{\Sigma}_Y \boldsymbol{\Delta}_{\hat{f}} \boldsymbol{\Sigma}_I^\top\right)^{-1} \boldsymbol{\Sigma}_I \boldsymbol{\Delta}_{\hat{f}} \mathbf{R}_{\hat{f}} \boldsymbol{\nu}, \tag{4.5.1}$$

where $\boldsymbol{\Delta}_{\hat{f}} = diag(\hat{f})$, $\mathbf{R}_{\hat{f}} = \mathbf{I}_{TIY} - (\mathbf{I}_{TI} \otimes \mathbf{1}_{Y\times Y})\boldsymbol{\Delta}_{\hat{f}}$, and $\boldsymbol{\Delta}_{\hat{\pi}} = diag(\hat{\pi})$.

(i) Letting $\boldsymbol{\Delta}_f = diag(\exp(\boldsymbol{\Sigma}_I^\top U + \boldsymbol{\nu}\tau - \boldsymbol{\Sigma}_Y^\top u))$ and $\boldsymbol{\Delta}_0 = diag(\exp(-u))$, show that the differentials $\partial U/\partial \tau$ and $\partial u/\partial \tau$, respectively of U and u with respect to τ, are obtained by

$$\begin{cases} -\boldsymbol{\Sigma}_I \boldsymbol{\Delta}_f \boldsymbol{\Sigma}_I^\top \frac{\partial U}{\partial \tau} + \boldsymbol{\Sigma}_I \boldsymbol{\Delta}_f \boldsymbol{\Sigma}_Y^\top \frac{\partial u}{\partial \tau} = \boldsymbol{\Sigma}_I \boldsymbol{\Delta}_f \boldsymbol{\nu}, \\ -\boldsymbol{\Sigma}_Y \boldsymbol{\Delta}_f \boldsymbol{\Sigma}_I^\top \frac{\partial U}{\partial \tau} + \left(\boldsymbol{\Delta}_0 + \boldsymbol{\Sigma}_Y \boldsymbol{\Delta}_f \boldsymbol{\Sigma}_Y^\top\right) \frac{\partial u}{\partial \tau} = \boldsymbol{\Sigma}_Y \boldsymbol{\Delta}_f \boldsymbol{\nu}. \end{cases}$$

Show that if (u, U) is a solution to system (4.4.14), *then $\boldsymbol{\Sigma}_I \boldsymbol{\Delta}_f \boldsymbol{\Sigma}_I^\top = I\boldsymbol{\Delta}_{\hat{\pi}}$, where $\boldsymbol{\Delta}_{\hat{\pi}} = diag\left(\hat{\pi}\right)$ and $\boldsymbol{\Sigma}_Y \boldsymbol{\Delta}_f \boldsymbol{\Sigma}_Y^\top = \mathbf{I}_{TI} - \boldsymbol{\Delta}_0$. Using the Schur complement formula recalled in appendix A.3.2, deduce a proof of formula* (4.5.1).

(ii) Provide an alternative derivation of formula (4.5.1) *using the result of exercise 2.9.*

Exercise 4.8 (Sensitivity of cost inversion). *The goal of this exercise is to compute the differential of $\hat{c}^\theta$ with respect to θ in the setting of section 4.4. More specifically, one shall derive the formulas that appear in steps 14–16 in table 4.5.*

(i) *By differentiating the identity $\widehat{\mathbf{D}}(\hat{c}^{\theta}-p)=\hat{p}$ with respect to θ, verify that the differential of $\hat{c}^{\theta}$ with respect to $\theta=(\tau^{\top},\alpha^{\top})^{\top}$ is given by*

$$\frac{\partial \hat{c}^{\theta}}{\partial \theta}=-\hat{\mathbf{D}}^{-1}\frac{\partial \hat{\mathbf{D}}}{\partial \theta}\left(\mathbf{I}_N\otimes(\hat{c}-p)\right). \tag{4.5.2}$$

(ii) *The differential of $\widehat{\mathbf{D}}$ is obtained from the differential of its transpose by reshuffling the indices $(\partial\widehat{\mathbf{D}}/\partial\theta)_{ty,nt'y'}=(\partial\widehat{\mathbf{D}}^{\top}/\partial\theta)_{t'y',nty}$. Using the results of exercise 2.9, show that the differential of $\widehat{\mathbf{D}}^{\top}$ is given by*

$$\frac{\partial \hat{\mathbf{D}}^{\top}}{\partial \theta}=\left(\mathbf{1}_N^{\top}\otimes\mathbf{H}\right)\odot\left(\frac{1}{I}\mathbf{\Sigma}_I\mathbf{\Delta}_{\hat{f}}\mathbf{R}_{\hat{f}}\left(\left(\mathbf{R}_{\hat{f}}\frac{\partial \hat{V}}{\partial \theta}\circledast\mathbf{R}_{\hat{f}}\frac{\partial V}{\partial p}\right)+\frac{\partial^2 V}{\partial\theta\partial p}\right)\right), \tag{4.5.3}$$

where $\circledast$ is the row-wise Kronecker product, and where $\hat{V}=\mathbf{\Sigma}_I^{\top}U^{\tau}+\boldsymbol{\nu}\tau$ and $V=\mathbf{\Sigma}_I^{\top}(\mathbf{\Phi}\lambda+\kappa)+\boldsymbol{\nu}\tau$.

(iii) *Verify that $\partial V/\partial p$ is given by* (4.4.19), *that*

$$\frac{\partial \hat{V}}{\partial \theta}=\left(\frac{\partial \hat{V}}{\partial \tau},\ \mathbf{0}_{TIY,K^e}\right)\ \text{where}\ \frac{\partial \hat{V}}{\partial \tau}=\mathbf{\Sigma}_I^{\top}\frac{\partial \hat{U}^{\tau}}{\partial \tau}+\boldsymbol{\nu}, \tag{4.5.4}$$

and that

$$\frac{\partial^2 V}{\partial\theta\partial p}=\left(\frac{\partial^2 V}{\partial\tau\partial p},\ \frac{\partial^2 V}{\partial\alpha\partial p}\right). \tag{4.5.5}$$

(iv) *Show that*

$$\left(\frac{\partial^2 V}{\partial\tau\partial p}\right)_{tiy,mt'y'}=\frac{\partial \boldsymbol{\nu}_{tiy,m}}{\partial p_{t'y'}}\ \text{and}\ \left(\frac{\partial^2 V}{\partial\alpha\partial p}\right)_{tiy,k^et'y'}=\left(\mathbf{\Sigma}_I^{\top}\frac{\partial \mathbf{\Phi}^e}{\partial p}\right)_{tiy,t'y'k^e}. \tag{4.5.6}$$

4.6 Problems

Problem 4.1. *The code in appendix B.4.8 loads the automobile dataset from Berry, Levinsohn, and Pakes using the library* `PyBLP` *by Conlon and Gortmaker. Based on that dataset, estimate the BLP model in a specification of your choice.*

Problem 4.2 (BLP using automatic differentiation). *Provide an alternative to the code provided in appendix B.4.7 using* `PyTorch`.

Problem 4.3 (MPEC formulation of BLP). *The BLP procedure described in table 4.5 has the following equivalent formulation, called* mathematical programming under equilibrium constraints *or MPEC formulation, which consists of*

$$\min_{U,u,\theta,\beta,\gamma} \|\mathbf{Z}^\top(\mathbf{Y}^{U,\theta} - \mathbf{X}(\beta^\top, \gamma^\top)^\top)\|_W^2 \tag{4.6.1}$$

subject to

$$\begin{cases} \hat{\pi} = \frac{1}{I}\boldsymbol{\Sigma}_I \exp(\boldsymbol{\Sigma}_I^\top U + \nu\tau - \boldsymbol{\Sigma}_Y^\top u) \\ \mathbf{1}_{TI} = \exp(-u) + \boldsymbol{\Sigma}_Y \exp(\boldsymbol{\Sigma}_I^\top U + \nu\tau - \boldsymbol{\Sigma}_Y^\top u), \end{cases} \tag{4.6.2}$$

where

$$\mathbf{Y}^{U,\theta} = ((U - \boldsymbol{\Phi}^e\alpha)^\top, (p + \hat{\mathbf{D}}^{-1}\hat{\pi})^\top)^\top$$

and where, in contrast, $\mathbf{X}$ *and* $\mathbf{Z}$ *are still defined by their expressions in* (4.4.22). *Write down the Lagrangian for that formulation of the problem, and implement it as an alternative to the code provided in appendix B.4.7 using the augmented Lagrangian algorithm described in appendix A.1.7.*

4.7 References and Notes

Expository texts covering random utility models include Train (2009), Wooldridge (2010), and Cameron and Trivedi (2005), among others.

A good reference for the characteristics approach is Lancaster (1966); see also Gorman (1980) and Heckman and Scheinkman (1987). The demand model developed in section 4.1 is the "pure characteristics model" of Berry and Pakes (2007). The Geweke–Hajivassiliou–Keane (GHK) algorithm is due to Geweke (1989), Hajivassiliou and McFadden (1998), and Keane (1994). The connection with optimal transport and Laguerre diagrams appeared in Bonnet, Galichon, Hsieh, O'Hara, and Shum (2022). A reference for the library `sdot` is Leclerc and Mérigot (2024), and one for the library `pypoman` is Caron (2023).

The random coefficient logit model is due to Berry, Levinsohn, and Pakes (1995); see also McFadden and Train (2000). The model was investigated under various aspects in Nevo (2001), while Dubé, Fox, and Su (2012) focus on computational aspects. Proposition 4.2.1 and the reinterpretation of the BLP contraction mapping algorithm (4.2.7) in terms of the IPFP/Sinkhorn's algorithm originate in Bonnet, Galichon, Hsieh, O'Hara, and Shum (2022). The fact that that algorithm is a contraction for Hilbert's metric was proven in Franklin and Lorenz (1989).

The central reference for sections 4.3 and 4.4 is the seminal paper Berry, Levinsohn, and Pakes (1995). The idea of performing an instrumental variable regression after demand inversion was first proposed in Berry (1994). The GMM method is due to Hansen and Singleton (1982), and heteroskedasticity-robust approaches were pioneered by Newey and West (1987). Other references include other works by Berry, Levinsohn, and Pakes (1999 and 2004), Brenkers and Verboven (2006), Andrews, Gentzkow, and Shapiro (2017), Knittel and Metaxoglou (2014), Nevo (2000 and 2001), and Petrin (2002). Problem 4.1 asks the reader to replicate the original study of Berry, Levinsohn, and Pakes. The `PyBLP` library by Conlon and Gortmaker has become a computational reference in that field; see the companion paper Conlon and Gortmaker (2020). An instructive exercise of BLP replication is carried out in the Python project by Lin (2025a), hosted on GitHub. The MPEC formulation of the BLP procedure introduced in problem 4.3 was proposed in Su and Judd (2012).

4.8 Summary

SUMMARY OF CHAPTER 4

In this chapter, the so-called characteristics-based approach to demand is studied.

Section 4.1 focuses on characteristics-based models, moving beyond the limitations of models that assume independent utility shocks for each alternative. It describes how these models, by considering the utility derived from the attributes of choices, allow for a more nuanced analysis of consumer preferences. The section covers the Lancasterian approach to utility, where utility is derived from goods' characteristics, and discusses computational aspects, including the use of computational geometry for estimating models and the challenges of modeling with characteristics-based approaches.

Section 4.2 delves into the random coefficient logit model (also known as the mixed logit model), which combines the flexibility of the logit model with random coefficients to account for heterogeneity in consumer preferences. This section explains how the model incorporates both characteristics-based utility and idiosyncratic taste variations, allowing for a richer depiction of consumer choice. It highlights the deep connection with entropic regularized optimal transport. It discusses the estimation and computation of the random coefficient logit model, highlighting its ability to capture preference heterogeneity and address some of the IIA limitations.

Section 4.3 is a gentle introduction to the challenges posed by demand estimation, in particular, the problem of endogeneity. This section introduces the main concept of the Berry, Levinsohn, and Pakes (BLP) estimation framework in a very simple model of binomial choice with only the price as an observed characteristic. In order to explain the price variations, one needs to assume that there is an unobserved characteristic that agents perceive, but that the analyst does not observe. It is natural to recover this unobservable characteristic using the market shares by an inversion of the market share map and then regressing on other characteristics, but we may then experience an endogeneity issue, as a number of observable characteristics (including the price) are most likely determined at equilibrium, and hence are endogenous. For that reason, one uses an instrumental variable approach: by determining an appropriate set of instruments, one can write down the adequate orthogonality restrictions that allow for the parametric estimation. This approach is very general and allows for models that are more sophisticated than the logit model, such as the random coefficient logit model. The methodology is presented in full generality in section 4.4. For the supply side, a imperfect price competition à la Bertrand is assumed. The assumption of profit maximization allows us to write down first-order conditions that predict the markups, leading to the possibility of guessing the costs based on the prices; these costs are regressed on the appropriate regressors with the relevant instruments.

CHAPTER 5

Generalized Linear Models for Allocation and Equilibrium Problems

In this chapter we will see that the logistic regression framework introduced in chapter 3, and, more generally, the addition of fixed effects in generalized linear models, can be used to capture a wide range of econometric models, with a large number of applications covering international trade, labor economics, family economics, and demand estimation, among others.

5.1 Macro Logistic Regression

Consider the conditional logit model of section 3.2, and assume that the individuals i are clustered into types $x \in [X]$ so that x_i denotes the type of i, and x_i is sufficient to explain any variation in the systematic part of the utility of i, so that $\phi_{iyk} = \phi_{x_i yk}$. (We reuse the notation ϕ indexed by types x instead of individuals i, but that should not lead to any confusion.) Given a type $x \in [X]$ and an option $y \in [Y]$, we define $\hat{\mu}_{xy} = \sum_i \hat{\pi}_{iy} 1\{x_i = x\}$ as the number of individuals of type x who choose option y, and let $n_x = \sum_y \hat{\mu}_{xy}$ and $m_y = \sum_x \hat{\mu}_{xy}$ be the number of individuals in type x and the number of options y chosen in the sample, respectively.

The primal formulation (3.2.3) of the multinomial logistic regression becomes

$$\max_{\lambda \in \mathbb{R}^K} \left\{ \hat{\mu}^\top \Phi \lambda - \sum_{x \in [X]} n_x \log \left(\sum_{y \in [Y]} \exp(\Phi \lambda)_{xy} \right) \right\}, \tag{5.1.1}$$

with the understanding that in expression (5.1.1) and in its sequel, Φ stands for the matrix of shape $XY \times K$ with term $\phi_{xy,k} = \phi_{xyk}$, and $\hat{\mu}$ stands for the vector of size XY with term $\hat{\mu}_{xy}$. This can be estimated as a Poisson regression with x-fixed effects, where the set of observations is $[\Omega] = [X \times Y]$, the dependent variables are $\hat{\mu}_{xy}$, and the design matrix $R = \begin{pmatrix} \Phi & -\mathbf{I}_X \otimes \mathbf{1}_Y \end{pmatrix}$ is made of two blocks: the block of regressors Φ of the logistic regression, and $\mathbf{I}_X \otimes \mathbf{1}_Y$, the matrix of x-dummies. With the addition of these fixed effects, problem (5.1.1) expands into

$$\max_{\lambda,\tilde{u}} \left\{ \hat{\mu}^\top \Phi\lambda - n^\top \tilde{u} - \sum_{xy\in[X\times Y]} \exp((\Phi\lambda)_{xy} - \tilde{u}_x) \right\}, \tag{5.1.2}$$

so that for any λ, the optimal value of $\tilde{u}_x$ is $\tilde{u}_x = \log(\sum_y \exp(\Phi\lambda)_{xy}) - \log n_x$. As a result, the macro logistic regression can be interpreted as a Poisson regression plus x-fixed effects, and the dual to expression (5.1.2) is

$$\begin{aligned} \min_{\mu_{xy}\geq 0} & \sum_{xy\in[X\times Y]} \mu_{xy} \log \mu_{xy} \\ s.t. & \sum_{y\in[Y]} \mu_{xy} = n_x, \\ & \sum_{xy\in[X\times Y]} \mu_{xy}\phi_{xyk} = \sum_{xy\in[X\times Y]} \hat{\mu}_{xy}\phi_{xyk}. \end{aligned} \tag{5.1.3}$$

We see from both the primal and the dual forms (5.1.1) and (5.1.3) that the macro logistic regression is a slight generalization of the "micro" logistic regression introduced in section 3.2, to which it reduces if $n_x = 1$ for all $x \in [X]$.

5.2 Entropic Optimal Transport

We first consider an instance of the macro logistic regression described in section 5.1 with only y-fixed effects v_y as regressors, and no others. In that case, we have $\lambda = v$ and $\Phi = -\mathbf{1}_X \otimes \mathbf{I}_Y$, and dropping the tilde from $\tilde{u}$ and denoting it by u instead, equation (5.1.2) boils down to

$$\min_{u,v} \left\{ \sum_{x\in[X]} n_x u_x + \sum_{y\in[Y]} m_y v_y + \sum_{xy\in[X\times Y]} \exp\left(-u_x - v_y\right) \right\}, \tag{5.2.1}$$

where n_x and m_y have been defined above. After dividing by $I = \sum_x n_x = \sum_y m_y$, and replacing n_x by n_x/I, m_y by m_y/I, and u_x by $u_x + \log I$, and leaving v_y unchanged, one may assume without loss of generality that $\sum_x n_x = \sum_y m_y = 1$. The dual to (5.2.1) is then

$$\max_{\mu\geq 0} \left\{ - \sum_{xy\in[X\times Y]} \mu_{xy} \log \mu_{xy} \right\} \tag{5.2.2}$$

$$\text{s.t.} \sum_{y\in[Y]} \mu_{xy} = n_x,$$
$$\sum_{x\in[X]} \mu_{xy} = m_y.$$

The solution μ to the dual problem can easily be found to be the independent coupling of marginal distributions n and m, that is, $\mu_{xy} = n_x m_y$; the primal solutions can be easily deduced.

Problem (5.2.2) is a particular instance of an entropic optimal transport problem, as introduced in section 7.3 of *Optimal Transport Methods in Economics*, but it is a very particular instance: its objective function essentially consists of maximizing the entropy $-\sum_{xy} \mu_{xy} \log \mu_{xy}$ subject to margin constraints, which leads the dual solution μ to be the independent coupling of the margins n and m. More general optimal transport problems with entropic regularization consist of minimizing $\sum_x n_x u_x + \sum_y m_y v_y + \sum_{xy} \exp\left(\phi_{xy} - u_x - v_y\right)$. In order to achieve this, we need to consider weighted Poisson regression, discussed in appendix A.2.9, where observation xy gets weight w_{xy}. The weighted Poisson log-likelihood of the sample is then $\sum_{xy} w_{xy}\hat{\mu}_{xy}(-u_x - v_y) - \sum_{xy} w_{xy} \exp\left(-u_x - v_y\right)$, from which we see that if $w_{xy} = \exp \phi_{xy}$ and $\hat{\mu}_{xy} = n_x m_y \exp\left(-\phi_{xy}\right)$, the maximum likelihood problem boils down to

$$\min_{u,v} \left\{ \sum_x n_x u_x + \sum_y m_y v_y + \sum_{xy} \exp\left(\phi_{xy} - u_x - v_y\right) \right\}, \tag{5.2.3}$$

whose dual formulation, as given by expression (A.2.53), is now the classical entropy-regularized optimal transport problem, that is,

$$\max_{\mu_{xy}\geq 0} \left\{ \sum_{xy\in[X\times Y]} \mu_{xy}\phi_{xy} - \sum_{xy\in[X\times Y]} \mu_{xy} \log \mu_{xy} \right\} \tag{5.2.4}$$
$$\text{s.t.} \sum_{y\in[Y]} \mu_{xy} = n_x,$$
$$\sum_{x\in[X]} \mu_{xy} = m_y,$$

which we summarize into:

Proposition 5.2.1. *The entropic regularized optimal transport problem can be computed as a weighted Poisson regression of* $\hat{\mu}_{xy} = n_x m_y \exp(-\phi_{xy})$ *on regressors solely composed of x- and y-fixed effects, with weights* $w_{xy} = \exp \phi_{xy}$.

This way to recast the well-known problem of optimal transport as a Poisson regression, or, equivalently, as a generalized linear model, may be useful for computing optimal transport problems using tools to estimate generalized linear models. Appendix B.5.1 shows code doing exactly this.

Computing problem (5.2.3) can be done directly by gradient descent, as discussed in section 3.3. However, there is a powerful alternative method called *coordinate descent*, which allows for very fast computation of problem (5.2.3). The principle of the method is simple: at time t, fix a guess u^t of u, and minimize the objective function of problem (5.2.3) with respect to v only, holding $u = u^t$ fixed; call the optimal solution v^{t+1}. By inspection of the first-order conditions, it is easy to see that one can obtain v^{t+1} as an explicit function of u^t: $v_y^{t+1} = \log(m_y^{-1} \sum_x \exp(\phi_{xy} - u_x^t))$. Similarly, the partial maximization of the objective function with respect to u_x only, and fixing $v_y = v_y^{t+1}$, yields $u_x^{t+1} = \log(n_x^{-1} \sum_y \exp(\phi_{xy} - v_y^{t+1}))$. This procedure is called the *iterated proportional fitting procedure* (IPFP), or *Sinkhorn's algorithm.* Setting $K_{xy} = \exp(\phi_{xy})$, $A_x = \exp(-u_x)$, and $B_y = \exp(-v_y)$ yields its formal description as algorithm 5.2.1.

Algorithm 5.2.1 IPFP algorithm for solving problem (5.2.3)

Input : K_{xy}, B_y^0
Parameter: η (tolerance)
while $\max_{y\in[Y]} \|B_y^t - B_y^{t-1}\| \geq \eta$ **do**
  $A_x^t = n_x / \sum_{y\in[Y]} B_y^{t-1} K_{xy}$
  $B_y^t = m_y / \sum_{x\in[X]} A_x^t K_{xy}$
end
Return : A^t, B^t

Code for the IPFP is provided in appendix B.5.2. The method is guaranteed to decrease the objective function, and as seen in theorem A.1.9 in appendix A.1.8, a classical argument due to Birkhoff applies to show that this procedure is a contraction mapping in the so-called "Hilbert metric," which implies the convergence of the algorithm. This implies:

Theorem 5.2.1 (Convergence of the IPFP). *Consider the sequence* (A_x^t, B_y^t) *defined in algorithm 5.2.1. Then:*

(i) The quantities $(-\log A_x^t, -\log B_y^t)$ *converge (up to a constant) toward a solution* (u_x, v_y) *to* (5.2.3).
(ii) The quantity $\mu_{xy}^t = A_x^t B_y^t K_{xy}$ *converges toward the unique solution* μ_{xy} *to* (5.2.4).

Although we don't provide rates of convergence here, the interested reader will find them in the appendix in theorem A.1.9.

5.3 Gravity Equation

We have seen in section 5.1 that the macro logistic regression can be computed as a Poisson regression with x-fixed effects, and we have investigated in section 5.2 the case of a macro logistic regression with y-fixed effects as sole regressors. We now consider the case of a macro logistic regression not only with y-fixed effects, but also with other regressors. This is best motivated in the context of international trade, in connection with the "gravity equation." There, i is an individual trade, x_i is the country where the trade originates from, and y_i is the destination country of the trade.

Example 5.1. In the book Yotov, Piermartini, Monteiro, and Larch (2016), data on trade flows between pairs of countries is provided. In that case, $X = Y$, x is an exporter country, y is an importer country, and the value of the trade flow from country x to country y is provided for every exporter-importer pair xy. We seek to explain these trade flows by various measures of proximity/distance between countries, such as geographic distance, contiguity, common language, or common colonial origins. Many other regressors are possible. In the notebook Galichon (2025f), we show how the gravity equation can be computed as a Poisson regression with exporter and importer fixed effects.

Let us then investigate the macro logistic regression where the categorical variables are the destinations of the trades $y \in [Y]$ and the regressors are various measures of distance from the origin to the destination plus the destination fixed effects, namely y-fixed effects. The macro logistic regression problem (5.1.2) now becomes

$$\max_{\lambda,u,v} \left\{ \sum_{xy\in[X\times Y]} \hat{\mu}_{xy}((\Phi\lambda)_{xy} - u_x - v_y) - \sum_{xy\in[X\times Y]} \exp((\Phi\lambda)_{xy} - u_x - v_y) \right\}, \tag{5.3.1}$$

and its dual expression (5.1.3) becomes

$$\min_{\mu_{xy}\geq 0} \sum_{xy\in[X\times Y]} \mu_{xy} \log \mu_{xy} \tag{5.3.2}$$

$$s.t. \sum_{y\in[Y]} \mu_{xy} = n_x, \quad [u_x]$$

$$\sum_{x\in[X]} \mu_{xy} = m_y, \quad [v_y]$$

$$\sum_{xy\in[X\times Y]} \mu_{xy}\phi_{xyk} = \sum_{xy\in[X\times Y]} \hat{\mu}_{xy}\phi_{xyk} \quad [\lambda_k],$$

where the quantities in brackets denote the Lagrange multipliers corresponding to each constraint. By the first-order conditions, the optimal dual solution μ is related with the primal solution (λ, u, v) by the *structural gravity equation* of Anderson and van Wincoop, namely,

$$\mu_{xy} = \exp\left((\Phi\lambda)_{xy} - u_x - v_y\right). \tag{5.3.3}$$

In the language of international trade, $\exp(u_x)$ and $\exp(v_y)$ are *multilateral resistances*, which are here to fit the right flow of exports and imports, that is, denoting $K^{\lambda}_{xy} = \exp(\Phi\lambda)_{xy}$,

$$\sum_{y\in[Y]} \frac{K^{\lambda}_{xy}}{e^{u_x}e^{v_y}} = n_x \qquad \text{and} \qquad \sum_{x\in[X]} \frac{K^{\lambda}_{xy}}{e^{u_x}e^{v_y}} = m_y.$$

Thus, following the insight of Santos Silva and Tenreyro, the macro logistic regression can be computed as a Poisson regression plus x- and y-fixed effects. In that correspondence, the observations are still $[\Omega] = [X \times Y]$ and the dependent variables are $\hat{\mu}_{xy}$, but the regressor matrix $R = \begin{pmatrix}\Phi & -\mathbf{I}_X \otimes \mathbf{1}_Y & -\mathbf{1}_X \otimes \mathbf{I}_Y\end{pmatrix}$ is now made of three blocks: the regressor matrix Φ plus the regressor matrices corresponding to the x- and the y-fixed effects, respectively. The code provided in appendix B.5.3 leverages the Poisson regression formulation in order to estimate the gravity equation. In that example we thus set $X = Y$, and we follow the use of the structural gravity literature to disregard the trade flows from country x to itself, so we set $K^{\lambda}_{xx} = 0$.

An algorithm exists that takes advantage of the structure of problem (5.3.1); we discuss it in a slightly more general formulation with an additional L1 penalty, which is

$$\max_{\lambda,u,v}\left\{\sum_{xy\in[X\times Y]} \hat{\mu}_{xy}((\Phi\lambda)_{xy} - u_x - v_y) - \sum_{xy\in[X\times Y]} e^{(\Phi\lambda)_{xy}-u_x-v_y} - \gamma\sum_{k\in[K]}|\lambda_k|\right\}, \tag{5.3.4}$$

and which, in the case when $\gamma = 0$, coincides with problem (5.3.1). The SISTA algorithm, which solves problem (5.3.4), is displayed as algorithm 5.3.2 below, where the soft-thresholding function was defined in 3.7.7. The

Algorithm 5.3.2 Sinkhorn iterative shrinkage-thresholding algorithm (SISTA)

Input : λ_k^0, B_y^0
Parameter: η (tolerance)
while $\max_k |\lambda_k^t - \lambda_k^{t-1}| \geq \eta$ **do**
$A_x^t = n_x / \sum_{y \in [Y]} B_y^{t-1} \exp(\Phi \lambda^{t-1})_{xy}$
$B_y^t = m_y / \sum_{x \in [X]} A_x^t \exp(\Phi \lambda^{t-1})_{xy}$
$\mu_{xy}^t = A_x^t B_y^t \exp(\Phi \lambda^{t-1})_{xy}$
$\lambda^t = \text{SoftThresholding}_{\alpha\gamma}(\lambda^{t-1} + \alpha \Phi^\top (\hat{\mu} - \mu^t))$
end
Return : A^t, B^t, λ^t

algorithm alternates between coordinate descent steps on the fixed effects u_x and v_y, which can be solved in closed form as in the IPFP algorithm 5.2.1 (these are the "S steps," the letter S standing for Sinkhorn), and proximal gradient steps on the parameter vector λ_k (which are the "ISTA steps").

Code is provided in appendix B.5.4. Carlier, Dupuy, Galichon, and Sun prove the convergence of the algorithm, and, in addition, they prove that the convergence is achieved at a linear rate:

Theorem 5.3.1. *The parameter vector λ^t in the SISTA algorithm 5.3.2 converges to the solution to problem* (5.3.1).

5.4 Empirical Models of Matching with Transferable Utility

In this section, we discuss how to estimate matching surplus in two-sided matching models with transferable utility using a logistic regression. More specifically, we will discuss *empirical models of matching*, which are traditional models of matching with the addition of a random utility component. To introduce these models, we will take the example of the heterosexual marriage "market," where it is assumed, following Gary Becker, that men and women pick their partners in order to maximize their individual utilities, while bargaining on how to allocate some joint matching surplus through additive transfers.

Example 5.2. The paper Choo and Siow (2006) used marriage data from the census before and after the 1973 Roe v. Wade decision by the United States Supreme Court imposing fundamental abortion rights in all states. One can postulate a parametric model where the surplus from marriage depends in various ways on the ages of the partners, and on the existence of abortion rights in the given state at the given time. In the notebook Galichon (2025e), we show how the Choo and Siow model can be estimated using generalized linear models.

Assume that men are clustered into X types and women into Y types. Anyone can decide to match or to remain single. If an individual man i of type $x \in [X]$ matches with an individual woman j of type $y \in [Y]$, then i and j respectively enjoy utility surpluses equal to $\alpha_{xy} + \tau_{xy} + \varepsilon_{iy}$ for the man, and $\gamma_{xy} - \tau_{xy} + \eta_{jx}$ for the woman, where these terms are:

- *matching affinities*: α_{xy} and γ_{xy} are marital enjoyment terms that depend only on the types of both partners—they are exogenous quantities, and their sum

$$\phi_{xy} = \alpha_{xy} + \gamma_{xy}, \tag{5.4.1}$$

  which is called the joint systematic surplus, is therefore a primitive of the model, which we are seeking to estimate;
- *transfers*: the quantity τ_{xy} is a utility transfer (from woman j to man i if positive, from the man to the woman if negative). It is determined at equilibrium by market-clearing conditions, and thus it is a price. In fact, one may have assumed that the transfers τ_{ij} depend on the individuals i and j, not only on their types; but it is possible to show (see exercise 5.5) that, at equilibrium, this quantity τ_{ij} will depend only on the observable types of i and j. One sees that if two partners are matched, the amount of utility given up by one partner is fully appropriated by the other one, so the model under study is called *matching with transferable utility* (TU);
- *utility shocks*: ε_{iy} and η_{jx} are random utility shocks that capture idiosyncratic tastes for some types of partners rather than others. They are not observed by the econometrician, who only knows their distribution, possibly under a parametric form.

If man i remains unmatched, then his surplus is simply his utility shock ε_{i0}, which means that the systematic part of the surplus of unmatched individuals has been normalized to zero. Similarly, if woman j remains unmatched, then her surplus is η_{j0}.

We shall make a couple of additional assumptions:

- **Large market:** The market is large, which means that there are an infinite number of individual men i of each type x and women j of each type y. We let n_x and m_y be the fractions of individuals in the population who happen to be men of type x and women of type y, respectively, and thus

$$\sum_{x\in[X]} n_x + \sum_{y\in[Y]} m_y = 1. \tag{5.4.2}$$

- **Logit random utility:** The random utility framework is of the logit type: the vector (ε_{iy}) of man i's random utility shocks, which has $Y+1$ dimensions (one for each of the Y marital options plus one for singlehood), is i.i.d. Gumbel. Similarly, the $(X+1)$-dimensional vector (η_{jx}) of woman j's random utility shocks is also i.i.d. Gumbel.

Given a vector of transfers τ_{xy}, a man i of type $x\in[X]$ thus picks his preferred marital option by solving a discrete choice problem where the options are $[0{:}Y]=\{0\}\cup[Y]$, 0 being associated with singlehood, and $y\in[Y]$ with the choice of a partner of type y. In this problem, the systematic utility associated with option $y\in[Y]$ is $U_{xy}:=\alpha_{xy}+\tau_{xy}$, and the one associated with option 0 (singlehood) is normalized to zero. The average welfare of a man of type x is thus $u_x=\log(1+\sum_y \exp U_{xy})$, and the probability that a man of type x picks a marital option $y\in[Y]$ is $\exp(U_{xy}-u_x)$. We introduce μ_{xy}, the mass of matches xy that are formed at equilibrium, which should satisfy the market clearing conditions

$$\begin{cases}\sum_{y\in[Y]}\mu_{xy}+\mu_{x0}=n_x,\\ \sum_{x\in[X]}\mu_{xy}+\mu_{0y}=m_y.\end{cases} \tag{5.4.3}$$

At equilibrium, the transfers adjust so that everyone receives their first choice. The probability that x picks y as his first choice, for which a formula is provided by the logit framework, is therefore equal to the ratio between the mass μ_{xy} of xy matches and the mass n_x of men of type x, so one has

$$\begin{gathered}\frac{\mu_{xy}}{n_x}=\exp(U_{xy}-u_x) \qquad \text{and} \qquad \frac{\mu_{x0}}{n_x}=\exp(-u_x),\\ \text{where} \quad u_x=\log\Big(1+\sum_{y\in[Y]}\exp U_{xy}\Big).\end{gathered} \tag{5.4.4}$$

Similarly, a woman j of type $y\in[Y]$ picks her preferred marital option $x\in[0{:}X]$ by solving a discrete choice problem with systematic utility $V_{xy}:=\gamma_{xy}-\tau_{xy}$ associated with option $x\in[X]$, and assigning 0 for singlehood. On the female side of the market, one therefore has symmetric relations to (5.4.4), that is,

$$\frac{\mu_{xy}}{m_y} = \exp(V_{xy} - v_y) \quad \text{and} \quad \frac{\mu_{0y}}{m_y} = \exp(-v_y),$$
$$\text{where} \quad v_y = \log\Big(1 + \sum_{x\in[X]} \exp V_{xy}\Big). \tag{5.4.5}$$

Note that the systematic utilities U_{xy} that appear in the discrete choice problems on the men's side and the analog quantities V_{xy} on the women's side are related by $U_{xy} + V_{xy} = \alpha_{xy} + \tau_{xy} + \gamma_{xy} - \tau_{xy} = \phi_{xy}$. The market equilibrium adjusts τ_{xy} to allocate the joint utility surplus ϕ_{xy} between the two sides of the market, based on the competition between agents, which depends on the relative scarcity of each man's and each woman's type, and their desirability to the other side of the market.

By multiplying both sides of the two expressions of μ_{xy} implied by equations (5.4.4) and (5.4.5), and using the fact that $U_{xy} + V_{xy} = \phi_{xy}$, we arrive at an expression of the matching patterns μ as a function of the conditional welfare vectors u and v, as follows:

$$\begin{cases} \mu_{xy} = \sqrt{n_x m_y} \exp\left(\dfrac{\phi_{xy} - u_x - v_y}{2}\right), \\ \mu_{x0} = n_x \exp\left(-u_x\right), \\ \mu_{0y} = m_y \exp\left(-v_y\right), \end{cases} \tag{5.4.6}$$

which yields the celebrated *Choo and Siow equation* relating the equilibrium number of xy pairs μ_{xy} to the number of unmatched individuals μ_{x0} and μ_{0y} of the corresponding types:

$$\mu_{xy} = \sqrt{\mu_{x0}\mu_{0y}} \exp\left(\frac{\phi_{xy}}{2}\right). \tag{5.4.7}$$

The quantities u_x and v_y in (5.4.6) are determined at equilibrium by the market-clearing conditions, which express that the matching patterns μ account for the right masses of men of type x and women of type y in the population, that is,

$$\begin{cases} \sum_{y\in[Y]} \sqrt{n_x m_y} \exp\left(\dfrac{\phi_{xy} - u_x - v_y}{2}\right) + n_x \exp\left(-u_x\right) = n_x, \\ \sum_{x\in[X]} \sqrt{n_x m_y} \exp\left(\dfrac{\phi_{xy} - u_x - v_y}{2}\right) + m_y \exp\left(-v_y\right) = m_y. \end{cases} \tag{5.4.8}$$

Galichon and Salanié showed that the nonlinear system of equations (5.4.8) can be reformulated as an optimization problem. Indeed, these equations are the first-order conditions associated with the following convex optimization problem:

$$\min_{u,v} \left\{ \begin{array}{c} \sum_{x\in[X]} n_x u_x + \sum_{y\in[Y]} m_y v_y \\ +2\sum_{xy\in[X\times Y]} \sqrt{n_x m_y} e^{\frac{\phi_{xy}-u_x-v_y}{2}} + \sum_{x\in[X]} n_x e^{-u_x} + \sum_{y\in[Y]} m_y e^{-v_y} \end{array} \right\}, \tag{5.4.9}$$

whose objective function is strictly convex and coercive, which implies that there is a unique such solution.

As announced at the beginning of this section, we are seeking to estimate the matching surplus ϕ_{xy} based on the observation of empirical matching patterns $\hat{\mu}_{xy}$. To do this, we shall introduce a linear parametrization of that matching surplus, and we shall set

$$\phi^{\lambda}_{xy} = (\Phi\lambda)_{xy} = \sum_{k\in[K]} \phi_{xyk}\lambda_k, \tag{5.4.10}$$

where we again view Φ as the $XY \times K$ matrix whose entry (xy, k) is ϕ_{xyk}. As the systematic utilities of unmatched agents have been normalized to zero, we set $\phi_{x0k} = 0$ and $\phi_{0yk} = 0$. Given a parameter vector λ, we define the *predicted matching patterns* μ^{λ}_{xy} as the solution to (5.4.7) where the fixed effects u_x and v_y that appear are solutions to (5.4.8), and hence are optimal for problem (5.4.9). We shall review two methods of estimation: *moment matching estimation* and the standard maximum likelihood estimation method, and we shall compare them.

Moment matching estimation. In the moment matching estimation method, we seek the estimator $\hat{\lambda}$ of the parameter vector such that the predicted masses of matches fit the margins of the empirical sample, and the moments of Φ associated with the predicted matching patterns coincide with the observed ones, that is, $\sum_{xy} \mu^{\lambda}_{xy}\phi_{xyk} = \sum_{xy} \hat{\mu}_{xy}\phi_{xyk}$. To summarize, the estimation problem consists of finding (λ, u, v) such that

$$\begin{cases} \sum_{y\in[Y]} \sqrt{n_x m_y} \exp\left(\frac{(\Phi\lambda)_{xy}-u_x-v_y}{2}\right) + n_x \exp(-u_x) = n_x, \\ \sum_{x\in[X]} \sqrt{n_x m_y} \exp\left(\frac{(\Phi\lambda)_{xy}-u_x-v_y}{2}\right) + m_y \exp(-v_y) = m_y, \\ \sum_{xy\in[X\times Y]} \mu^{\lambda}_{xy}\phi_{xyk} = \sum_{xy\in[X\times Y]} \hat{\mu}_{xy}\phi_{xyk}, \end{cases} \tag{5.4.11}$$

where the first two equations impose that μ^{λ}_{xy} is an equilibrium matching, and the last one imposes that the predicted moments coincide with the observed moments. Galichon and Salanié showed that the system of equations (5.4.11) arises as the first-order conditions associated with the following convex optimization problem:

$$\min_{\lambda,u,v}\left\{\begin{array}{c}\sum_{x\in[X]} n_x u_x + \sum_{y\in[Y]} m_y v_y - \sum_{xyk\in[X\times Y\times K]} \hat{\mu}_{xy}\phi_{xyk}\lambda_k \\ +2\sum_{xy\in[X\times Y]}\sqrt{n_x m_y}e^{\frac{(\Phi\lambda)_{xy}-u_x-v_y}{2}} + \sum_{x\in[X]} n_x e^{-u_x} + \sum_{y\in[Y]} m_y e^{-v_y}\end{array}\right\}. \tag{5.4.12}$$

Note that the solution to the problem does not change if we replace $(\hat{\mu}_{xy}, \hat{\mu}_{x0}, \hat{\mu}_{0y})$ and the margins n_x and m_y by the same quantities divided by a scalar. This is related to the *constant return to scale* property of the Choo and Siow model: the behavior and the welfare of the market participants do not depend on the overall number of individuals, provided the frequency of each type remains constant. Indeed, what matters to the individuals are the types of their potential partners, and not the individual partners themselves. Hence the overall size of the population is irrelevant to the matching estimation procedure. As a result, we shall therefore introduce $I=\sum_{xy}\hat{\mu}_{xy}+\sum_x \hat{\mu}_{x0}+\sum_y \hat{\mu}_{0y}$, the total number of matches observed, and instead of $\hat{\mu}$ we shall rather work with

$$\hat{\pi}=\frac{\hat{\mu}}{I}, f_x=\frac{n_x}{I}, \text{ and } g_y=\frac{m_y}{I}. \tag{5.4.13}$$

Again, one can reformulate problem (5.4.12) as a generalized linear model with a logarithmic link function (i.e., a Poisson regression). We denote by $B=XY+X+Y$ the number of household types, so that $b\in[B]$ indexes the household types, listed (in that order) as $xy\in[X\times Y]$ (married households), $x0$ for $x\in[X]$ (single male households), and $0y$ for $y\in[Y]$ (single female households). We shall then take as dependent variables the masses $\hat{\mu}_b$ of households of type $b\in[B]$. The regressor matrix R includes the previously introduced regressors ϕ_{xyk}, plus the categorical dummies $\mathbf{1}\{x\in b\}$ for $x\in[X]$ and $\mathbf{1}\{y\in b\}$ for $y\in[Y]$; hence we have

$$R=\begin{pmatrix}\Phi & -\mathbf{I}_X\otimes\mathbf{1}_Y & -\mathbf{1}_X\otimes\mathbf{I}_Y \\ \mathbf{0}_{X\times K} & -\mathbf{I}_X & \mathbf{0}_{X\times Y} \\ \mathbf{0}_{Y\times K} & \mathbf{0}_{Y\times X} & -\mathbf{I}_Y\end{pmatrix}.$$

The matrix R therefore has shape $B\times P$, where $P=K+X+Y$. Setting $\tilde{u}_x=u_x-\log f_x$ and $\tilde{v}_y=v_y-\log g_y$, the Poisson regression parameter $\theta=(\lambda^\top,\tilde{u}^\top,\tilde{v}^\top)^\top$ is of size P, and, denoting $\mathbf{\Delta}_s=diag(s)$, problem (5.4.12) rewrites more compactly as

$$\max_{\theta\in\mathbb{R}^P}\left\{\hat{\pi}^\top R\theta - s^\top\exp(\mathbf{\Delta}_s^{-1}R\theta)\right\}, \tag{5.4.14}$$

where the vector of household sizes s is defined by $s_b = 2$ if b is a matched household, which is when $b = xy \in [X \times Y]$, and $s_b = 1$ if it is a single household, which is when $b = x0$ or $b = 0y$. Problem (5.4.14) is thus a weighted Poisson regression with dependent variable $\hat{\pi}$, regressor matrix $\mathbf{\Delta}_s^{-1}R$, and weights (s_b), which has optimality conditions

$$R^\top \exp(\mathbf{\Delta}_s^{-1}R\theta) = R^\top \hat{\pi}, \tag{5.4.15}$$

from which, letting $\mathbf{V}_\pi = \mathbf{\Delta}_\pi - \pi\pi^\top$ be the asymptotic covariance matrix of $\hat{\pi}$, the asymptotic distribution of $\hat{\theta}$ can easily be deduced from theorem A.2.8 in appendix A.2.9. We are able to state the following result, which is illustrated with Python code in appendix B.5.5:

Theorem 5.4.1 (Galichon and Salanié). *The parametric matching surplus estimator $\hat{\lambda}$ defined by* (5.4.11) *can be obtained as the first component of the regression coefficient $\hat{\theta}$ of a weighted Poisson regression with dependent variable $(\hat{\mu}_b)$, regressor matrix $\mathbf{\Delta}_s^{-1}R$, and weights (s_b). Letting $B = R^\top \mathbf{\Delta}_\pi \mathbf{\Delta}_s^{-1} R$ and $C = R^\top \mathbf{V}_\pi R$, the convergence $\hat{\theta} \to \theta_0$ holds almost surely as $N \to +\infty$, and*

$$\sqrt{N}(\hat{\theta} - \theta_0) \xrightarrow{d} \mathcal{N}\left(0, B^{-1}CB^{-1}\right). \tag{5.4.16}$$

Note that, unlike the models discussed in the previous section, the estimation method just presented formulates as a Poisson regression, but not as a logistic regression: indeed, the y-fixed effects are missing in the $x0$ observations, and the x-fixed effects are missing in the $0y$ observations.

Maximum likelihood estimation. We now investigate the maximum likelihood estimator procedure in this model. The likelihood of observation $b \in [B]$ in the sample is given by expression $\pi_b^\theta = \exp((\mathbf{\Delta}_s^{-1}R\theta)_b)/\mathbf{1}_B^\top \exp\left(\mathbf{\Delta}_s^{-1}R\theta\right)$, and the maximum likelihood estimator is

$$\max_\theta \left\{\hat{\pi}^\top \mathbf{\Delta}_s^{-1}R\theta - \log\left(\mathbf{1}_B^\top \exp\left(\mathbf{\Delta}_s^{-1}R\theta\right)\right)\right\}, \tag{5.4.17}$$

which is this time a logistic regression, whose first-order conditions are

$$R^\top \mathbf{\Delta}_s^{-1}\pi^{\hat{\theta}} = R^\top \mathbf{\Delta}_s^{-1}\hat{\pi}, \tag{5.4.18}$$

or, more explicitly,

$$\begin{cases} \sum_{y\in[Y]} \frac{1}{2}\pi_{xy}^\theta + \pi_{x0}^\theta = \sum_{y\in[Y]} \frac{1}{2}\hat{\pi}_{xy} + \hat{\pi}_{x0}, \\ \sum_{x\in[X]} \frac{1}{2}\pi_{xy}^\theta + \pi_{0y}^\theta = \sum_{x\in[X]} \frac{1}{2}\hat{\pi}_{xy} + \hat{\pi}_{0y}, \\ \sum_{xy\in[X\times Y]} \pi_{xy}^\theta \phi_{xyk} = \sum_{xy\in[X\times Y]} \hat{\pi}_{xy}\phi_{xyk}. \end{cases} \tag{5.4.19}$$

This is reminiscent of equations (5.4.11), but differs by a factor ½ in front of π^{θ}_{xy} and $\hat{\pi}_{xy}$. This has a straightforward interpretation: while in the moment matching method we match the number of individuals in the sample with the predicted number of individuals of each type, in the maximum likelihood estimation procedure, we see from (5.4.19) that we are matching the average share of the household represented by individuals of each type: indeed, if x is in a matched household of type xy, then he stands for half of that household, hence the weight one-half, and if he is in a single household of type $x0$, he stands for the entire household.

The reader can verify that the linearization of equation (5.4.19) recovers standard results on maximum likelihood estimation. We summarize all of this into:

Theorem 5.4.2 (Maximum likelihood estimation in the Choo and Siow model). *The maximum likelihood estimator $\hat{\lambda}$ defined by* (5.4.17) *can be obtained as a multinomial logistic regression with dependent variable $\hat{\mu}$ and regressor matrix $\mathbf{\Delta}_s^{-1}R$. Assuming that the model is well specified, that is, the population π coincides with its predicted counterpart π^{θ_0} at the true value of the parameter, and letting*

$$B_0 = R^{\top}\mathbf{\Delta}_s^{-1}\mathbf{\Delta}_{\pi}\mathbf{\Delta}_s^{-1}R \text{ and } C_0 = R^{\top}\mathbf{\Delta}_s^{-1}\mathbf{V}_{\pi}\mathbf{\Delta}_s^{-1}R \tag{5.4.20}$$

where we recall that $\mathbf{V}_{\pi} = \mathbf{\Delta}_{\pi} - \pi\pi^{\top}$, the convergence $\hat{\theta} \to \theta_0$ holds almost surely as $N \to +\infty$, and

$$\sqrt{N}(\hat{\theta} - \theta_0) \xrightarrow{d} \mathcal{N}\left(0, B_0^{-1}C_0B_0^{-1}\right). \tag{5.4.21}$$

5.5 Coalition Formation Models

In this section, we generalize the setting of section 5.4 by allowing agents to form any type of coalition. We are thus moving away from bipartite, one-to-one models to more general models. The matching models analyzed in this section, which were investigated in the bipartite case in Pauline Corblet's doctoral thesis, are no longer two-sided: the market participants need not be drawn from two distinct populations, so the models developed in this section can apply to same-sex marriage, for example. They need not be one-to-one either: in these models, more than two agents may form a coalition, and thus the tools developed in this section apply to labor market situations where one firm can hire several workers. However, the models we consider in this section are still of the transferable utility kind, as they will assume that members of a coalition can split the total surplus in an additive manner.

Let $[Z]=\{1,\ldots,Z\}$ be the set of agents' types, and consider a large population with an exogenous mass q_z of agents of type z. A *coalition* is a group of agents who come in various numbers and types. If B is the total number of possible coalitions, a coalition $b \in [B]$ is thus represented by the number Σ_{zb} of agents of type z in the coalition, for every $z \in [Z]$. Let $\mathbf{\Sigma}$ be the $Z \times B$ matrix of term Σ_{zb}. The size s_b of a coalition b is the total number of its members, that is,

$$s_b = \sum_{z\in[Z]} \Sigma_{zb}, \text{ so } s = \mathbf{\Sigma}^\top \mathbf{1}_Z. \tag{5.5.1}$$

Given a coalition $b \in [B]$, $\mu_b \geq 0$ denotes the mass of coalitions of type b formed at equilibrium, which is subject to the population counting condition

$$q_z = \sum_{b\in\mathbb{N}^Z} \mu_b \Sigma_{zb}, \text{ so } q = \mathbf{\Sigma}\mu, \tag{5.5.2}$$

which expresses that μ_b correctly accounts for the total mass of individuals of type z in the population. Assume that the total systematic part of the surplus to be shared between members of a coalition b is ϕ_b, an exogenous quantity called the *coalition surplus* associated with coalition b. We postulate that, at equilibrium, this surplus is split in an endogenous manner, in such a way that U_{zb} is the share of the surplus that goes to z if $\Sigma_{zb} > 0$, and we make sure to set $U_{zb} = -\infty$ if z is not represented in coalition b, that is, if $\Sigma_{zb} = 0$. We have thus

$$\phi_b = \sum_{z\in[Z]} \Sigma_{zb} U_{zb}, \tag{5.5.3}$$

with the convention that $\Sigma_{zb} U_{zb} = 0$ when $\Sigma_{zb} = 0$.

Before we go further into describing the equilibrium, it is a good time to take a few examples to map these notations to familiar concepts.

Example 5.3 (The Choo and Siow model as a coalition model). In the Choo and Siow model described in section 5.4, $Z = X + Y$ is the number of types of men and women; $z \in [1:X]$ will index men's types and $z \in [X+1:X+Y]$ will index women's types. The masses of individuals of each type are $q_z = n_z$ for $z \in [1:X]$ (in which case z is a man's type), and $q_z = m_{z-X}$ for $z \in [X+1:X+Y]$ (in which case z is a woman's type). Consistent with our convention of normalizing the surplus of unmatched individuals to zero, we set $\phi_b = 0$ for coalitions of a single individual, that is, when $s_b = 1$. We set $\phi_b = \phi_{xy}$ when coalition b contains exactly one man of type x and one woman of type y, that is, when $s_b = 2$ and $\Sigma_{xb} = \Sigma_{(y+X)b} = 1$. We set $\phi_b = -\infty$ in all other cases. Then, one shall see that $\mu_b = 0$ unless b is a coalition of the form xy, $x0$, or $0y$, and (5.5.2) rewrites in this context as the classical margin conditions $\sum_y \mu_{xy} + \mu_{x0} = n_x$ and $\sum_x \mu_{xy} + \mu_{0y} = m_y$, which we saw as conditions (5.4.3) in section 5.4. If b is the coalition of type xy, U_{zb} is what we denoted in section 5.4 as U_{xy} if $z = x$, and V_{xy} if $z = y$, so that equation (5.5.3) boils down to $U_{xy} + V_{xy} = \phi_{xy}$, which we saw in equation (5.4.1) in section 5.4.

Example 5.4 (A model of same-sex marriage). Ciscato, Galichon, and Goussé estimate a model of same-sex marriage, which can be recast in the present framework. Consider a homosexual population, and let $[Z]$ be the set of individuals' types. We rule out coalitions b unless $s_b \leq 2$. If $s_b = 1$, then the surplus is zero; if $s_b = 2$, then either there is an index z such that $\Sigma_{zb} = 2$, in which case the coalition will represent a match between two individuals with the same observable type; or there are two $z \neq z'$ such that $\Sigma_{zb} = \Sigma_{z'b} = 1$, which means that the coalition is a match between two partners of type z and z'. The quantity $(\Phi\lambda)_b$ is interpreted as the joint surplus to be shared between partners. Code for this specific example is provided in appendix B.5.6.

Example 5.5 (A model of the labor market). Corblet proposes a model of one-to-many matching where one firm hires several workers. A simplistic presentation of her model is as follows. Assume the type of a firm is its industry, and it is denoted by $z \in [X]$. Each firm can hire a certain number of executives who all have the same type denoted by $z = X + 1$; a certain number of engineers whose unique type is $z = X + 2$; and a certain number of salespeople whose unique type is $z = X + 3$. As a result, ϕ_b will be finite if and only if $\sum_{z \in [1:X]} \Sigma_{zb} = 1$ (one firm and an arbitrary set of workers) or $\sum_{z \in [1:X]} \Sigma_{zb} = 0$ and $\sum z \in [X + 1 : Z] = 1$ (unassigned worker), and the surplus will depend on the industry of the firm, and the numbers b_{X+1} of its executives, b_{X+2} of its engineers, and b_{X+3} of its salespeople. The industry-specific production function may be, for instance, a constant elasticity of substitution (CES) function of the various groups of employees.

Equilibrium. If an agent i of type z chooses a coalition b in which it belongs (i.e., such that $\Sigma_{zb} \geq 1$), they get utility $U_{zb} + \varepsilon_{ib}$. The agent therefore solves the following discrete choice problem:

$$u_i = \max_{b \in [B]} \{U_{zb} + \varepsilon_{ib}\}, \tag{5.5.4}$$

where we assume that the utility shocks ε_{ib} are i.i.d. with a centered Gumbel distribution, so that we are in the logit framework. Taking the average of u_i over the individuals i of type z, one obtains the classical log-sum-exp formula for u_z, the mean welfare of type z, namely, $u_z = \log(\sum \exp U_{zb})$. The probability that an individual of type z picks a coalition $b \in [B]$ is thus $\exp(U_{zb} - u_z)$. As the overall mass of individuals of type z is q_z, the mass of individuals of type z appearing in coalition b is $q_z \exp(U_{zb} - u_z)$; this number should coincide with $\mu_b \Sigma_{zb}$, and one can thus summarize these findings into:

$$u_z = \log\Big(\sum_{b \in [B]} \exp U_{zb}\Big) \quad \text{and} \quad \mu_b \Sigma_{zb} = q_z \exp\left(U_{zb} - u_z\right). \tag{5.5.5}$$

Taking the logarithm of the second equation displayed in (5.5.5), multiplying by Σ_{zb}, and summing over z yields

$$s_b \log \mu_b = \sum_{z \in [Z]} \Sigma_{zb} \log \frac{q_z}{\Sigma_{zb}} + \phi_b - (\mathbf{\Sigma}^\top u)_b = -s_b \ell_b + \phi_b - (\mathbf{\Sigma}^\top \tilde{u})_b, \tag{5.5.6}$$

where we have defined $\ell_b := (\sum_z \Sigma_{zb} \log \Sigma_{zb}) / s_b$ and $\tilde{u}_z := u_z - \log q_z$, which leads to

$$\mu_b = \exp(-\ell_b) \exp(\frac{\phi_b - (\mathbf{\Sigma}^\top \tilde{u})_b}{s_b}), \tag{5.5.7}$$

and the price-like vector $\tilde{u}_z$ is adjusted at equilibrium by the population equation (5.5.2).

Estimation. The estimation of the coalition surplus ϕ_b is our focus in this section. We assume that we observe the vector $\hat{\mu}_b$ of empirical frequencies of coalitions b. Taking, as before, a linear parametrization $\phi_b^\lambda = (\Phi\lambda)_b = \sum_k \phi_{bk} \lambda_k$, where Φ is a $B \times K$ matrix, we shall look for the estimator $\hat{\lambda}$ called the *coalition surplus estimator* such that the equilibrium vector μ of masses of coalitions predicted by the parameter λ has the same moments as those observed in the sample. In other words, we shall seek λ that leads to a predicted vector μ such that the population counting and the moment matching conditions hold:

$$\begin{cases} \sum_{b \in \mathbb{N}^Z} \mu_b \Sigma_{zb} = \sum_{b \in \mathbb{N}^Z} \hat{\mu}_b \Sigma_{zb}, \\ \sum_{b \in [B]} \mu_b \phi_{bk} = \sum_{b \in [B]} \hat{\mu}_b \phi_{bk}. \end{cases} \tag{5.5.8}$$

Calling $\theta = (\lambda^\top, \tilde{u}^\top)^\top$ the parameter of size $P = K + Z$, the relation becomes

$$\mu_b^\theta = \exp\left(-\ell_b\right) \exp\left(\frac{(\Phi\lambda)_b - (\mathbf{\Sigma}^\top \tilde{u})_b}{s_b}\right).$$

The parameter θ satisfies the population counting and the moment matching conditions (5.5.8), yielding the system of equations

$$\begin{cases} \sum_{b\in[B]} \mu_b^\theta \Sigma_{zb} = \sum_{b\in[B]} \hat{\mu}_b \Sigma_{zb}, \\ \sum_{b\in[B]} \mu_b^\theta \phi_{bk} = \sum_{b\in[B]} \hat{\mu}_b \phi_{bk}, \end{cases}$$

which, as usual, can be interpreted as the first-order conditions associated to the following optimization problem over $(\lambda, \tilde{u})$:

$$\max_{\lambda,\tilde{u}} \left\{ \sum_{b\in[B]} \hat{\mu}_b \left((\Phi\lambda)_b - (\mathbf{\Sigma}^\top \tilde{u})_b \right) - \sum_{b\in[B]} s_b e^{-\ell_b} \exp\left(\frac{(\Phi\lambda)_b - (\mathbf{\Sigma}^\top \tilde{u})_b}{s_b} \right) \right\}. \tag{5.5.9}$$

We see that problem (5.5.9) can be interpreted as a Poisson regression. We introduce the vector of weights $w_b = s_b \exp(-\ell_b)$, the regressor matrix

$$R = diag(s)^{-1} \left(\Phi, \quad -(\mathbf{\Sigma})^\top \right), \tag{5.5.10}$$

and the dependent variable $\hat{\mu}_b \exp(\ell_b)$, and we obtain the following claim that constitutes an extension of theorem 5.4.1:

Theorem 5.5.1. *The coalition surplus estimator $\hat{\lambda}$ defined above can be estimated using a weighted Poisson regression of dependent variable $\hat{\mu}_b \exp(\ell_b)$ on regressor matrix R defined by* (5.5.10), *with weights $w_b = s_b \exp(-\ell_b)$.*

Python code is provided in appendix B.5.6; it is written for the case of surplus estimation in same-sex marriage of example 5.4, but can easily be adapted for other applications.

5.6 Quasilinear Hedonic Pricing Models

In a hedonic model, producers and consumers may produce and consume one unit of a good that comes in several varieties or qualities. There is heterogeneity among producers and consumers, and, as a result, the unitary production cost to a producer depends on the interaction of the characteristics of the producer and of the good; and, similarly, on the consumer's side. Each quality has its own unit price, called *hedonic price*, which is determined at equilibrium by the adjustment of supply and demand. Both producers and consumers have utilities that are quasilinear in the price.

As an introductory example, consider the market for a car passenger service where the city's surface is divided into Z areas. There is a mass n_x of drivers with characteristics $x \in [X]$, where X represents the number of all possible characteristics of drivers, including their current location. Each driver of type x incurs an (exogenous) systematic cost C_{xz} to pick up a passenger in area $z \in [Z]$; this cost is dependent not only on the driver's current location, but also on other factors such as their willingness to travel longer distances. The surplus of a driver i of type x picking up a passenger in area z is given by

$$p_z - C_{xz} + \varepsilon_{iz}, \tag{5.6.1}$$

where p_z is the (endogenous) payment received for service in area z, and ε_{iz} is a random utility shock that accounts for unobserved preferences. We assume full participation. Similarly, there is a mass m_y of passengers with characteristics $y \in [Y]$. A passenger j of type y derives a surplus from the service, represented as

$$U_{yz} - p_z + \eta_{jz}, \tag{5.6.2}$$

where U_{yz} is the (exogenous) utility a passenger of type y derives from being serviced in area z, p_z is the price paid for the service, and η_{jz} is a random utility shock for passengers. All passengers participate, and their total number equals the total number of drivers.

Assuming that the random utility terms ε_{iz} and η_{jz} are i.i.d. Gumbel, the average welfare of a driver of type x and a passenger of type y can be respectively expressed as

$$u_x = \log\Big(\sum_{z \in [Z]} e^{p_z - C_{xz}}\Big) \quad \text{and} \quad v_y = \log\Big(\sum_{z \in [Z]} e^{U_{yz} - p_z}\Big), \tag{5.6.3}$$

and the mass of drivers x choosing to pick up at location z is given by $\mu_{xz} = n_x \exp\left(p_z - C_{xz} - u_x\right)$, while the mass of passengers of type y choosing location z is $\mu_{yz} = m_y \exp\left(U_{yz} - p_z - v_y\right)$.

The market clearing condition requires that the supply for any service area z equals the demand for that area, which can be formalized as

$$\sum_{x \in [X]} n_x \exp(p_z - C_{xz} - u_x) = \sum_{y \in [Y]} m_y \exp(U_{yz} - p_z - v_y), \tag{5.6.4}$$

where the left-hand side represents the total supply of drivers for area z, and the right-hand side represents the total demand from passengers for area z. This implies the logit hedonic pricing formula

$$p_z = \frac{1}{2} \log \frac{\sum_{y \in [Y]} m_y \exp(U_{yz} - v_y)}{\sum_{x \in [X]} n_x \exp(-C_{xz} - u_x)}, \tag{5.6.5}$$

which provides the value of the equilibrium price as a function of the vectors u and v, which are other equilibrium quantities. In exercise 5.6, the student is asked to show that u and v are solutions to an interesting optimization problem.

The reader should by now be used to recognizing equations (5.6.3) combined with equations (5.6.4) as the first-order conditions to a convex optimization problem. Setting $\tilde{u}_x = u_x - \log n_x$ and $\tilde{v}_y = v_y - \log m_y$, the vectors $\tilde{u}$ and $\tilde{v}$ are the solutions to

$$\min_{p,\tilde{u},\tilde{v}} \left\{ \sum_{x\in[X]} n_x \tilde{u}_x + \sum_{y\in[Y]} m_y \tilde{v}_y + \sum_{xz\in[X\times Z]} e^{p_z - C_{xz} - \tilde{u}_x} + \sum_{yz\in[Y\times Z]} e^{U_{yz} - p_z - \tilde{v}_y} \right\},$$

which again can be interpreted as the Poisson regression of the vector of dimension $(XZ + YZ)$ obtained by the concatenation of the vector of term $n_x \exp(C_{xz})$ for $xz \in [X \times Z]$ and the vector of term $m_y \exp\left(-U_{yz}\right)$ for $yz \in [Y \times Z]$, on the regressor matrix R defined by

$$R = \begin{pmatrix} \mathbf{1}_X \otimes \mathbf{I}_Z & -\mathbf{I}_X \otimes \mathbf{1}_Z & \mathbf{0} \\ -\mathbf{1}_Y \otimes \mathbf{I}_Z & \mathbf{0} & -\mathbf{I}_Y \otimes \mathbf{1}_Z \end{pmatrix},$$

which is associated with the vector of coefficients $\theta = (p, \tilde{u}, \tilde{v})$, and with a vector of weights w that is the concatenation of the vector of term $\exp\left(-C_{xz}\right)$ for $xz \in [X \times Z]$ and the vector of term $\exp(U_{yz})$ for $yz \in [Y \times Z]$. The Python code for this pricing problem is provided in appendix B.5.7.

5.7 Exercises

Exercise 5.1. *Show that formulation* (5.3.1) *of the gravity equation has the following equivalent formulation:*

$$\begin{aligned} &\max_{\lambda,u,v} \sum_{xy\in[X\times Y]} \hat{\mu}_{xy}\big((\Phi\lambda)_{xy} - u_x - v_y\big) \\ &s.t. \sum_{xy\in[X\times Y]} \exp\left((\Phi\lambda)_{xy} - u_x - v_y\right) \leq 1, \end{aligned} \tag{5.7.1}$$

which, after normalizing $\sum_x n_x = \sum_y m_y = 1$, *is in turn equivalent with*

$$\max_{\lambda,u,v} \sum_{xy\in[X\times Y]} \hat{\mu}_{xy}\big((\Phi\lambda)_{xy} - u_x - v_y\big) - \log\Big(\sum_{xy\in[X\times Y]} \exp\left((\Phi\lambda)_{xy} - u_x - v_y\right)\Big). \tag{5.7.2}$$

Exercise 5.2. *Show that the roles played by x and y in the gravity equation are symmetric: that is, the same formulation of the gravity equation would have been obtained by picking x as a categorical dependent variable, and including an x-fixed effect into the macro logistic regression.*

Exercise 5.3. *What is the dual formulation of problem* (5.4.12) *in section 5.4? Interpret.*

Exercise 5.4. *Consider a model of one-to-one matching with linear taxes. If a worker of type $x \in [X]$ works with a firm $y \in [Y]$ and they agree on a wage w_{xy}, then their respective utilities are*

$$\alpha_{xy} + (1-\tau) w_{xy} + \varepsilon_y \qquad \text{and} \qquad \gamma_{xy} - w_{xy} + \eta_x,$$

where α_{xy} is the monetary valuation by the worker of the job's amenities, $\tau \in (0,1)$ is the (flat) tax rate, γ_{xy} is the output of worker x appropriated by the firm, measured in monetary units, and ε_y and η_x are random utility shocks. It is assumed that the payoff of an unmatched worker (respectively, firm) is ε_0 (respectively, η_0). We assume that $(\varepsilon_y)_{y \in [0:Y]}$ and $(\eta_x)_{x \in [0:X]}$ are i.i.d. Gumbel random vectors of respective dimensions $X+1$ and $Y+1$.

(i) Show that the equilibrium in this market maximizes an objective function similar to expression (5.4.12) *in section 5.4. Comment on the differences.*

(ii) Do the two quantities

$$A = \sum_{xy \in [X \times Y]} \mu_{xy} \alpha_{xy} - \sum_{xy \in [X \times Y]} \mu_{xy} \log \mu_{xy} - \sum_{x \in [X]} \mu_{x0} \log \mu_{x0},$$

$$\Gamma = \sum_{xy \in [X \times Y]} \mu_{xy} \gamma_{xy} - \sum_{xy \in [X \times Y]} \mu_{xy} \log \mu_{xy} - \sum_{y \in [Y]} \mu_{0y} \log \mu_{0y}$$

increase or decrease when the tax rate τ increases? Same question for their sum $A + \Gamma$. Interpret.

Exercise 5.5. *The goal of this exercise is to motivate the assumption we made in section 5.4 when we assumed that the transfer τ between i and j depended on the observable types x_i and y_j only, rather than being idiosyncratic and depending on i and j. Assume that the transfers depend on i and j, denote it by τ_{ij}, and consider two men i and i' of type x and two women j and j' of type y, and assume i is matched with j and i' with j'. Show that $\tau_{ij} \geq \tau_{ij'}$, and that $\tau_{i'j'} \geq \tau_{i'j}$, and, similarly, that $\tau_{ij} \leq \tau_{i'j}$ and $\tau_{i'j'} \leq \tau_{ij'}$. Conclude that all these inequalities* *are equalities. Hence, at equilibrium, the transfers are not individual-specific but type-specific.*

Exercise 5.6. *In the context of the hedonic equilibrium pricing seen in section 5.6, define the* $Z \times X$ *matrix* A *by* $A_{zx} = \exp(-C_{xz})$, *the* $Z \times Y$ *matrix* B *by* $B_{zy} = \exp U_{yz}$, *the* X*-dimensional vector* a *by* $a_x = \exp(-\tilde{u}_x)$, *and the* Y*-dimensional vector* b *by* $b_y = \exp\left(-\tilde{v}_y\right)$. *Show that the vector of equilibrium prices* p *is such that*

$$p_z = \frac{1}{2} \log \frac{(Bb)_z}{(Aa)_z},$$

where (a_x) *and* (b_y) *are the solutions to*

$$\max_{a_x \geq 0, b_y \geq 0} \sum_{x \in [X]} n_x \log a_x + \sum_{y \in [Y]} m_y \log b_y - 2 \sum_{z \in [Z]} \sqrt{(Aa)_z \, (Bb)_z}.$$

Exercise 5.7. *We set ourselves in the context of the matching model of section 5.4. Use appendix A.2.10 on exponential families of distributions to show that the model can be recast as an exponential family with* $\theta^\top = (\lambda^\top, a^\top, b^\top)$ *and*

$$\Lambda^*(\theta) = \log\Big(\sum_{xy \in [X \times Y]} e^{\frac{(\Phi\lambda)_{xy} - a_x - b_y}{2}} + \sum_{x \in [X]} e^{-a_x} + \sum_{y \in [Y]} e^{-b_y}\Big),$$

and interpret the optimality conditions in (A.2.58) *in this context.*

5.8 Problems

Problem 5.1 (An econometric model of the marriage market). *Our data is Choo and Siow's (2006) marriage data. The data is made of the marriages between men and women of various age categories* x *and* y. $x = 0$ *is 16 year old (yo) male,* $x = 1$ *is 17 yo male, etc., until* $x = 24$, *which is 40 yo male. Similarly,* $y = 0$ *to* $y = 25$ *denote females from age 16 to age 40. You should run the code in appendix B.5.8 to load* $\mu_{xy}, \mu_{x0}, \mu_{0y}$ *before answering the following questions:*

(i) *Assume that the joint surplus is zero for any pair* xy, *i.e.,* $\phi_{xy} = 0$. *What fraction of individuals is married according to the model?*

(ii) *We now consider a model of the surplus given by* $(\Phi\lambda)_{xy} = \lambda_0 - \lambda_1 |x - y| - \lambda_2 \mathbf{1}\{x > y\} - \lambda_3 \mathbf{1}\{x < y\}$ *for* $xy \in [X \times Y]$ *and* $(\Phi\lambda)_{x0} = (\Phi\lambda)_{0y} = 0$.

Compute the estimator of λ *matching the moments* ϕ_{xyk} *for* $k = 0, \ldots, 3$.

(iii) *Provide the confidence interval for the previous estimator.*

Problem 5.2 (Estimating the gravity equation). *The code provided in appendix B.5.9 loads trade data from the book by Yotov, Piermartini, Monteiro, and Larch (2016) for years 1986, 1990, 1994, 1998, 2002, and 2006. Based on that dataset:*

(i) *Estimate the gravity equation with regressors* `'ln_DIST'`, `'CNTG'`, `'LANG'`, `'CLNY'` *for year 1986 (corresponding to index $t=0$).*

(ii) *Now include the full set of years, and time-country fixed effects u_{tx} and v_{ty}. How does one include these fixed effects in the regressor matrix using Kronecker products? Run the regression and report the results.*

Problem 5.3 (Welfare in a Ricardian model of trade). *In this problem, we investigate Pareto efficient allocations in a Ricardian model of trade. There are Ω types of traded goods, and X countries. The total mass of the labor force in country $x \in [X]$ is n_x. One unit of the labor force in country $x \in [X]$ yields $\zeta_{x\omega}$ units of good $\omega \in [\Omega]$, so that $\zeta_{x\omega}$ is the matrix controlling the comparative advantages.*

We classically model "iceberg costs" as follows: for any type of good, one unit of good (when measured at departure) exported from country $x \in [X]$ to country $y \in [X]$ yields $1/\tau_{xy}$ units of the same good (when measured at destination), so $\tau_{xy} \geq 1$ is some kind of measure of the losses occurring during transit. We define $\mu_{xy\omega}$ as the mass of goods $\omega \in [\Omega]$ consumed in country $y \in [X]$ that originates from country $x \in [X]$. We define $q_{y\omega}$ as the total mass of goods $\omega \in [\Omega]$ consumed in country $y \in [X]$.

The utility of the representative agent in country $y \in [X]$ is $Q_y(q)$, which is assumed not to depend on the entries $q_{y'\omega}$ for $y' \neq y$, and which is a concave and nondecreasing function of $(q_{y\omega})_\omega$.

(i) *Show that Pareto efficient outcomes are solutions to the welfare maximization problem*

$$\max_{q \in \mathbb{R}_+^{X\Omega}, \mu \in \mathbb{R}_+^{X^2\Omega}} \sum_{y \in [X]} P_y Q_y(q) \tag{5.8.1}$$

$$s.t.\ n_x = \sum_{y\omega \in [X \times \Omega]} \frac{\tau_{xy}\mu_{xy\omega}}{\zeta_{x\omega}}, \ \forall x \in [X], \tag{5.8.2}$$

$$q_{y\omega} = \sum_{x \in [X]} \mu_{xy\omega}, \ \forall y\omega \in [X \times \Omega] \tag{5.8.3}$$

for a choice $P_y \geq 0$ of the Pareto weight of each country $y \in [X]$.

(ii) *Show that the dual to problem* (5.8.1) *in the previous question is*

$$\min_{w\in\mathbb{R}^X, p\in\mathbb{R}^{X\Omega}} \sum_{x\in[X]} n_x w_x + \sum_{y\in[Y]} P_y S_y\left(\left(\frac{p_{y\omega}}{P_y}\right)_{\omega\in[\Omega]}\right) \quad (5.8.4)$$

$$s.t.\ p_{y\omega} \leq \frac{\tau_{xy} w_x}{\zeta_{x\omega}} \quad \forall xy\omega \in [X \times X \times \Omega], \quad (5.8.5)$$

where for $\varphi \in \mathbb{R}^\Omega$, *we have defined* $S_y(\varphi) = \max_{q\in\mathbb{R}_+^\Omega}\{Q_y(q) - \sum_{\omega\in[\Omega]} \varphi_\omega q_\omega\}$.

(iii) *Interpret the variables* w_x *and* $p_{y\omega}$ *in the dual problem* (5.8.4) *and the function* $S_y(\varphi)$ *defined just above. Show that if* w *and* p *are solutions to the dual problem, then* $p_{y\omega} = \min_x\{\tau_{xy} w_x/\zeta_{x\omega}\}$ *and* $w_x = \max_{y\omega}\{p_{y\omega}\zeta_{x\omega}/\tau_{xy}\}$, *and show that if* $\mu_{xy\omega} > 0$, *then constraint* (5.8.5) *is binding. Show that one has* $\mu_{xy\omega} = q_{y\omega}\partial p_{y\omega}/\partial w_x$ *whenever the right-hand side exists.*

(iv) *Using the interpretation given in the previous question of the variables* w_x *and* $p_{y\omega}$, *express country* x*'s trade imbalance, defined as the country's labor income minus the total value of its imports, as a function of the aforementioned variables and* $q_{y\omega}$. *Deduce that, in general, all Pareto efficient allocations are not Walrasian equilibria.*

(v) *Assume only for this question that* $Q_y(q) = \sum_\omega M_{y\omega} q_{y\omega}$ *for* M *a vector of size* $X\Omega$. *Show that problem (5.8.1) is now a linear programming problem and write its dual. Compute numerically using Gurobi based on data generated in appendix B.5.10. Show that the problem has a solution in closed-form and compare the numerical expression given by Gurobi with the one given by the closed-form expression.*

Problem 5.4 (The Eaton–Kortum model of trade). *This problem is a sequel to problem* (5.3), *where we investigate the Walrasian equilibrium. We now assume that country* $x \in [X]$ *maximizes utility subject to the "balance of trade" constraint, thus solving the program*

$$\max_{(q_{x\omega})_\omega \in\mathbb{R}^\Omega} Q_x(q) \quad (5.8.6)$$

$$s.t.\ n_x w_x = \sum_{\omega\in[\Omega]} q_{x\omega} p_{x\omega}. \quad (5.8.7)$$

(i) *Show that, at equilibrium, the allocations* (q, μ) *are Pareto efficient, and, more specifically, that they solve program* (5.8.1) *for some Pareto weights*

(P_x), which one shall relate to the Lagrange multiplier associated with the constraint (5.8.7).

In the sequel, we assume that the utility functions $Q_y(q)$ are all of the constant elasticity of substitution (CES) form with constant return to scale and elasticity of substitution $\eta > 1$, that is,

$$Q_y(q) = \left(\frac{1}{\Omega} \sum_{\omega \in [\Omega]} q_{y\omega}^{\frac{\eta-1}{\eta}}\right)^{\frac{\eta}{\eta-1}} \text{ for all } y \in [X]. \tag{5.8.8}$$

(ii) Show that, at equilibrium, the Pareto weight of country $x \in [X]$ is given by

$$P_x = \left(\frac{1}{\Omega^\eta} \sum_\omega p_{x\omega}^{1-\eta}\right)^{\frac{1}{1-\eta}}, \tag{5.8.9}$$

and deduce that the quantity of good $\omega \in [\Omega]$ consumed by country $x \in [X]$ is given by

$$q_{x\omega} = \frac{n_x w_x p_{x\omega}^{-\eta}}{\sum_{\omega \in [\Omega]} p_{x\omega}^{1-\eta}}. \tag{5.8.10}$$

In the remainder of the problem, we assume in addition that $\zeta_{x\omega}$ are i.i.d. Fréchet variables, that is, given $X\Omega$ i.i.d. draws $\varepsilon_{x\omega}$ from a standard Gumbel distribution,

$$\zeta_{x\omega} = T_x^{1/\theta} \exp\left(\frac{\varepsilon_{x\omega}}{\theta}\right) \tag{5.8.11}$$

for some positive numbers T_x and θ.

(iii) Letting $A_y = (\sum_x T_x \left(\tau_{xy} w_x\right)^{-\theta})^{-1/\theta}$, show that $p_{y\omega}$ is equal in distribution to $A_y \exp(-\epsilon_\omega/\theta)$, where ϵ_ω are Ω draws from an i.i.d. Gumbel distribution over $\mathbb{R}$.

(iv) Defining $f_\Omega(t) = \Omega^{-1} \sum_\omega e^{t\epsilon_\omega}$, and letting $\mu_{xy} = \sum_\omega \mu_{xy\omega}$ be the total trade flows from country x to country y, deduce that

$$\begin{cases} q_{y\omega} = \dfrac{n_y w_y}{A_y} \dfrac{\Omega^{-1} \exp(\eta\epsilon_\omega/\theta)}{f_\Omega((\eta-1)/\theta)}, \\ \mu_{xy\omega} = \dfrac{n_y w_y}{A_y^{1-\theta}} \left(\tau_{xy} w_x\right)^{-\theta} T_x \dfrac{\Omega^{-1} \exp(\eta\epsilon_\omega/\theta)}{f_\Omega((\eta-1)/\theta)}, \\ \mu_{xy} = \dfrac{n_y w_y}{A_y^{1-\theta}} \left(\tau_{xy} w_x\right)^{-\theta} T_x \dfrac{f_\Omega(\eta/\theta)}{f_\Omega((\eta-1)/\theta)}. \end{cases} \tag{5.8.12}$$

(v) *When $\Omega \to +\infty$, one has $f_\Omega(t) \to f(t) = \mathbb{E}\left[e^{t\epsilon}\right] = \Gamma(1-t)$, where Γ is Euler's gamma function. Show that μ_{xy} tends to*

$$\mu_{xy} = \frac{n_y w_y}{A_y^{1-\theta}} \left(\tau_{xy} w_x\right)^{-\theta} T_x \frac{\Gamma(1-\frac{\eta}{\theta})}{\Gamma(1+\frac{1-\eta}{\theta})}. \tag{5.8.13}$$

5.9 References and Notes

Optimal transport with entropic regularization refers to a modification of the classical optimal transport problem, where an entropy term is added to the optimization objective to encourage smoother, more computationally tractable solutions. This is an old idea, of which historical details are given in Idel (2016). As discussed there, algorithm 5.2.1 is known under many names, including the *iterative proportional fitting procedure, matrix scaling, Sinkhorn–Knopp algorithm, Kruithof double-factor model, RAS method, matrix balancing,* and *Furness iteration procedure,* among others. In the transportation engineering literature, the role of entropy was highlighted by the work of Alan Wilson in the 1960s; see Wilson (1967) and his more recent book Wilson (2011). An early reference in the optimal transport literature is Ruschendorf (1995). In economics, the method was first applied to matching models by Galichon and Salanié (2009); subsequently, Cuturi (2013) popularized the method among the machine learning community in an influential paper, followed by the book Peyré and Cuturi (2019). The proof of the convergence of the method using Hilbert's metric is due to Franklin and Lorenz (1989), based on an important result by Birkhoff (1957) on positive matrices.

The structural gravity equation is due to Anderson and Van Wincoop (2003). The estimation using a generalized linear model was first proposed in Santos Silva and Tenreyro (2006). A comprehensive survey of the topic is in the handbook chapter by Head and Mayer (2014). Numerical methods for the estimation of the gravity equation are given in Guimarães and Portugal (2010), and in Correia, Guimarães, and Zylkin (2020). The data discussed in example 5.1 is taken from the book Yotov, Piermartini, Monteiro, and Larch (2016). The SISTA algorithm displayed as algorithm 5.3.2 was proposed by Carlier, Dupuy, Galichon, and Sun (2023), building on earlier work on "inverse optimal transport" by Dupuy and Galichon (2014), and by Dupuy, Galichon, and Sun (2019).

The Choo and Siow model originates from a 2006 paper by these authors, who incorporate logit random utility in a Beckerian model, following Becker's theory of marriage proposed in Becker (1973). Galichon and Salanié (2009 and 2022) proposed a method-of-moments estimator of a parametric version of the model, and they extended it beyond the logit model to allow for arbitrary distributions of random utilities. The estimation using a generalized linear model is worked out in Galichon and Salanié (2024).

The extension of the Choo and Siow model to more general models of coalition formation, presented in section 5.5, builds on the doctoral thesis of Corblet (2022b) and the paper by Corblet (2022a), which propose a one-to-many matching problem with general random utilities and apply it to the labor market. Section 5.5 also presents an extension of the Choo–Siow model in a different direction, based on the work of Ciscato, Galichon, and Goussé (2020), who develop a model of unipartite one-to-one matching markets with logit random utilities and apply it to the estimation of marital preferences among same-sex partners.

The hedonic model with logit heterogeneity presented in section 5.6 was proposed by Dupuy, Galichon, and Henry (2014), building on a very large literature on hedonic models. These originated with Tinbergen (1956), but were popularized with Lancaster's (1966) characteristic approach to consumer demand. Rosen (1974) investigated how competition and differentiation affect hedonic pricing. In a series of paper in the 2000s, Heckman, with Ekeland, Matzkin, and Nesheim, laid down a rigorous approach to hedonic models, and worked out various econometric approaches immune from any endogeneity bias; see Ekeland, Heckman, and Nesheim (2004) and Heckman, Matzkin, and Nesheim (2010). Chiappori, McCann, and Nesheim (2010) and Ekeland (2010) established a close connection between hedonic models and the theory of optimal transport. Chernozhukov, Galichon, Henry, and Pass (2021) generalized quantile identification methods to the multivariate case using optimal transport in a different context.

Problems 5.3 and 5.4 are based on Eaton and Kortum's (2002) model of international trade, which expands upon the classic Ricardian model of Dornbusch, Fischer, and Samuelson (1977). Alvarez and Lucas Jr. (2007) provide an analysis of equilibrium in Eaton and Kortum's model.

5.10 Summary

SUMMARY OF CHAPTER 5

In this chapter, the logistic regression framework introduced in chapter 3 demonstrates its versatility through its applicability across a wide range of econometric models. This framework captures a wide range of applications in fields such as international trade, labor economics, family economics, and demand estimation.

The "macro logistic regression model," defined in section 5.1, slightly extends the logistic regression to cluster individuals into types, allowing for a generalized form of logistic regression that aggregates individuals of the same type x. This model is particularly useful for capturing variations in utility across different types without needing to model each individual separately. The primal formulation of this model maximizes a function related to the utility of types across different choices, and it can be estimated using a Poisson regression with x-fixed effects.

The entropic optimal transport problem discussed in section 5.2 is a specialized instance of the macro logistic regression with appropriately chosen weights and y-fixed effects as sole regressors. This offers a convenient way to compute the optimal transport problem with entropic regularization using the formulation as a generalized linear model with x- and y-fixed effects.

In section 5.3, one adds other regressors beyond y-fixed effects in the macro logistic regression; this yields the gravity equation, an important econometrics framework used to model trade flows between entities, and to predict them based on various measures of distance between them. Again, this approach allows for the computation of trade flows as a Poisson regression problem, incorporating exporter- and importer-fixed effects.

The logistic regression framework is shown to apply to two-sided matching markets with transferable utility, which are discussed in section 5.4, and have application to the heterosexual marriage market and the labor market, among others. The model is a Becker model with logit heterogeneity that assumes that matches are formed to maximize a joint surplus that is then divided among the participants. One can reformulate the equilibrium conditions and moment conditions as an optimization problem that is a weighted Poisson regression, in order to estimate the surplus generated from matches between different types of agents.

Extending beyond two-sided markets, coalition formation models discussed in section 5.5 consider the formation of groups (coalitions) among agents of different types. These models generalize the concept of matching by allowing for more complex arrangements beyond pairwise matches, addressing scenarios such as labor markets where firms may hire multiple workers. The surplus from forming a coalition is distributed among its members, and, again, the equilibrium and estimation conditions for these models can be analyzed using generalized linear models, emphasizing the framework's adaptability to a variety of economic settings.

The chapter ends with section 5.6 on hedonic equilibrium pricing models, in which demand and supply form in a quality space. The hedonic equilibrium is reformulated as a generalized linear model with product, seller, and buyer fixed effects, and the computation of the hedonic prices can be computed in that way.

CHAPTER 6

Dynamic Discrete Choice

Until now, we have considered models that were static, in the sense that the decision-maker was faced with a one-time, one-off choice problem, based on a payoff associated with every option. But in many economic settings, the decision-maker needs to make a series of choices, where choices made in one period affect the options available in later periods, as well as the payoffs associated with these options. Examples such as saving decisions, career choices, maintenance decisions, and retirement decisions illustrate scenarios where the optimal long-run decision does not coincide with the "myopic" decision based only on short-term considerations. The present chapter will therefore explore dynamic models, which incorporate a dynamic programming component into models. The basic idea is to define a state space in which a decision-maker finds herself at the beginning of every period. There are Markovian transitions between states from one period to the next, and these transitions are affected by the choices agents make in each period. Under certain assumptions about the probabilistic structure of the random shocks, we will demonstrate that it is possible to associate a value with every state the decision-maker occupies at any given period. The probability of transitioning from one state to another necessitates that the agent consider the expected value of the state in the next period, conditional on the current period's choice. Consequently, choosing an option in the current period involves resolving a trade-off between the immediate payoff and the conditional expectation of the value of the next period state, appropriately discounted.

Example 6.1. In his landmark paper, Rust (1987) tells the story of an engineer who operates a fleet of buses and must decide when to send the buses into maintenance, resulting in a trade-off between operational costs and maintenance costs. At each period and for each bus, the engineer decides either to send the bus out, which will increase the mileage of its engine (and, with it, its operational costs), or to maintain it, which will reset its mileage to zero.

See the notebook Galichon (2025b) for an illustration loosely based on Rust's setting, and another one, Galichon (2025a), for a partial replication of Rust's analysis, based on the paper's original dataset.

In this chapter, we will start by examining the simplest form of a dynamic choice model, characterized by a finite state space, absence of heterogeneity, discrete time steps, and a finite horizon. We will discover that this model can be expressed through a linear programming formulation, and we will delve into the interpretation of the resulting duality. Still in the finite-horizon case, we will introduce heterogeneity by assuming that the random utility comprises i.i.d. Gumbel components, thus yielding a logit structure. Subsequently, we will transition to the infinite horizon scenario, demonstrating how the equations of the stationary model can be derived from a singular potential function called the Z function. In the limit of no discounting, that is, when the discount factor approaches 1, we will show that the model can be estimated once more by a generalized linear model. The chapter concludes with an exploration of dynamic matching problems, showing how much of the formalism introduced earlier can be straightforwardly extended to this context.

6.1 Finite Time Horizon, No Heterogeneity

We assume that each unit within the population can occupy one of X possible states. For instance, in the bus maintenance example, $x \in [X]$ could represent a mileage bracket, indicating the age of the engine. Each unit is in a specific state at any given period but will typically change states from one period to the next. The scenario unfolds over T periods of time. At the initial period, $t=1$, there is a mass N_x of units in state $x \in [X]$. We let n_{tx} be the mass of units that are at time t in state x, so that the initial condition has

$$n_{1x} = N_x. \tag{6.1.1}$$

We shall implicitly make a large market assumption, and suppose that there are initially a large number of units per state x. This is why we refer to the "mass" of units in state x, as opposed to the "number" of these units.

We define $[Y]$ as the set of decisions that can be applied to units. In the context of the bus maintenance example, there are only two possible decisions ($Y=2$): to continue operating the bus ($y=1$) or to perform maintenance ($y=2$). More generally, if we define μ_{txy} as the mass of units in state x at time t for which choice $y \in [Y]$ is applied, we will sometimes call this vector *policy variable.* As one decision is applied per unit, it follows that

$$n_{tx} = \sum_{y \in [Y]} \mu_{txy}, \quad t \in [T]. \tag{6.1.2}$$

In a vectorized manner, as explained in appendix A.3.1, we stack the entries μ_{txy} into a single-column vector μ by listing the indices in the lexicographic order (which extends the row-major order to multidimensional arrays), to get:

$$n = (\mathbf{I}_T \otimes \mathbf{I}_X \otimes \mathbf{1}_Y^\top)\mu. \tag{6.1.3}$$

The decisions can undergo constraints, such as a limit on the total number of units for which decision y is applied. We shall assume that such constraints on μ are given by

$$\mathbf{A}\mu \leq c, \tag{6.1.4}$$

where c is a vector of size L, and $\mathbf{A}$ is a matrix of size $L \times (TXY)$. A plausible instance of constraints is $\sum_x \mu_{txy} \leq c$ for some y; this will be the case in the bus example if one assumes that the capacity of the maintenance workshop is limited, thus setting a constraint on the total number of units to which decision $y = 2$ can be applied at a given time t. In that case, $\mathbf{A} = \mathbf{I}_T \otimes \mathbf{1}_X^\top \otimes (0\ 1)$. Throughout the sequel, we shall always assume that there are feasible μ such that $(\mathbf{A}\mu)_l < c_l$ for all l, hence the capacity constraint is not a threat to the feasibility of the programs we shall write.

A decision $y \in [Y]$, made at time $t \in [T]$ for a unit in state $x \in [X]$, has two effects.

- The decision has a short-term effect, represented by the current period payoff ϕ_{txy} associated with that decision. In the bus maintenance example, ϕ_{txy} for $y = 1$ (operating) is expected to be positive at lower mileage values and negative at higher mileage values, whereas for $y = 2$ (maintaining), it is anticipated to reflect a significant negative value, as the cost of an overhaul is likely to be higher by one order of magnitude than the operating profit or loss.
- However, the decision also influences the future state of the unit. We assume that if a unit is in state x and a decision y is made at time t, the probability that the unit will find itself in state x' at time $t + 1$ is $P_{x'xy}$. In the bus example, operating the vehicle (decision $y = 1$) carries a positive probability of the bus remaining in its current state, as well as a probability of transitioning to a state representing higher mileage. Conversely, if the bus undergoes maintenance (decision $y = 2$), it will transition back to the zero-mileage state with probability 1.

A careful planner therefore needs to take into account not only the short-term payoff ϕ_{txy}, but also the impact on the future stream of payoffs through the expectation of future transitions to other states. These transitions are modeled using Markov chains on the state space conditional on the decision. Letting $P_{x'xy}$ be the probability of transitioning from state x to state

x' if option y has been chosen, one has

$$n_{tx'} = \sum_{xy \in [X \times Y]} P_{x'xy} \mu_{(t-1)xy}, \quad t \in [2:T]. \tag{6.1.5}$$

Putting together (6.1.5) with (6.1.1) and (6.1.2) allows one to substitute out n in order to obtain the *evolution equations* of μ:

$$\begin{cases} \sum_{y \in [Y]} \mu_{1xy} = N_x, & t = 1, \\ \sum_{y \in [Y]} \mu_{txy} - \sum_{\tilde{x}y \in [X \times Y]} P_{x\tilde{x}y} \mu_{(t-1)\tilde{x}y} = 0, & t \in [2:T]. \end{cases} \tag{6.1.6}$$

This vectorizes into

$$\left(\mathbf{I}_T \otimes \mathbf{I}_X \otimes \mathbf{1}_Y^\top - \mathbf{J}_T \otimes \mathbf{P} \right) \mu = \mathbf{e}_T^1 \otimes N, \tag{6.1.7}$$

where $\mathbf{J}_T$ is the $T \times T$ lower shift matrix with 1s on the subdiagonal and 0s elsewhere, i.e., $(\mathbf{J}_T)_{tt'} = \mathbf{1}\{t = t' + 1\}$, and $\mathbf{P}$ is the $X \times (XY)$ matrix whose (x', xy) term is $P_{x'xy}$. Recall that $\mathbf{e}_T^1$ is the first vector of the canonical basis of $\mathbb{R}^T$.

We have defined ϕ_{txy} as the immediate payoff associated with making decision y for one unit in state x at time t. These quantities are measured in a single numéraire, so that it makes sense to sum them; for example, if one assumes geometric discounting at rate β, one could denominate all payoffs into the time $t = 0$ unit by setting $\phi_{txy} = \beta^t \phi_{xy}$. The decision-making problem will therefore be the maximization of the *intertemporal objective function*

$$\sum_{txy \in [T \times X \times Y]} \mu_{txy} \phi_{txy} \tag{6.1.8}$$

subject to the evolution equations (6.1.6) and the policy constraints (6.1.4). Clearly, this is a linear programming problem, for which an example of Python code is provided in appendix B.6.1; the program can be elegantly written in vectorized terms. Using the reminders on linear programming provided in appendix A.1.5, we can state:

Proposition 6.1.1. *The intertemporal decision-maker's problem, whose primal linear programming formulation is*

$$\max_{\mu \geq 0} \mu^\top \phi \tag{6.1.9}$$

$$s.t. \quad (\mathbf{I}_T \otimes \mathbf{I}_X \otimes \mathbf{1}_Y^\top - \mathbf{J}_T \otimes \mathbf{P})\mu = \mathbf{e}_T^1 \otimes N, \tag{6.1.10}$$

$$\mathbf{A}\mu \leq c, \tag{6.1.11}$$

has the dual formulation

$$\min_{u\in\mathbb{R}^{TX},\tau\in\mathbb{R}^L_+} \quad (\mathbf{e}^1_T \otimes N)^\top u + c^\top \tau \tag{6.1.12}$$

$$\text{s.t.} \quad (\mathbf{I}_T \otimes \mathbf{I}_X \otimes \mathbf{1}_Y - \mathbf{J}_T^\top \otimes \mathbf{P}^\top)u + \mathbf{A}^\top \tau \geq \phi. \tag{6.1.13}$$

Back to the coordinates notation, the previous dual formulation is expressed as

$$\min_{(u_{tx}),(\tau_l)\geq 0} \quad \sum_{x\in[X]} N_x u_{1x} + \sum_{l\in[L]} c_l \tau_l \tag{6.1.14}$$

$$\text{s.t.} \quad u_{tx} \geq \phi_{txy} - \sum_{l\in[L]} A_{ltxy}\tau_l + \sum_{x'\in[X]} P_{x'xy} u_{(t+1)x'}, \ \forall txy \in [T\times X\times Y],$$

$$u_{(T+1)x} = 0, \ \forall x \in [X],$$

whose interpretation calls for some remarks. Since τ is the vector of Lagrange multipliers associated with the scarcity constraints (6.1.11), $\sum_l A_{ltxy}\tau_l$ is interpreted as the shadow cost of option y for a bus in state x at time t. In the bus example, if there is an upper limit on the capacity of the maintenance warehouse, maintaining a bus will create an externality on the other buses as spots in the warehouse may run out. If $\tau^\star$ is the optimal vector (τ_l) appearing in the solution to (6.1.14), in the socially efficient solution, agents will internalize these externalities and act as if their short-term payoff ϕ_{txy} were modified into $\phi^\star_{txy} = \phi_{txy} - \sum_l A_{ltxy}\tau^\star_l$.

Conditional on knowing $\tau^\star$, the optimal value of τ, it is easy to see that the vector u can be solved in closed form by backward induction, namely,

$$\begin{cases} u_{tx} = \max_{y\in[Y]}\{\phi^\star_{txy} + \sum_{x'\in[X]} P_{x'xy} u_{(t+1)x'}\}, \ \forall tx \in [T\times X], \\ u_{(T+1)x} = 0, \ \forall x \in [X], \end{cases} \tag{6.1.15}$$

which are *Bellman equations* allowing one to interpret u_{tx} as the value associated with being in state x at time t—“value” here meaning the maximum of the expected total payoff starting at t in state x. The Bellman equation is a dynamic programming principle relating the value of each state at time t to the value of each state at time $t+1$. It provides a practical algorithm to compute the value function starting with $t=T$ and iteratively backward until $t=1$; this is called *backward induction* or *value iteration*. Once the value function u_{tx} has been computed, one may compute the policy μ_{txy} by *forward induction*, or *policy iteration*. The backward-forward induction algorithm is displayed as algorithm 6.1.1, and implemented in Python in appendix B.6.2.

Note that the backward-forward algorithm is of no help in computing the vector τ of Lagrange multipliers associated with the scarcity constraint

Algorithm 6.1.1 Backward-forward induction algorithm

Input : Transition probabilities $P_{x'xy}$, payoff function $\phi^{\star}_{txy}$
Output : Value function u_{tx}, policy μ_{txy}

Backward Phase (value iteration):
$u_{Tx} \leftarrow \max_{y \in [Y]} \{\phi^{\star}_{Txy}\}$
for $t \leftarrow T-1$ **to** 1 **by** -1 **do**
  $u_{tx} \leftarrow \max_{y \in [Y]} \left\{\phi^{\star}_{txy} + \sum_{x' \in [X]} P_{x'xy} u_{(t+1)x'}\right\}$
end

Forward Phase (policy iteration):
Set $n_{1x} = N_x$, $u_{(T+1)x} = 0 \; \forall x \in [X]$
for $t \leftarrow 1$ **to** T **do**
  forall $x \in [X]$ **do**
    Pick one $y^{\circ} \in [Y]$ such that $\phi^{\star}_{txy^{\circ}} + \sum_{x'} P_{x'xy^{\circ}} u_{(t+1)x'} = u_{tx}$
    Set $\mu_{txy} = n_{tx}$ for $y = y^{\circ}$, $\mu_{txy} = 0$ otherwise
  end
  Compute $n_{(t+1)x} = \sum_{x',y'} P_{xx'y'} \mu_{tx'y'} \forall x \in [X]$
end
return μ_{txy}, u_{tx}

$\mathbf{A}\mu \leq c$; it computes u and μ assuming that τ is known, which is the case in particular when there is no scarcity constraint, in which case τ is zero.

Remark 6.1.1. In the absence of the capacity constraint $\mathbf{A}\mu \leq c$, there is no externality across the treatment of the various units, and therefore the dynamic associated with a population with N_x units in state x in the first period coincides with the superposition of the dynamic of the sum over x of N_x agents in state x in the first period. This configuration is sometimes called the "single-agent dynamic."

6.2 Adding Heterogeneity

We now assume that in addition to the short-term payoff ϕ_{txy}, for each unit $i \in [I]$, there is an additional random utility shock at time t associated with every decision $y \in [Y]$, denoted by ε_{ity}. While two units i and i' in the same state x and the same period t have independent random utility shocks, we want to allow for dependence between these shocks for a given unit i across alternatives y; that is, we would like to allow for dependence between ε_{iyt} and $\varepsilon_{iy't}$. Therefore, we denote by $\mathcal{P}_{tx}$ the distribution of the Y-dimensional random vector $\varepsilon_t = (\varepsilon_{ty})_y$ from which the draws ε_{ity} are made. The following assumption is crucial:

Assumption 6.1. *We assume that the sequence $x_t, (x_t, \varepsilon_t), (x_t, y_t), x_{t+1}, \ldots$ constitutes a Markov chain. Conditional on x_t, a Y-dimensional random utility vector ε_t is drawn from a distribution $\mathcal{P}_{tx_t}$; based on x_t and the vector ε_t, the optimal choice y_t is selected; and, finally, the next state, x_{t+1}, is drawn from the transition probability $P_{x'xy}$ given $xy = x_t y_t$.*

What this assumption means is that, at any stage, only the current state matters for predicting the next one. In particular, the distribution of ε_t conditional on ε_{t-1} is completely explained by x_t. This rules out "persistence" of the random utility shocks. Similarly, y_t is fully explained by (x_t, ε_t), and x_{t+1} by (x_t, y_t).

The primal problem becomes the maximization of $\sum_{txy} \mu_{txy}(\phi_{txy} + \mathbb{E}_{\mathcal{P}}[\varepsilon_{ty}])$ over the policy vectors μ that are subject to the evolution equations (6.1.6) and the policy constraints (6.1.4). Note that here we are choosing not to discount the random part of the utility, as the terms $\sum_{xy} \mu_{txy}\mathbb{E}[\varepsilon_{ty}]$ will appear each with a weight of 1 in the intertemporal objective function of the decision-maker. However, when we consider the problem with infinite horizon in section 6.5, we shall apply a geometric discounting to both the systematic and the random parts of the utility. For now, we do not discount the random part of the utility, and we have:

Theorem 6.2.1. *In the model with logit heterogeneities, that is, when $\mathcal{P}$ is the distribution of i.i.d. standard Gumbel terms, the primal problem has formulation*

$$\max_{\mu_{txy} \geq 0} \left\{ \sum_{txy \in [T \times X \times Y]} \mu_{txy}\phi_{txy} - \sum_{txy \in [T \times X \times Y]} \mu_{txy} \log \frac{\mu_{txy}}{\sum_{y \in [Y]} \mu_{txy}} \right\} \quad (6.2.1)$$

$$\text{s.t.} \sum_{y \in [Y]} \mu_{1xy} = N_x, \quad x \in [X], \quad (6.2.2)$$

$$\sum_{y \in [Y]} \mu_{txy} = \sum_{\tilde{x}y \in [X \times Y]} P_{x\tilde{x}y}\mu_{(t-1)\tilde{x}y}, \quad tx \in [(2{:}T) \times X], \quad (6.2.3)$$

$$\mathbf{A}\mu \leq c, \quad (6.2.4)$$

which has the dual formulation

$$\min_{u_{tx}, \tau_l} \left\{ \sum_{x \in [X]} N_x u_{1x} + \sum_{l \in [L]} c_l \tau_l \right\} \quad (6.2.5)$$

$$\text{s.t. } u_{tx} = \log\left(\sum_{y\in[Y]} \exp\left(\phi - \mathbf{A}^\top \tau + (\mathbf{J}_T^\top \otimes \mathbf{P}^\top)u\right)_{txy}\right), \quad tx \in [T \times X]. \tag{6.2.6}$$

Further, if $\mu \geq 0$ is the solution to the primal problem and (u, τ) to the dual problem, then defining $n_{tx} = \sum_y \mu_{txy}$ to be the number of units in state x at time t, the conditional choice probability of choosing option y in state x at time t, denoted by $\pi_{txy} := \mu_{txy}/n_{tx}$, is expressed by $\pi = \exp(\phi - \mathbf{A}^\top \tau + (\mathbf{J}_T^\top \otimes \mathbf{P}^\top - \mathbf{I}_{TX} \otimes \mathbf{1}_Y)u)$.

Before giving the proof of this duality result, it is worth making explicit that the expression of the entry txy of the vector $\phi - \mathbf{A}^\top \tau + (\mathbf{J}_T^\top \otimes \mathbf{P}^\top)u$ that appears in formula (6.2.6) is $\phi_{txy} - (\mathbf{A}^\top \tau)_{txy} + \sum_{x'} P_{x'xy} u_{(t+1)x'}$ for $t \leq T-1$ and $\phi_{Txy} - (\mathbf{A}^\top \tau)_{Txy}$ for $t = T$. In words, the intertemporal surplus associated with option y for a unit in state x at time t is the sum of the short-term payoff, the shadow cost associated with the constraints, and, if t is not the last period, the expected continuation payoff. Similarly, the explicit expression of the conditional choice probability vector π is

$$\pi_{txy} = \exp\left(\phi_{txy} - (\mathbf{A}^\top \tau)_{txy} + \sum_{x'\in[X]} P_{x'xy} u_{(t+1)x'} - u_{tx}\right) \tag{6.2.7}$$

for $t \leq T-1$, and a similar expression, without the continuation term, for $t = T$.

Proof. It follows from proposition 1.3.2 and the expression of the generalized entropy of choice associated with the logit model from example 1.2 that $\mathbb{E}[\mu_{txy}\varepsilon_{ty}] = -\mu_{txy}\log(\mu_{txy}/n_{tx})$ and therefore that the expression of the primal problem is (6.2.1). Letting u_{1x} be the Lagrange multipliers associated with constraints (6.2.2), u_{tx} for $t \geq 2$ be the Lagrange multipliers associated with constraints (6.2.3), and τ be the L-dimensional vector of Lagrange multipliers associated with constraints (6.2.4), the Lagrangian $\mathcal{L}(\mu; u, \tau)$ of the problem is thus

$$\begin{aligned}\mathcal{L}(\mu; u, \tau) := &\sum_{txy\in[T\times X\times Y]} \mu_{txy}\phi_{txy} - \sum_{txy\in[T\times X\times Y]} \mu_{txy} \log \frac{\mu_{txy}}{\sum_y \mu_{txy}} \\ &+ \sum_{x\in[X]} N_x u_{1x} - \sum_{xy\in[X\times Y]} \mu_{1xy} u_{1x} - \sum_{txy\in[(2:T)\times X\times Y]} \mu_{txy} u_{tx} \\ &+ \sum_{tx\tilde{x}y\in[(2:T)\times X\times X\times Y]} P_{x\tilde{x}y}\mu_{(t-1)\tilde{x}y} u_{tx} + \sum_{l\in[L]} c_l \tau_l - \sum_{ltxy\in[L\times T\times X\times Y]} A_{ltxy}\mu_{txy}\tau_l,\end{aligned}$$

and the primal formulation is $\max_{\mu\geq 0} \min_{u,\tau} \mathcal{L}(\mu; u, \tau)$, which, by virtue of theorem A.1.5, coincides with the value of its dual, $\min_{u,\tau} \max_{\mu\geq 0} \mathcal{L}(\mu; u, \tau)$.

But one may rearrange $\mathcal{L}(\mu; u, \tau)$ in order to isolate terms that do not depend on μ, and we obtain:

$$\mathcal{L}(\mu; u, \tau) = \sum_{x\in[X]} N_x u_{1x} + \sum_{l\in[L]} c_l \tau_l + \sum_{txy\in[(T-1)\times X\times Y]}$$
$$\times \mu_{txy}\left(\phi_{txy} - (\mathbf{A}^\top \tau)_{txy} + \sum_{x'\in[X]} P_{x'xy} u_{(t+1)x'} - u_{tx} - \log \frac{\mu_{txy}}{\sum_{y\in[Y]} \mu_{txy}}\right)$$
$$+ \sum_{txy\in[X\times Y]} \mu_{txy}\left(\phi_{Txy} - (\mathbf{A}^\top \tau)_{Txy} - u_{Tx} - \log \frac{\mu_{Txy}}{\sum_{y\in[Y]} \mu_{Txy}}\right).$$

A similar calculation to the one conducted in exercise 1.7 shows that the value of $\max_{\mu\geq 0} \mathcal{L}(\mu; u, \tau)$ is $\sum_x N_x u_{1x} + \sum_l c_l \tau_l$ if each of the constraints

$$\begin{cases} \sum_{y\in[Y]} \exp(\phi_{txy} - (\mathbf{A}^\top \tau)_{txy} + \sum_{x'\in[X]} P_{x'xy} u_{(t+1)x'} - u_{tx}) \\ \qquad \leq 1,\ \forall tx \in [(T-1) \times X], \\ \sum_{y\in[Y]} \exp(\phi_{Txy} - (\mathbf{A}^\top \tau)_{Txy} - u_{Tx}) \leq 1,\ \forall x \in [X] \end{cases} \tag{6.2.8}$$

is met and is $\max_{\mu\geq 0} \mathcal{L}(\mu; u, \tau) = +\infty$ otherwise. As a result, the dual formulation of the problem becomes the minimization of $\sum_x N_x u_{1x} + \sum_l c_l \tau_l$ over the set of vectors u that satisfy (6.2.8), which is the claimed result. ■

In light of the above expressions, in the sequel we shall use the notation $(\mathbf{P}^\top u_{t+1})_{xy}$ to denote $((\mathbf{J}_T \otimes \mathbf{P}^\top) u)_{txy}$. Note then that we can rewrite the equality constraint (6.2.6) of the dual problem in theorem 6.2.1 as $u_{tx} \geq \log \sum_{y\in[Y]} \exp(\phi_{txy} - (\mathbf{A}^\top \tau)_{txy} - (\mathbf{P}^\top u_{t+1})_{xy})$, which is an inequality constraint. Indeed, we are minimizing an objective function (6.2.5) that depends positively on the entries of (u_{1x}), and therefore one can easily see that if the inequality constraints were to hold strictly for some feasible solution (u_{tx}), it would be possible to decrease the objective function, and (u_{tx}) would not be optimal. By taking exponentials and dividing by $\exp u_{tx}$, one can rewrite the dual problem (6.2.5) in theorem (6.2.1) as

$$\min_{u_{tx},\tau_l} \sum_{x\in[X]} N_x u_{1x} + \tau^\top c \tag{6.2.9}$$
$$s.t. \sum_{y\in[Y]} \exp\left(\phi_{txy} - (\mathbf{A}^\top \tau)_{txy} - (\mathbf{P}^\top u_{t+1})_{xy} - u_{tx}\right) \leq 1,\ tx \in [T \times X],$$

where we have taken the convention that $u_{(T+1)x} = 0$ for all $x \in [X]$. Denoting by $(n_{tx}) \geq 0$ the vector of Lagrange multipliers, and introducing the

Lagrangian $\mathcal{S}(u, \tau; n)$, one has

$$\mathcal{S}(u, \tau; n) := \sum_{x\in[X]} N_x u_{1x} + \tau^\top c \qquad (6.2.10)$$

$$+ \sum_{txy\in[T\times X\times Y]} n_{tx} e^{\phi_{txy} - (\mathbf{A}^\top \tau)_{txy} + (\mathbf{P}^\top u_{t+1})_{xy} - u_{tx}} - \sum_{tx\in[T\times X]} n_{tx}.$$

Note that this Lagrangian $\mathcal{S}(u, \tau; n)$ is not quite the same as the Lagrangian $\mathcal{L}(\mu; u, \tau)$ that appears in the proof of theorem 6.2.1; indeed, the latter stems from a different way to express the constraint on the dual program. In any case, problem (6.2.5) consists of determining a saddle point[1] $(u, \tau; n)$ in the minimax problem $\min_{u,\tau} \max_n \mathcal{S}(u, \tau; n)$, which we state as:

Proposition 6.2.1. *Problem* (6.2.5) *takes the following minimax formulation*

$$\min_{u_{tx},\tau_l} \max_{n_{tx}\geq 0} \left\{ \begin{array}{c} \sum_{x\in[X]} N_x u_{1x} + \tau^\top c \\ + \sum_{txy\in[T\times X\times Y]} n_{tx} e^{\phi_{txy} - (\mathbf{A}^\top \tau)_{txy} + (\mathbf{P}^\top u_{t+1})_{xy} - u_{tx}} - \sum_{tx\in[T\times X]} n_{tx} \end{array} \right\}, \qquad (6.2.11)$$

with the convention that $u_{(T+1)x} = 0$. *Further, if one lets*

$$\mu_{txy} = n_{tx} \exp\left(\phi_{txy} - (\mathbf{A}^\top \tau)_{txy} + (\mathbf{P}^\top u_{t+1})_{xy} - u_{tx}\right), \qquad (6.2.12)$$

then (u_{tx}, τ_l, n_{tx}) *is a saddle point of problem* (6.2.11) *if and only if* (μ_{txy}) *and* (u_{tx}, τ_l) *are respectively solutions to the primal and dual problems* (6.2.1) *and* (6.2.5), *and* n_{tx} *is the mass of units in state* x *at time* t.

6.3 Inference, Logit Model with Finite Horizon

In the dynamic model with logit heterogeneities and finite horizon introduced in the previous section, we consider a linear parametrization of the surplus $(\Phi\lambda)_{txy} = \sum_k \phi_{txyk}\lambda_k$. Assume that the observations $t_i x_i y_i$, $i \in [I]$ are sampled in an i.i.d. fashion, to form a sample of size I. We denote by $\hat{\mu}_{txy}$ the number of units that are in state x at time t and for which option y is chosen. This model is therefore a dynamic version of the static model described in chapter 3.

Let u^λ be the optimal vector of utilities given λ. Then the conditional likelihood of choosing option y for a unit in state x at time t is $\pi^\lambda_{txy} = \mu^\lambda_{txy}/n_{tx} = \exp((\Phi\lambda)_{txy} + (\mathbf{P}^\top u^\lambda_{t+1})_{xy} - u^\lambda_{tx})$. The conditional log-likelihood

[1] We refer the reader to appendix A.1.4 for background on saddle points and minimax theory.

is thus $\log(\pi^{\lambda}_{txy}) = (\Phi\lambda)_{txy} + (\mathbf{P}^{\top} u^{\lambda}_{t+1})_{xy} - u^{\lambda}_{tx}$, and the log-likelihood of the sample is thus

$$\ell(\lambda) := \sum_{txy} \hat{\mu}_{txy}\big((\Phi\lambda)_{txy} + (\mathbf{P}^{\top} u^{\lambda}_{t+1})_{xy} - u^{\lambda}_{tx}\big) = \hat{\mu}^{\top}(\Phi\lambda + \mathbf{\Psi} u^{\lambda}),$$

where we have set $\mathbf{\Psi} := \mathbf{J}_T^{\top} \otimes \mathbf{P}^{\top} - \mathbf{\Sigma}_Y^{\top}$ and $\mathbf{\Sigma}_Y := \mathbf{I}_{TX} \otimes \mathbf{1}_Y^{\top}$. Recall that $\mathbf{J}_T$ is the $T \times T$ matrix such that $(\mathbf{J}_T)_{t,t'} = 1\{t = t' + 1\}$, and also that $\mathbf{P}$ is the $X \times (XY)$ matrix whose (x', xy) term is $P_{x'xy}$. Finally, u^{λ} is defined as the solution to the Bellman equation $u^{\lambda}_{tx} = \log \sum_y \exp((\Phi\lambda)_{txy} + (\mathbf{P}^{\top} u^{\lambda}_{t+1})_{xy})$ for all $tx \in [T \times X]$, and $u^{\lambda}_{(T+1)x} = 0$. But this Bellman equation rewrites $\sum_y \exp((\Phi\lambda)_{txy} + (\mathbf{P}^{\top} u^{\lambda}_{t+1})_{xy} - u^{\lambda}_{tx}) = 1$, that is,

$$\mathbf{\Sigma}_Y \exp\left(\Phi\lambda + \mathbf{\Psi} u^{\lambda}\right) = \mathbf{1}_{TX}. \tag{6.3.1}$$

We have therefore:

Proposition 6.3.1. *The maximum likelihood in the dynamic model with finite horizon and logit heterogeneities is*

$$\max_{u,\lambda} \hat{\mu}^{\top} (\Phi\lambda + \mathbf{\Psi} u) \tag{6.3.2}$$

$$s.t.\ \mathbf{\Sigma}_Y \exp(\Phi\lambda + \mathbf{\Psi} u) = \mathbf{1}_{TX}. \tag{6.3.3}$$

This result calls for a series of remarks.

Remark 6.3.1 (Interpretation of optimality). Let n_{tx} be the Lagrange multiplier associated with constraint (6.3.3), and let $\mu_{txy} = n_{tx} \exp((\Phi\lambda + \mathbf{\Psi} u)_{txy})$ be the predicted policy variable. Then the optimality conditions with respect to λ yield

$$\Phi^{\top}\mu = \Phi^{\top}\hat{\mu},$$

which means that the predicted moments of the policy variable match with the observed ones. Likewise, the optimality conditions with respect to u yield $\mathbf{\Psi}^{\top}\mu = \mathbf{\Psi}^{\top}\hat{\mu}$, that is,

$$\left(\mathbf{J}_T \otimes \mathbf{P} - \mathbf{\Sigma}_Y\right)\hat{\mu} = \left(\mathbf{J}_T \otimes \mathbf{P} - \mathbf{\Sigma}_Y\right)\mu,$$

which implies that $\sum_y \mu_{1xy} = N_x$, or, in words, in the initial period, the number of units in each state implied by $\hat{\mu}$ and μ coincides; and in subsequent periods the change in the number of units in each state implied by $\hat{\mu}$ and μ coincide.

Remark 6.3.2 (Representative panel). We say that we have a *representative panel* if $\sum_y \hat{\mu}_{(t+1)xy} = n_{(t+1)x} = \sum_{x'y} P_{x,x'y}\hat{\mu}_{tx'y}$. In that case, the log-likelihood of the sample simplifies into

$$\ell(\lambda) = \sum_{txyk} \hat{\mu}_{txy}\phi_{txyk}\lambda_k - \sum_{x\in[X]} N_x u_{1x} = \hat{\mu}^\top \Phi\lambda - N^\top u_1.$$

Remark 6.3.3 (Numerical computation of the value function). Recall the constraint (6.3.3) of the problem above: $u_{tx} = \log \sum_y \exp((\Phi\lambda)_{txy} + (\mathbf{P}^\top u_{t+1})_{xy})$. If the value of the argument of one of the exponentials in the sum over y is relatively large (say, above several dozens), the exponentiation will blow up, causing numerical instability. As a remedy for that, it is advised to use the fact that $\log \sum_y \exp U_y = c + \log \sum_y \exp(U_y - c)$ for any scalar c. Taking $c = -\max_y U_y$ ensures that no argument of the exponentials is positive, thus restoring numerical stability. The value function iteration will therefore embed an intermediate step, as follows:

$$\begin{cases} \bar{u}_{tx} = \max_{y\in[Y]}\{(\Phi\lambda)_{txy} + (\mathbf{P}^\top u_{t+1})_{xy}\}, \\ u_{tx} = \bar{u}_{tx} + \log \sum_{y\in[Y]} \exp((\Phi\lambda)_{txy} + (\mathbf{P}^\top u_{t+1})_{xy} - \bar{u}_{tx}). \end{cases}$$

This is the "log-sum-exp trick," which we encountered at section 3.8.

In order to compute $\max_\lambda \ell(\lambda)$ by gradient descent, we need to compute the gradient of the objective function, $\nabla\ell(\lambda)$. For that, we need to compute the differential of u with respect to λ. In order to do this, we take the differential of constraint (6.3.3) with respect to λ in order to obtain

$$\mathbf{\Sigma}_Y \mathbf{\Delta}_\pi (\Phi + \mathbf{\Psi} du) = 0, \tag{6.3.4}$$

where du denotes the differential of u with respect to λ (see appendix A.3.1), and where we have set $\mathbf{\Delta}_\pi := diag(\pi)$, and we recall that $\pi_{txy} = \mu_{txy}/n_{tx} = \exp((\Phi\lambda)_{txy} + (\mathbf{P}^\top u_{t+1})_{xy} - u_{tx})$. As a result, we have

$$du = -(\mathbf{\Sigma}_Y \mathbf{\Delta}_\pi \mathbf{\Psi})^{-1} \mathbf{\Sigma}_Y \mathbf{\Delta}_\pi \Phi. \tag{6.3.5}$$

Using expression (A.3.12) for the second differentiation of (6.3.3), we have

$$\mathbf{\Sigma}_Y \mathbf{\Delta}_\pi \mathbf{\Psi} d^2u + \mathbf{\Sigma}_Y \mathbf{\Delta}_\pi (\Phi + \mathbf{\Psi} du)^{\circledast 2} = 0,$$

where $M^{\circledast 2}$ stands for $M \circledast M$, where the row-wise Kronecker product $\circledast$ has been defined in A.3.6, and where the second differential d^2u is a $TX \times K^2$ matrix whose definition is provided in appendix A.3.1. Therefore

$$d^2u = -(\mathbf{\Sigma}_Y \mathbf{\Delta}_\pi \mathbf{\Psi})^{-1} \mathbf{\Sigma}_Y \mathbf{\Delta}_\pi (\Phi + \mathbf{\Psi} du)^{\circledast 2}. \tag{6.3.6}$$

Thanks to the expressions of du and d^2u provided in (6.3.5) and (6.3.6), we are finally able to compute the first and second derivatives of $\ell(\lambda)$. We have

$$d\ell\,(\lambda) = \hat{\mu}^\top(\Phi + \Psi du) \quad \text{and} \quad d^2\ell\,(\lambda) = \hat{\mu}^\top \Psi d^2 u.$$

As a result, the sample estimator $\hat{\lambda}$ of the true parameter λ is determined by the optimality condition $\hat{\mu}^\top(\Phi + \Psi du(\hat{\lambda})) = 0$, and therefore, letting $D^2\ell\,(\lambda)$ be the Hessian matrix of ℓ, which is simply the $K \times K$ matrix that is obtained by reshaping $d^2\ell\,(\lambda)$ in the row-major order, one has $\hat{\lambda} - \lambda = -(D^2\ell(\lambda))^{-1}(\Phi^\top + du^\top\Psi^\top)(\hat{\mu} - \mu) + o_P(\sqrt{I})$. This intuition can be made rigorous using the theory of M-estimators recalled in appendix A.2.5, and we are led to the following.

Theorem 6.3.1. *Assume that the observations $t_i x_i y_i$ are drawn from the population in an i.i.d. fashion and collected in a sample of size I. Letting $\hat{\mu}_{txy}$ be the number of observations of type txy in the sample, the following convergence holds in distribution*

$$\sqrt{I}(\hat{\lambda} - \lambda) \xrightarrow{d} H^{-1}CH^{-1},$$

where H is estimated by $\hat{H} = D^2\ell(\lambda)/I$ and $C = BVB^\top$, where B is estimated by $\hat{B} = (\Phi^\top + Du(\hat{\lambda})^\top\Psi^\top)$ and V by $\hat{V} = diag(\hat{\mu})/I - \hat{\mu}\hat{\mu}^\top/I^2$.

Appendix B.6.3 provides an implementation of the problem in `NumPy`, while appendix B.6.4 solves the same problem in `PyTorch`.

6.4 Potential Function Approach

In this section, we will introduce the potential function approach, which is an important tool to help reformulate the models seen so far, allowing us to formulate various extensions.

We study the dynamic model with finite horizon and logit heterogeneities introduced in section 6.2, and assume for now that no capacity constraint is imposed. We consider the dual program (6.2.11) in proposition 6.2.1, and focus on how $u_1 = (u_{1x})$, the payoff vector in period 1, depends on $u_2 = (u_{2x})$, the corresponding vector in the next period. Given the vector u_2, the vector u_1 is determined by

$$\min_{u_{1x}} \max_{n_x \geq 0} \left\{ \sum_{x \in [X]} N_x u_{1x} + Z\left(u_1, u_2', n\right) \right\}, \tag{6.4.1}$$

where the function Z, a fundamental tool for the analysis developed in this chapter, is defined here by

$$Z\left(u, u', n\right) := \sum_{xy \in [X \times Y]} n_x \exp(\phi_{1xy} + (\mathbf{P}^\top u')_{xy} - u_x) - \sum_{x \in [X]} n_x \tag{6.4.2}$$

if $n \geq 0$, and $Z(u, u', n) = +\infty$ otherwise. Recall that for now we are considering the case without capacity constraints; as we shall see, the function Z will take different forms when capacity constraints are imposed.

At equilibrium, $u_x = \log \sum_y \exp(\phi_{1xy} + (\mathbf{P}^\top u')_{xy})$, and therefore the quantity $\pi_{xy} = \exp(\phi_{1xy} + (\mathbf{P}^\top u')_{xy} - u_x)$ is the probability that option y is chosen for a unit in state x. Hence $\sum_y \exp(\phi_{1xy} + (\mathbf{P}^\top u')_{xy} - u_x) = 1$, and thus at equilibrium, $Z(u, u', n) = 0$. It is not hard to see that the function Z is convex in (u, u') and linear (hence concave) in $n \geq 0$. Letting $\mu_{xy} = n_x \exp(\phi_{1xy} + (\mathbf{P}^\top u')_{xy} - u_x)$, one has

$$\frac{\partial Z}{\partial n_x} = \sum_{y \in [Y]} \frac{\mu_{xy}}{n_x} - 1, \qquad \frac{\partial Z}{\partial u_x} = - \sum_{y \in [Y]} \mu_{xy}, \qquad \text{and} \qquad \frac{\partial Z}{\partial u'_x} = (\mathbf{P}\mu)_x. \tag{6.4.3}$$

Formulas (6.4.3) highlight the interest of the function Z: its partial derivatives encompass all the quantities of interest of the model. In particular, if we want to interpret μ_{xy}/n_x as the conditional choice probability of option y conditional on being in state x, we need to have $\sum_y \mu_{xy}/n_x = 1$, and thus $\partial Z/\partial n_x = 0$. In that case, we have $\sum_y \mu_{xy} = n_x$, and thus

$$\frac{\partial Z}{\partial u_x} = -n_x,$$

meaning that the partial derivative of Z with respect to u_x, the current-period utility in state x, is equal to minus the number of units in state x in the current period, while if n'_x denotes the number of units that are in state x in the next period, we have $n'_x = (\mathbf{P}\mu)_x$, and thus, because of the last relation displayed in (6.4.3), one has

$$\frac{\partial Z}{\partial u'_x} = n'_x,$$

which means that the partial derivative of Z with respect to u'_x, the next-period utility in state x, is equal to the number of units in state x in the next period.

We now discuss how to incorporate capacity constraints into this analysis. The capacity constraints in period 1 are $\sum_{xy} A_{l,xy}\mu_{xy} \leq c_l$ for $l \in [L]$, and we let $\tau_l \geq 0$ be the respective Lagrange multipliers. It follows from proposition 6.2.1 that given the vector u_2, the vector u_1 is obtained by

$$\min_{u_{1x}} \max_{n_x \geq 0} \left\{ \sum_{x\in[X]} N_x u_{1x} + Z\left(u_1, u_2', n\right) \right\}, \tag{6.4.4}$$

where, this time, the expression for the function Z is

$$Z\left(u, u', n\right) := \min_{\tau_l \geq 0} \left\{ \begin{array}{c} \sum_{xy\in[X\times Y]} n_x \exp\left(\phi_{1xy} - (\mathbf{A}\tau)_{xy} + (\mathbf{P}^\top u')_{xy} - u_x\right) \\ - \sum_{x\in[X]} n_x + \sum_{l\in[L]} c_l \tau_l \end{array} \right\}. \tag{6.4.5}$$

We provide several examples to start illustrating the usefulness of the function Z, even in the static case.

Example 6.2 (Static logit model using the function Z)**.** Consider the static logit discrete choice problem discussed in example 1.2 in chapter 1, where there are N_x agents of type $x \in [X]$. An agent of type x choosing option $y \in [Y]$ gets utility $\phi_{xy} + \varepsilon_y$, where (ε_y) is a Y-dimensional vector of i.i.d. Gumbel random variables. The problem is static, so $u_x' = 0$. Then $u_x = \mathbb{E}[\max_{y\in[Y]}\{\phi_{xy} + \varepsilon_y\}]$ is the solution to

$$\min_u \left\{ \sum_{x\in[X]} N_x u_x + Z\left(u, 0, N\right) \right\},$$

with the expression for the function Z given by (6.4.2). Indeed, by first-order conditions, one has $N_x = \sum_y N_x \exp(\phi_{xy} - u_x)$, and thus $u_x = \log \sum_y \exp \phi_{xy}$.

Next, we show that the function Z can be very helpful in parametric estimation problems, and, in particular, in the logistic regression problem—here again, still in the static case.

Example 6.3 (Static logistic regression using the function Z)**.** Consider a parametric version of the model discussed above, where it is assumed that an agent of type x choosing option $y \in [Y]$ gets utility $(\Phi\lambda)_{xy} + \varepsilon_y$, where $(\Phi\lambda)_{xy} = \sum_k \phi_{kxy}\lambda_k$ and (ε_y) is still a Y-dimensional vector of i.i.d. Gumbel random variables. As before, the problem is static, so $u_x' = 0$. One observes $\hat{\mu}_{xy}$, the mass of agents x who have chosen option y, and we denote $N_x = \sum_{xy} \hat{\mu}_{xy}$. The logistic regression estimator consists of

$$\min_{u,\lambda} \left\{ \sum_{xy\in[X\times Y]} \hat{\mu}_{xy}(u_x - (\Phi\lambda)_{xy}) + Z\left(\lambda, u, 0, N\right) \right\},$$

where the function Z is given by

$$Z\left(\lambda, u, u', n\right) := \sum_{xy\in[X\times Y]} n_x \exp\left((\Phi\lambda)_{xy} + (\mathbf{P}^\top u')_{xy} - u_x\right) - \sum_{x\in[X]} n_x. \tag{6.4.6}$$

Our next example shows that one can estimate a static matching model with transferable utility, which is the Choo and Siow model introduced in section 5.4, using the function Z.

Example 6.4 (Estimation of a static TU matching model using the function Z). Consider the static Choo and Siow model of section 5.4. Then the estimation described in theorem 5.4.1 can be performed by the following problem:

$$\min_{u,v,\lambda} \max_{n,m} \left\{ n^\top u + m^\top v - \sum_{xy\in[X\times Y]} \hat{\mu}_{xy}(\Phi\lambda)_{xy} + Z\left(\lambda, u, v, n, m\right) \right\},$$

where the function Z is given by

$$Z\left(\lambda, u, v, n, m\right) = 2 \sum_{xy\in[X\times Y]} \sqrt{n_x m_y} \exp\left(\frac{(\Phi\lambda)_{xy} - u_x - v_y}{2}\right)$$
$$+ \sum_{x\in[X]} n_x \exp(-u_x) + \sum_{y\in[Y]} m_y \exp(-v_y) - \sum_{x\in[X]} n_x - \sum_{y\in[Y]} m_y.$$

One sees that the function Z is equal to the difference between two terms that should coincide at equilibrium with the total number of individuals in the population. The first such term counts the individuals at the level of matches. It records twice the number of matched pairs, plus the total number of unassigned agents. The second of these two terms is simply the total number of x-individuals, plus the total number of y-individuals. These two quantities should be equal, hence Z should be zero at equilibrium.

Moving on to the dynamic case, the next example is simply a reexpression of proposition 6.2.1 in terms of the function Z.

Example 6.5 (Discrete choice estimation, logit and finite horizon using Z). Consider the dynamic discrete choice problem with finite horizon and logit heterogeneity of section 6.3, and assume that the constraints are expressed period by period by $\sum_{xy} \mathbf{A}_{tl,cy}\mu_{xy} \leq c_{tl}$. It directly follows from proposition 6.2.1 that the solution of the problem is such that μ is given by expression (6.2.12), where $(u, \tau; n)$ is a saddle point for

$$\min_{u} \max_{n \geq 0} \left\{ N^{\top} u_1 + \sum_{t \in [T]} Z_t \left(c_t, u_t, u_{t+1}, n_t \right) \right\},$$

where the expression for the function Z_t is the same as (6.4.5) after replacing ϕ_{1xy} by ϕ_{txy}.

Lastly, we show that the estimation of the previous dynamic discrete choice model with logit heterogeneity and finite horizon—thus combining the features of examples 6.3 and 6.5—also has a nice formulation using the function Z:

Example 6.6. The finite-horizon dynamic discrete choice estimation problem, which formulates as problem (6.3.2) in proposition 6.3.1, rewrites as the following saddle point problem involving the function Z:

$$\max_{u,\lambda} \min_{n \geq 0} \left\{ \hat{\mu}^{\top} \Phi \lambda + N^{\top} u - \sum_{t \in [T]} Z_t \left(\lambda, u_t, u_{t+1}, n_t \right) \right\}, \tag{6.4.7}$$

where the expression for Z_t is the same as (6.4.6) after replacing $(\Phi\lambda)_{xy}$ by $(\Phi\lambda)_{txy}$. The saddle point conditions for Lagrangian (6.6.9) are first-order conditions that express that the partial derivatives of the objective function with respect to n, u, and λ are each zero. The first-order conditions with respect to n_{tx} yield $\partial Z / \partial n_{tx} = 0$, that is, $u_{tx} = \log \sum_y \exp((\Phi\lambda)_{txy} + (\mathbf{P}^{\top} u_{t+1})_{xy})$. The first-order conditions with respect to u_{tx} imply $n_1 = N$ (for $t = 1$) and $n_{t+1} = \mathbf{P}\mu^{\lambda}$ (for $t > 1$). Hence, the first-order conditions with respect to u and n express that the system is at equilibrium. Lastly, the first-order conditions with respect to λ yield $\Phi^{\top} \hat{\mu} = \Phi^{\top} \mu^{\lambda}$, which expresses that the predicted moments match with the observed moments.

The five examples just presented are simple reformulations of models we previously introduced and solved without recourse to the function Z. In light of this, the reader may legitimately question the usefulness of introducing the function Z if it merely serves to paraphrase existing results and methods. But as we shall see subsequently, the function Z will be very useful to extend our setting to dynamic discrete choice problems with infinite horizon (which we will cover in section 6.5), and to more general dynamic models of coalition formation (which we will explore in section 6.9).

6.5 The Model with Infinite Horizon

We now consider the model introduced in section 6.3, but with infinite horizon, so $T=+\infty$. In that case, we assume a constant geometric discounting $\beta \in (0,1)$, so that, measured in period-zero equivalent, the systematic payoff associated with option y chosen at time t for a unit in state x is $\phi_{txy}=\beta^{t-1}\tilde{\phi}_{xy}$. For consistency, we should also use the same geometric discounting for the random part of the utility, and we shall assume that the logit random utility term is discounted with the same rate β, so that the total payoff associated with option y chosen at time t for a unit in state x is $\beta^{t-1}\tilde{\phi}_{xy}+\beta^{t-1}\varepsilon_y)$, where the components ε_y of the vector ε are i.i.d. draws from a Gumbel distribution. This is a noticeable departure from the assumptions adopted previously in sections 6.2 and 6.3, where we allowed the systematic payoff to vary arbitrarily with t, but where we assumed that the logit utility shock was undiscounted; with infinite horizon, that assumption is not sustainable, as, if we discounted the systematic part of the utility but not its random component, the latter would soon dwarf the former, which is why we need to discount both terms. As a result, the intertemporal objective of the central planner is now

$$\sum_{txy\in[\infty\times X\times Y]} \mu_{txy}\beta^{t-1}\tilde{\phi}_{xy} - \sum_{txy\in[\infty\times X\times Y]} \beta^{t-1}\mu_{txy}\log\left(\frac{\mu_{txy}}{\sum_{y\in[Y]}\mu_{txy}}\right).$$

We further assume that the constraint does not depend on time, so that the capacity constraints rewrite

$$\sum_{xy\in[X\times Y]} \tilde{A}_{l,xy}\mu_{txy} \leq \tilde{c}_l \text{ for each } tl \in [\infty \times L],$$

and we denote by τ_{tl} the Lagrange multiplier associated with each constraint. As before, the dual will consist of minimizing the objective $\sum_x N_x u_{1x} + \sum_{tl}\tau_{tl}\tilde{c}_l$ subject to the Bellman equation, which is now $u_{tx}=\mathbb{E}[\max_y\{\beta^{t-1}\tilde{\phi}_{xy} - (\tilde{\mathbf{A}}\tau_t)_{xy} + (\mathbf{P}^\top u_{t+1})_{xy} + \beta^{t-1}\varepsilon_y\}]$, and therefore, it makes sense to introduce $\tilde{u}_{xt}=u_{tx}/\beta^{t-1}$ and $\tilde{\tau}_{tl}=\tau_{tl}/\beta^{t-1}$, which express the value function at time t and the shadow price of the capacity constraints at time t in their current period denomination, and no longer in their period-zero equivalent. The Bellman equation becomes

$$\tilde{u}_{tx} = \log\sum_y \exp\left(\tilde{\phi}_{xy} - (\tilde{\mathbf{A}}^\top \tilde{\tau}_t)_{xy} + \beta(\mathbf{P}^\top \tilde{u}_{t+1})_{xy}\right),$$

and the objective of the dual problem becomes $\sum_x N_x \tilde{u}_{1x} + \sum_{tl}\beta^{t-1}\tilde{\tau}_{tl}\tilde{c}_l$. To ensure readability, we shall from now on drop all tildes, and we will replace $\tilde{\phi}_{xy}$ by ϕ_{xy}, $\tilde{\mathbf{A}}_{xy,l}$ by $\mathbf{A}_{xy,l}$, $\tilde{c}_l$ by c_l, $\tilde{u}_{tx}$ by u_{tx}, and $\tilde{\tau}_{tl}$ by τ_{tl}. We get:

Theorem 6.5.1. *The intertemporal decision program of the decision-maker in the problem with logit heterogeneities and infinite horizon has primal formulation*

$$\max_{\mu_{txy}\geq 0}\left\{\sum_{txy\in[\infty\times X\times Y]}\mu_{txy}\beta^{t-1}\phi_{xy}-\sum_{txy\in[\infty\times X\times Y]}\beta^{t-1}\mu_{txy}\log\left(\frac{\mu_{txy}}{\sum_{y\in[Y]}\mu_{txy}}\right)\right\} \tag{6.5.1}$$

$$\text{s.t.}\ \sum_{y\in[Y]}\mu_{1xy}=N_x,\ x\in[X], \tag{6.5.2}$$

$$\sum_{x'y\in[X\times Y]}P_{x,x'y}\mu_{tx'y}=\sum_{y\in[Y]}\mu_{(t+1)xy},\ tx\in[\infty\times X], \tag{6.5.3}$$

$$\sum_{xy\in[X\times Y]}A_{l,xy}\mu_{txy}\leq c_l,\ tl\in[\infty\times L], \tag{6.5.4}$$

whose dual formulation is

$$\min_{u,\tau}\sum_{x\in[X]}N_x u_{1x}+\sum_{tl\in[\infty,L]}\beta^{t-1}\tau_{tl}c_l \tag{6.5.5}$$

$$\text{s.t.}\ u_{tx}=\log\sum_{y\in[Y]}\exp(\phi_{xy}-(\mathbf{A}^\top\tau_t)_{xy}+\beta(\mathbf{P}^\top u_{t+1})_{xy}). \tag{6.5.6}$$

Assume that there are no capacity constraints. In that case, we incorporate a discount rate into expression (6.4.2) so that the function Z is given by:

$$Z\left(u,u',n\right)=\sum_{xy\in[X\times Y]}n_x\exp(\phi_{xy}+\beta(\mathbf{P}^\top u')_{xy}-u_x)-\sum_{x\in[X]}n_x. \tag{6.5.7}$$

Let μ_{txy} and u_{tx} be the solutions to the primal and dual problems (6.5.1) and (6.5.5) (there is no vector τ as we are abstracting away from capacity constraints). Let $n_{tx}=\sum_y\mu_{txy}$. If we set $\mu_{txy}=n_{tx}\exp(\Phi_{xy}+\beta(\mathbf{P}^\top u_{t+1})_{xy}-u_{tx})$, we can reexpress all the equations of the model using the function Z. First, the market clearing equation is $n_{tx}=\sum_y\mu_{txy}$, which can be rewritten as $n_{tx}=-\partial Z(u_t,u_{t+1},n_t)/\partial u_x$. The value iteration equation is $u_{tx}=\log\sum_{y\in[Y]}\exp(\phi_{xy}+\beta(\mathbf{P}^\top u_{t+1})_{xy})$, which can be reexpressed as $\partial Z(u_t,u_{t+1},n_t)/\partial n_x=0$. Finally, the Markov transition equations are $n_{(t+1)x}=\sum_{x'y}P_{x|x'y}\mu_{tx'y}$, which can be expressed as $n_{(t+1)x}=\beta^{-1}\partial Z(u_t,u_{t+1},n_t)/\partial u'_x$. This leads to the *dynamic equations of the model*:

Definition 6.5.1. *The dynamic equations of the model with logit heterogeneity and infinite horizon are given using the expression of Z in* (6.5.7) *by*

$$\begin{cases} n_{1x} = N_x, \ \forall x \in [X], \\ 0 = \frac{1}{\beta}\frac{\partial Z}{\partial u'_x}\left(u_t, u_{t+1}, n_t\right) + \frac{\partial Z}{\partial u_x}(u_{t+1}, u_{t+2}, n_{t+1}), \ \forall tx \in [\infty \times X], \\ 0 = \frac{\partial Z}{\partial n_x}\left(u_t, u_{t+1}, n_t\right), \ \forall tx \in [\infty \times X]. \end{cases} \tag{6.5.8}$$

Remark 6.5.1 (Various generalizations). Note that these equations can easily be modified to incorporate various features. If we need to deal with capacity constraints $\mathbf{A}\mu \leq c$, we change the expression of Z into (6.4.5), and equations (6.5.8) prevail.

In another generalization where we take into account births and deaths, that is, if we have a flow of arrival b_{tx} of units in state x at time t, and a flow d_{tx} of units in state x leaving the system at time t, we can denote by $e_{tx} = d_{tx} - b_{tx}$ the net exiting flow, and equations (6.5.8) are modified into

$$\begin{cases} n_{1x} = N_x, \ \forall x \in [X], \\ e_{tx} = \frac{1}{\beta}\frac{\partial Z}{\partial u'_x}\left(u_t, u_{t+1}, n_t\right) + \frac{\partial Z}{\partial u_x}(u_{t+1}, u_{t+2}, n_{t+1}), \ \forall tx \in [\infty \times X], \\ 0 = \frac{\partial Z}{\partial n_x}\left(u_t, u_{t+1}, n_t\right), \ \forall tx \in [\infty \times X]. \end{cases} \tag{6.5.9}$$

Returning to the case without capacity constraint, we have:

Proposition 6.5.1. *Consider the dynamic discrete choice problem with logit heterogeneity, infinite horizon, and no capacity constraint. Then the primal and dual solutions (μ_{txy}) and (u_{tx}) to* (6.5.1) *and* (6.5.5) *are such that the dual solutions u_{tx} do not depend on t, and the primal solutions μ_{txy} are given by*

$$\mu_{txy} = n_{tx} \exp\left(\phi_{xy} + \beta(\mathbf{P}^\top u)_{xy} - u_x\right),$$

where the quantities n and u are the solution to:

$$\begin{cases} n_{1x} = N_x \ \forall x \in [X], \\ 0 = \frac{1}{\beta}\frac{\partial Z}{\partial u'_x}\left(u, u, n_t\right) + \frac{\partial Z}{\partial u_x}\left(u, u, n_{t+1}\right) \ \forall tx \in [\infty \times X], \\ 0 = \frac{\partial Z}{\partial n_x}\left(u, u, n_t\right) \ \forall tx \in [\infty \times X], \end{cases} \tag{6.5.10}$$

and the function Z is given by expression (6.5.7).

Idea of the proof. The idea of the proof is based on the fact that the value function u_{tx} remains bounded, as its absolute value cannot exceed a constant

K equal to $(1-\beta)^{-1}$ times the maximum short-term per-unit expected payoff, which in the logit model is $|\phi|_\infty + \log Y$. From theorem A.3.1, the value iteration operator is a contraction mapping with constant β, which implies that given two solutions u_{tx} and u'_{tx} to the value iteration equation, we have for any integer $s \geq 0$

$$\left|u_t - u'_t\right|_\infty \leq \beta^s \left|u_{(t+s)} - u'_{(t+s)}\right|_\infty \leq 2\beta^s K,$$

which, after letting $s \to +\infty$, implies that in infinite horizon there is a unique bounded solution to the value iteration equation. Taking $u'_{t+1} = u_t$, we see that this solution is a fixed point of the value iteration operator. ■

Note that although the primal solution μ_{txy} is time-dependent because n_{tx} depends in general on t, the choice probabilities, which are given by $\pi_{txy} = \exp(\phi_{xy} + \beta(\mathbf{P}^\top u)_{xy} - u_x)$, are stationary. In other words, the nonstationarity of the policy function only comes from the marginals n_{xt}. One may write down the equation associated with a fully stationary solution, i.e., a solution such that both the primal and dual solutions are time-independent. Because the total size of the population $\sum_x n_{tx}$ is conserved, we need to impose it; for instance, one may normalize it to 1, and we are led to the stationarity equations:

Definition 6.5.2 (Stationary solution). *A stationary solution is such that* $\mu_{xy} = n_x \exp(\Phi_{xy} + \beta(\mathbf{P}^\top u)_{xy} - u_x)$, *where, taking* Z *as in (6.5.7),* n *and* u *solve*

$$\begin{cases} \sum_{x\in[X]} n_x = 1, \\ 0 = \frac{1}{\beta}\frac{\partial Z}{\partial u'_x}(u, u, n) + \frac{\partial Z}{\partial u_x}(u, u, n), \ \forall x \in [X], \\ 0 = \frac{\partial Z}{\partial n_x}(u, u, n), \ \forall x \in [X]. \end{cases}$$

Theorem 6.5.2. *Consider the dynamic discrete choice problem with logit heterogeneity, infinite horizon, and no capacity constraint. Assume that for every pair of states* x *and* x' *in* $[X]$, *there is an action* $y \in [Y]$ *such that* $\mathbf{P}_{x',xy} > 0$. *Then a stationary solution* $n^\star$ *exists, and it is the limit as* $t \to +\infty$ *of the solution* n_t *to* (6.5.10), *irrespective of the starting point* $n_1 = N$.

Proof. Let $Q_{x'x} = \sum_y P_{x'xy}\pi_{xy}$ with $\pi_{xy} = \exp(\Phi_{xy} + \beta(\mathbf{P}^\top u)_{xy} - u_x)$, and denote by $\mathbf{Q}$ the matrix of term (x', x), $Q_{x'x}$. We have $Q_{x'x} > 0$ for every x and x' in $[X]$. As $\mathbf{1}_X^\top \mathbf{Q} = \mathbf{1}_X^\top$, the Perron–Frobenius theorem (theorem A.3.4) implies that the Perron eigenvalue of $\mathbf{Q}$ is 1, and $\mathbf{1}_X$ is its left Perron eigenvector. Letting (n_x) be its right Perron eigenvector, one has $\mathbf{Q}n = n$, which implies that

$$n_{x'} = \sum_{xy \in [X \times Y]} \pi_{xy} n_x P_{x',xy}.$$

Therefore, setting $\mu_{xy} = n_x \exp(\Phi_{xy} + \beta(\mathbf{P}^\top u)_{xy} - u_x)$, we have $\mathbf{P}\mu = n$, which implies that $n_{t+1} = n_t$ for all t. ■

When the agent becomes infinitely patient, the discounting rate β goes to 1. In that case, one can express the previous conditions as the solution to a minimax problem, as developed below:

Remark 6.5.2 (Unit discounting limit). When $\beta = 1$, which is the limit of unit discount rate, and still with Z given by expression (6.5.7), the equations associated with the fully stationary solution become

$$\begin{cases} \sum_{x \in [X]} n_x = 1, \\ 0 = \frac{\partial Z}{\partial u'_x}(u, u, n) + \frac{\partial Z}{\partial u_x}(u, u, n), \ \forall x \in [X], \\ 0 = \frac{\partial Z}{\partial n_x}(u, u, n), \ \forall x \in [X], \end{cases}$$

which are the first-order conditions associated with the saddle point problem

$$\min_{u \in \mathbb{R}^X} \max_{n \in \Delta_X} Z(u, u, n),$$

where $\Delta_X = \{n \in \mathbb{R}_+^X : \sum_x n_x = 1\}$.

6.6 Inference, Logit Model with Infinite Horizon

We seek to estimate the dynamic discrete choice model with logit heterogeneity and without capacity constraints in the case of infinite horizon. Let $\beta \in [0, 1)$ and assume that we have a sample of size I of observations of the form $x_i y_i$, $i \in [I]$. Assume that the observations are independently sampled. The conditional log-likelihood of observation i is therefore $\log \pi^\lambda_{x_i y_i}$, where the vector of conditional choice probabilities π^λ_{xy} is given by

$$\pi^\lambda_{xy} = \exp\left((\Phi\lambda + \beta \mathbf{P}^\top u^\lambda)_{xy} - u^\lambda_x\right), \tag{6.6.1}$$

where u^λ_x is the unique solution to the *nested fixed-point equation*,

$$u^\lambda_x = \log \sum_{y \in [Y]} \exp(\Phi\lambda + \beta \mathbf{P}^\top u^\lambda)_{xy}, \tag{6.6.2}$$

which can be written in a vectorized manner as

$$\mathbf{\Sigma}_Y \exp\left(\Phi\lambda + \mathbf{\Psi} u^{\lambda}\right) = \mathbf{1}_X, \tag{6.6.3}$$

where we have let $\mathbf{\Sigma}_Y := \mathbf{I}_X \otimes \mathbf{1}_Y^{\top}$ and $\mathbf{\Psi} := \beta \mathbf{P}^{\top} - \mathbf{\Sigma}_Y^{\top}$.

The maximum likelihood estimator. As the log-likelihood of the sample is the sum of the log-likelihood terms associated with the observations, we have $\ell_I(\lambda) = \sum_i (\Phi\lambda)_{x_i y_i} + \beta \sum_i (\mathbf{P}^{\top} u^{\lambda})_{x_i y_i} - \sum_i u^{\lambda}_{x_i}$, which, letting $\hat{\mu}_{xy} = \sum_i 1\{x_i = x, y_i = y\}$, can be written as $\ell(\lambda) = \hat{\mu}^{\top}(\Phi\lambda + (\beta\mathbf{P}^{\top} - \mathbf{I}_X \otimes \mathbf{1}_Y) u^{\lambda})$. Therefore the maximum likelihood estimator of λ is given by

$$\max_{\lambda \in \mathbb{R}^K} \ell_I(\lambda) = \max_{\lambda \in \mathbb{R}^K} \left\{ \hat{\mu}^{\top}\left(\Phi\lambda + \mathbf{\Psi} u^{\lambda}\right) \right\}. \tag{6.6.4}$$

Problem (6.6.4) as written is called the "nested fixed-point" formulation of the MLE, because the computation of u^{λ} involves solving the nested fixed-point equation (6.6.2), given the value of the parameter vector λ. Problem (6.6.4) has the following equivalent MPEC formulation, where the acronym stands for "mathematical program under equilibrium constraint":

$$\begin{aligned} &\max_{\lambda \in \mathbb{R}^K, u \in \mathbb{R}^X} \left\{ \hat{\mu}^{\top}(\Phi\lambda + \mathbf{\Psi} u) \right\} \\ &\quad s.t.\ \mathbf{\Sigma}_Y \exp(\Phi\lambda + \mathbf{\Psi} u) = \mathbf{1}_X. \end{aligned} \tag{6.6.5}$$

While formulations (6.6.4) and (6.6.5) may seem trivially equivalent, and are indeed mathematically the exact same thing, they hint at different computational approaches. The name "nested fixed-point" suggests computing a fixed point (up to numerical precision) after each update of the parameter λ. However, it may not be necessary to carry out this calculation up to machine precision for each value of the parameter; indeed, the value of λ in the first steps is likely to be quite far from the optimal one, and updating it does not require a very precise knowledge of u. As a result, other methods from the optimization toolbox to compute the problem written as in (6.6.5) are generally advisable. We will review both techniques, and to start with, we will investigate the nested fixed-point algorithm. But first, it will be useful to compute the first- and second-order derivatives of the value of each state u^{λ}_x with respect to λ_k.

First and second differentials of the value function. We aim at computing the first and second differentials (with respect to λ) of u^{λ} as defined by the

Bellman equation (6.6.2). The implicit function theorem applies here without any difficulty, and will allow us to obtain these quantities. Denoting by du the differential of u with respect to λ as defined in appendix A.3.1, the differentiation of equation (6.6.3) yields $\mathbf{\Sigma}_Y \mathbf{\Delta}_\pi (\Phi + \mathbf{\Psi} du) = 0$, which yields the differential of u as

$$du = -(\mathbf{\Sigma}_Y \mathbf{\Delta}_\pi \mathbf{\Psi})^{-1} \mathbf{\Sigma}_Y \mathbf{\Delta}_\pi \Phi. \tag{6.6.6}$$

Turning to the second differential, and proceeding by a computation similar to the one carried out in section 6.3, we obtain, using indexed notation,

$$0 = \sum_y \pi_{xy} \left(\sum_{x'} \mathbf{\Psi}_{xy,x'} \frac{\partial^2 u_{x'}}{\partial \lambda_k \partial \lambda_l} \right) + \sum_y \pi_{xy} \frac{\partial \log \pi_{xy}}{\partial \lambda_l} \left(\phi_{xyk} + \sum_{x'} \mathbf{\Psi}_{xy,x'} \frac{\partial u_{x'}}{\partial \lambda_k} \right).$$

Therefore, as $d \log \pi = \Phi + \mathbf{\Psi} du$, and recalling the expression for the row-wise Kronecker product $\circledast$ given in expression (A.3.6), the equation can be rewritten in a vectorized form as $0 = \mathbf{\Sigma}_Y \mathbf{\Delta}_\pi \mathbf{\Psi} d^2 u + \mathbf{\Sigma}_Y \mathbf{\Delta}_\pi ((\Phi + \mathbf{\Psi} du)^{\circledast 2})$, where the notation $M^{\circledast 2}$ stands for $M \circledast M$. As a result, we obtain the following expression for the second differential of u:

$$d^2 u = -(\mathbf{\Sigma}_Y \mathbf{\Delta}_\pi \mathbf{\Psi})^{-1} \mathbf{\Sigma}_Y \mathbf{\Delta}_\pi ((\Phi + \mathbf{\Psi} du)^{\circledast 2}). \tag{6.6.7}$$

The nested fixed-point algorithm. With the knowledge of the differential du of u, we can deduce the first and second derivatives of the log-likelihood of the sample. Indeed, as the log-likelihood of the sample is $\ell(\lambda) = \hat{\mu}^\top (\Phi \lambda + \mathbf{\Psi} u^\lambda)$, it follows from expression (6.6.6) that its differential is

$$d\ell(\lambda) = \hat{\mu}^\top \left(\mathbf{I}_{XY} - \mathbf{\Psi} (\mathbf{\Sigma}_Y \mathbf{\Delta}_\pi \mathbf{\Psi})^{-1} \mathbf{\Sigma}_Y \mathbf{\Delta}_\pi \right) \Phi, \tag{6.6.8}$$

and from the formula for the second differential of u in (6.6.7), one can readily deduce the expression of $d^2 \ell(\lambda) = \hat{\mu}^\top \mathbf{\Psi} d^2 u$. In a gradient descent algorithm, such as the one we describe in algorithm 6.6.2 below, one shall need only the first differential of u; however, the second differential will be useful if we want to use a Newton method instead.

Close inspection of algorithm 6.6.2 reveals that the essential step of the method consists of updating the parameter value λ by an increment that is proportional to the gradient of $\ell(\lambda)$. The computation of that gradient requires the differential of the value function du; however, the computation of du in turn requires the computation of the value function u itself, which is performed by the nested fixed-point algorithm. A Python implementation of algorithm 6.6.2 is provided in appendix B.6.5.

Algorithm 6.6.2 Dynamic discrete choice estimation via nested fixed-point iteration

Input : λ^0 (initial guess)
Parameters: β (discount factor), ϵ (tolerance), α (learning rate), S (maximum iterations)
Output : Estimated parameters $\hat{\lambda}$ (estimator)
Initialize : $s \leftarrow 0$, $\lambda^s \leftarrow \lambda^0$
cont $\leftarrow$ **True**
while cont = **True do**
  Compute u_x with $\lambda = \lambda^s$ using fixed-point iteration (6.6.2);
  Compute $\nabla \ell(\lambda^s)$ using (6.6.8) and the value of u_x;
  Update $\lambda^{s+1} = \lambda^s - \alpha \nabla \ell(\lambda^s)$;
  $s \leftarrow s+1$;
  if $\|\lambda^{s+1} - \lambda^s\| < \epsilon$ ***or*** $s = S$ **then**
    cont $\leftarrow$ **False**;
  end
end
return $\hat{\lambda} = \lambda^s$

Minimax formulation. Problem (6.6.5) can be rewritten as $\max_{\lambda,u} \min_n \mathcal{L}(\lambda, u; n)$, where the Lagrangian $\mathcal{L}$ is defined by

$$\mathcal{L}(\lambda, u; n) := \sum_{xy} \hat{\mu}_{xy}((\Phi\lambda + \beta \mathbf{P}^\top u)_{xy} - u_x) + \sum_x n_x - \sum_{xy} n_x \exp(\Phi\lambda + \mathbf{\Psi} u)_{xy}, \tag{6.6.9}$$

that is, $\mathcal{L}(\lambda, u; n) = \hat{\mu}^\top(\Phi\lambda + \mathbf{\Psi} u) - Z(\lambda, u, n)$, where Z is the penalty function defined by $Z(\lambda, u, n) := \sum_{xy} n_x \exp((\Phi\lambda + \beta \mathbf{P}^\top u)_{xy} - u_x) - \sum_x n_x$ as in section 6.4. Equivalently, problem (6.6.5) can be formulated as $\max_{\lambda,u} \min_n \mathcal{L}(\lambda, u; n)$. Letting $\theta = (\lambda, u, n)$, and denoting

$$\mu^\theta_{xy} := n_x \exp((\Phi\lambda + \beta \mathbf{P}^\top u)_{xy} - u_x), \tag{6.6.10}$$

one can express the derivatives of Z from (6.5.7) with respect to λ, u, and n as:

$$\begin{cases} \frac{\partial Z}{\partial \lambda_k}(\lambda, u, n) = \sum_{xy \in [X \times Y]} \mu^\theta_{xy} \phi_{xyk}, \\ \frac{\partial Z}{\partial u_x}(\lambda, u, n) = \beta \sum_{x'y \in [X \times Y]} \mathbf{P}_{x,x'y} \mu^\theta_{x'y} - \sum_{y \in [Y]} \mu^\theta_{xy}, \\ \frac{\partial Z}{\partial n_x}(\lambda, u, n) = \sum_{y \in [Y]} \frac{\mu^\theta_{xy}}{n_x} - 1. \end{cases} \tag{6.6.11}$$

These quantities are relatively straightforward to interpret. $\partial Z/\partial\lambda_x$ is the kth moment of ϕ_k predicted by μ^θ_{xy}. $\partial Z/\partial u_x$ is β times the number of units that are transitioning to state x from the previous period minus the number of units currently in state x; in a stationary solution, this is equal to $\beta - 1$ times the number of units in state x. $\partial Z/\partial n_x$ is the ratio of the number of units that are in state x divided by n_x; at equilibrium, this should be equal to 1.

Interpretation of the optimality conditions. Consider the vector $\theta = (\lambda, u, n)$ that is the saddle point for the Lagrangian (6.6.9). By first-order conditions, one has the following relations:

- Optimality with respect to λ_k implies

$$\sum_{xy\in[X\times Y]} \mu^\theta_{xy}\phi_{xyk} = \sum_{xy\in[X\times Y]} \hat{\mu}_{xy}\phi_{xyk} \quad \text{for all } k\in[K],$$

  which means that the K predicted moments that Φ associates with the predicted policy vector μ^θ_{xy} match with the observed corresponding moments. This extends the property from the static case, seen in section 3.2, in particular, proposition 3.2.1, that the multinomial logistic regression ensures that the predicted moments will match with the observed moments.
- Optimality with respect to the value u_x yields

$$\beta \sum_{x'y\in[X\times Y]} \mathbf{P}_{x,x'y}\mu^\theta_{x'y} - \sum_{y\in[Y]} \mu^\theta_{xy} = \beta \sum_{x'y\in[X\times Y]} \mathbf{P}_{x,x'y}\hat{\mu}_{x'y} - \sum_{y\in[Y]} \hat{\mu}_{xy} \quad \text{for all } x\in[X],$$

  which expresses that the predictor μ^θ_{xy} is such that β times the total number of units transitioning to state x from the previous period minus the total number of units in state x matches its observed analog. When $\beta = 0$, this amounts to matching the total number of units in state x, which is also a feature of the static multinomial logistic regression.
- Optimality with respect to the marginal number n_x gives us

$$\sum_{y\in[Y]} \frac{\mu^\theta_{xy}}{n_x} = 1 \quad \text{for all } x\in[X],$$

  which expresses that $n_x = \sum_y \mu^\theta_{xy}$, allowing one to interpret n_x, which was introduced as the Lagrange multiplier associated with the constraint x in problem (6.6.5), as being indeed the number of units in state x.

Computation using the augmented Lagrangian method. We can use the augmented Lagrangian algorithm to compute the maximum likelihood estimator using the MPEC formulation (6.6.5). The augmented Lagrangian algorithm is described in display A.1.2 in appendix A.1.7. Given the expression of the Lagrangian provided in expression (6.6.9), we can formulate the augmented Lagrangian as

$$\mathcal{L}_A(\lambda, u; n; \gamma) = \hat{\mu}^\top(\Phi\lambda + \Psi u) - Z(\lambda, u, n) - \gamma \|\nabla_n Z(\lambda, u, n)\|^2, \quad (6.6.12)$$

and the particularization of the augmented Lagrangian algorithm to our setting consists of algorithm 6.6.3. We recall that if $\theta = (\lambda, u, n)$, the expression of μ^θ is given by expression (6.6.10) above. Python code for algorithm 6.6.3 is provided in appendix B.6.6.

Algorithm 6.6.3 Augmented Lagrangian algorithm for dynamic choice estimation

---

Input : Initial values λ^0, u^0, n^0
Parameter: Step size ϵ, tolerance η, maximum iterations S, sequence $\gamma^t \to +\infty$
Initialize iteration counter $s = 0$
while $\|Z(\lambda^s, u^s)\|_\infty \geq \eta$ **and** $s < S$ **do**
  $(\lambda^{s+1}, u^{s+1}) = \arg\max_{\lambda,u} \mathcal{L}_A(\lambda, u; n^s; \gamma^s)$
  $n_x^{s+1} = n_x^s + \gamma^s(1 - \sum_y \exp((\Phi\lambda^{s+1} + \beta \mathbf{P}^\top u^{s+1})_{xy} - u_x^{s+1})$
  $s = s + 1$
end
Return : λ^s, u^s, n^s

Asymptotic distribution of the estimator. One can obtain the Hessian matrix $D^2\ell(\lambda)$ by reshaping the entries of the K^2-dimensional vector $d^2\ell(\lambda)$ provided above in expression (6.6.7) into a $K \times K$ matrix, and a very similar argument to the one leading to theorem 6.3.1 now yields an asymptotic distribution result:

Theorem 6.6.1. *Assume that the observations $x_i y_i$ are drawn from the population in an i.i.d. fashion and collected in a sample of size I. Letting $\hat{\mu}_{xy}$ be the number of observations of type xy in the sample, the following convergence holds in distribution:*

$$\sqrt{I}(\hat{\lambda} - \lambda) \xrightarrow{d} \mathcal{N}(0, H^{-1}CH^{-1}),$$

where H is estimated by $\hat{H}=D^2\ell(\lambda)/I$ and $C=BVB^\top$, where B is estimated by $\hat{B}=\Phi^\top+Du(\hat{\lambda})^\top\Psi^\top$ and V by $\hat{V}=diag(\hat{\mu})/I-\hat{\mu}\hat{\mu}^\top/I^2$.

6.7 Nonparametric Identification, Infinite Horizon

Given the data—summarized by the vector $(\hat{\mu}_{xy})$, which represents the number of observations of each type xy—nonparametric identification concerns the uniqueness of the payoff vector ϕ that rationalizes the data. We discuss three key notions related to this question: set estimators, identified statistics, and identifying restrictions.

Set estimators. Given the conditional choice probabilities $\hat{\pi}_{xy}$ computed from the observations $\hat{\mu}_{xy}$, and a discounting rate $\beta\in[0,1]$, the set estimator of ϕ is the set of those vectors $\hat{\phi}$ that lead to predicting the conditional choice probabilities $\hat{\pi}_{xy}$. Recall from expression (6.6.1) that

$$\log\pi=\phi+(\beta\mathbf{P}^\top-\mathbf{\Sigma}_Y^\top)u, \tag{6.7.1}$$

where the expression of $\mathbf{\Sigma}_Y^\top$ was given by $\mathbf{\Sigma}_Y^\top=\mathbf{I}_X\otimes\mathbf{1}_Y$, and where π is subject to the restriction $\mathbf{\Sigma}_Y\pi=\mathbf{1}_X$. As a result, the set estimator of ϕ is given by

$$\{\log\hat{\pi}+(\mathbf{\Sigma}_Y^\top-\beta\mathbf{P}^\top)u : u\in\mathbb{R}^X\}. \tag{6.7.2}$$

In the static case ($\beta=0$), the set estimator of ϕ is simply the set of vectors $\hat{\phi}$ such that $\hat{\phi}_{xy}=\log\hat{\pi}_{xy}+u_x$. This means that $\log\hat{\pi}$ belongs in that set, as well as any quantity obtained by adding any x-fixed effects to it. In the general dynamic case, the set estimator of ϕ is the set of vectors $\hat{\phi}$ such that $\hat{\phi}_{xy}=\log\hat{\pi}_{xy}+u_x-\beta\sum_{x'}P_{x',xy}u_{x'}$, for any vector $u\in\mathbb{R}^X$. This means that for $\beta>0$, $\log\hat{\pi}$ still belongs in the set estimator; but in general, $\log\hat{\pi}$ plus an arbitrary x-fixed effect will not belong in the set estimator.

Identified statistics. Letting $F:\mathbb{R}^{XY}\to\mathbb{R}^P$, we say that F is an identified statistic if $\hat{\phi}\mapsto F(\hat{\phi})$ is constant on the set estimator of ϕ. In the current setting, this means that if we let for instance $\hat{\phi}:=\log\hat{\pi}$, then for any vector $u\in\mathbb{R}^X$, we have $F(\hat{\phi}+(\mathbf{\Sigma}_Y^\top-\beta\mathbf{P}^\top)u)=F(\log\hat{\pi})$. When F is differentiable, F is an identified statistic if and only if

$$dF(\hat{\phi})(\mathbf{\Sigma}_Y^\top-\beta\mathbf{P}^\top)=0$$

holds for all $\hat{\phi}$ in the set estimator of ϕ. For instance, in the static case, that is, when $\beta=0$, $F(\hat{\phi})=(\mathbf{I}_{XY}-\mathbf{\Sigma}_Y^\top\mathbf{\Sigma}_Y/Y)\hat{\phi}$ is an identified statistic since

$(\mathbf{I}_{XY} - \boldsymbol{\Sigma}_Y^\top \boldsymbol{\Sigma}_Y / Y)\boldsymbol{\Sigma}_Y^\top = 0$. This corresponds to taking $F_{xy}(\hat{\phi}) = \log \hat{\pi}_{xy} - Y^{-1} \sum_y \log \hat{\pi}_{xy}$, which is the deaveraging operation consisting of removing the conditional average of $\log \hat{\pi}_{xy}$ given x.

Identifying restrictions. Letting $\rho : \mathbb{R}^{XY} \to \mathbb{R}^X$, we say that the restrictions $\rho(\hat{\phi}) = 0$ are identifying if there is only one vector $\hat{\phi}$ in the set estimator of ϕ such that $\rho(\hat{\phi}) = 0$. In our setting, this means that if we set $\hat{\phi} := \log \hat{\pi} + (\boldsymbol{\Sigma}_Y^\top - \beta \mathbf{P}^\top) u$ and if we assume that $\rho(\hat{\phi}) = 0$, then the equation $\rho(\hat{\phi} + (\boldsymbol{\Sigma}_Y^\top - \beta \mathbf{P}^\top) r) = 0$ has no other solution than $r = 0_X$. We have:

Lemma 6.7.1. *If* $0 \leq \beta < 1$, *the matrix* $(\boldsymbol{\Sigma}_Y^\top - \beta \mathbf{P}^\top)$ *is of full rank.*

Proof. We shall show that $(\boldsymbol{\Sigma}_Y^\top - \beta \mathbf{P}^\top) u = 0$ implies $u = 0$. Assume $\boldsymbol{\Sigma}_Y^\top u - \beta \mathbf{P}^\top u = 0$. Letting $\mathbf{Q} = \mathbf{P} \boldsymbol{\Sigma}_Y^\top / Y$, which is an $X \times X$ matrix, we have $\boldsymbol{\Sigma}_Y \boldsymbol{\Sigma}_Y^\top u / Y - \beta \mathbf{Q}^\top u = 0$, but as $\boldsymbol{\Sigma}_Y \boldsymbol{\Sigma}_Y^\top / Y = \mathbf{I}_X$, we have $u = \beta \mathbf{Q}^\top u$, which implies that $u = \beta^k (\mathbf{Q}^\top)^k u$ for all integers k, which implies that $u = 0$. ■

For instance, requiring $\phi_{x1} = 0$ for all x is an identifying restriction. Indeed, in that case, $\rho(\phi) = (\mathbf{I}_X \otimes (\mathbf{e}_Y^1)^\top)\phi$, and we have $d\rho(\phi)(\boldsymbol{\Sigma}_Y^\top - \beta \mathbf{P}^\top) = (\mathbf{I}_X \otimes (\mathbf{e}_Y^1)^\top)(\boldsymbol{\Sigma}_Y^\top - \beta \mathbf{P}^\top) = \mathbf{I}_X - \beta \tilde{\mathbf{Q}}^\top$, where the matrix $\tilde{\mathbf{Q}}_{x',x}$ is defined by $\tilde{\mathbf{Q}}_{x',x} = \mathbf{P}_{x',x1}$, and a similar argument to the one given in lemma 6.7.1 implies that $\mathbf{I}_X - \beta \tilde{\mathbf{Q}}^\top$ is invertible.

6.8 The Conditional Choice Probability Approach

We are now ready to present Hotz and Miller's conditional choice probability (CCP) approach to nonparametric identification for the logit model of dynamic discrete choice with infinite horizon. The approach consists of solving the demand inversion problem, to derive the systematic utilities consistent with the conditional choice probabilities, and then using these systematic quantities to nonparametrically or parametrically estimate the vector of short-term payoffs ϕ_{xy}.

Nonparametric identification. Recall from the previous section that the restrictions $\phi_{x1} = 0$ for all $x \in [X]$ are identifying, and therefore we may assume that $\phi_{x1} = 0$ for all $x \in [X]$, where $1 \in [Y]$ is a particular option, without loss of generality of the one with index $y = 1$. This corresponds to setting $\rho(\phi) = \delta^1 \phi$, where $\delta^1 := \mathbf{I}_X \otimes (\mathbf{e}_Y^1)^\top$. If a is the column vector of size XY whose entries are a_{xy}, $xy \in [X \times Y]$, then $\delta^1 a$ is the column vector of size X whose entries are a_{x1}, $x \in [X]$. The identifying equation $\log \pi_{xy} = (\phi +$

$(\beta \mathbf{P}^\top - \boldsymbol{\Sigma}_Y^\top) u)_{xy}$ can be rewritten for $y=1$ as

$$\log \pi_{x1} = \beta(\mathbf{P}^\top u)_{x1} - u_x,\ \forall x \in [X],$$

which, as $\boldsymbol{\delta}^1 \log \pi$ is the vector $(\log \pi_{x1})$ and $\boldsymbol{\delta}^1 \phi = 0$, we can rewrite in vectorized terms as $\boldsymbol{\delta}^1 \log \pi = \boldsymbol{\delta}^1(\beta \mathbf{P}^\top - \boldsymbol{\Sigma}_Y^\top) u$, and therefore u is identified by Hotz and Miller's formula for the *conditional choice probability estimator*, which is

$$u = \left(\delta^1(\beta \mathbf{P}^\top - \boldsymbol{\Sigma}_Y^\top)\right)^{-1} \delta^1 \log \pi,$$

and $\phi = \log \pi + (\boldsymbol{\Sigma}_Y^\top - \beta \mathbf{P}^\top) u$ can readily be deduced. Therefore we can state:

Theorem 6.8.1. *Given the identifying restrictions $\phi_{x1} = 0$ for all $x \in [X]$, ϕ is identified from the data by*

$$\phi = \mathbf{A} \log \pi \quad \text{where} \quad \mathbf{A} = \mathbf{I}_{XY} + (\boldsymbol{\Sigma}_Y^\top - \beta \mathbf{P}^\top)\left(\delta^1(\beta \mathbf{P}^\top - \boldsymbol{\Sigma}_Y^\top)\right)^{-1} \delta^1.$$

If $\hat{\pi}$ is a consistent estimator of π that is asymptotically normal, so that the convergence $I^{1/2}\hat{\pi} \xrightarrow{d} \mathcal{N}(0, V_\pi)$ holds in distribution as $I \to \infty$, we let $\hat{\phi} = \mathbf{A} \log \hat{\pi}$, and we have

$$I^{1/2}\hat{\phi} \xrightarrow{d} \mathcal{N}\left(0, \mathbf{A}\boldsymbol{\Delta}_\pi^{-1} V_\pi \boldsymbol{\Delta}_\pi^{-1} \mathbf{A}^\top\right),$$

where $\boldsymbol{\Delta}_\pi$ is the diagonal square matrix of size XY with the vector π on the diagonal.

An implementation of the CCP estimator is provided in appendix B.6.7.

Parametric estimation using generalized least squares. We now consider a parametric version of the conditional choice probability estimator. Assume as before that ϕ has a linear parametrization $\phi^\lambda = \Phi\lambda$, where, as before, Φ is an $XY \times K$ matrix whose (xy, k) entry is ϕ_{xyk}. We rewrite equation (6.7.1) to get

$$\log \pi = \Phi\lambda + (\beta \mathbf{P}^\top - \boldsymbol{\Sigma}_Y^\top) u, \tag{6.8.1}$$

and letting $\mathbf{W}$ be a square matrix of weights of size XY that is symmetric, positive definite, we may estimate (λ, u) by generalized least squares using the following problem:

$$\min_{\lambda, u} |\log \hat{\pi} - \Phi\lambda - (\beta \mathbf{P}^\top - \boldsymbol{\Sigma}_Y^\top) u|_{\mathbf{W}}, \tag{6.8.2}$$

where $|z|^2_{\mathbf{W}} = z^\top \mathbf{W} z$. Methods exist to pick a matrix $\mathbf{W}$ with good statistical efficiency properties. Compared to the other approaches for estimating λ, the approach based on solving problem (6.8.2) is particularly simple. However, it is more sensitive to the zero cell problem, because is requires taking the logarithms of each entry π_{xy} of the vector of conditional choice probabilities, and therefore it requires that all of these quantities be positive.

6.9 Dynamic Matching Models

Corblet, Fox, and Galichon build a dynamic model of two-sided matching with transferable utility based on the static model of section 5.4, where workers of type $x \in [X]$ seek to match with positions, which we will call firms, of type $y \in [Y]$. At the current period, there are n_x workers of type x, and m_y firms of type y.

We assume that a matching game is played at every period, and that at each period, the systematic part of the current-period value created by a match of type xy is $\phi_{xy} = (\Phi\lambda)_{xy} = \sum_k \phi_{xyk}\lambda_k$, where λ is a K-dimensional parameter vector to be estimated. Likewise, if x or y remain unmatched, the systematic part of their current-period utility is respectively $\phi_{x0} = (\Phi\lambda)_{x0}$ or $\phi_{0y} = (\Phi\lambda)_{0y}$. We also assume that conditional on being matched with a firm of type y, an employee of type x has a probability $P_{x',xy}$ of transitioning to type x' in the next period. Similarly, conditional on being matched with an employee of type x, a firm of type y has a probability $Q_{y',xy}$ of transitioning to type y'. We denote by $\mathbf{P}$ the matrix of shape $X \times X(Y+1)$ and term $P_{x',xy}$, and by $\mathbf{Q}$ the matrix of shape $Y \times (X+1)Y$ and term $Q_{y',xy}$. Letting u'_x and v'_y be the average intertemporal payoffs of agents of respective types x and y in the next period, the intertemporal joint surplus at the current period is given by

$$(\Phi\lambda)_{xy} + \beta(\mathbf{P}^\top u')_{xy} + \beta(\mathbf{Q}^\top v')_{xy} + \varepsilon_y + \eta_x,$$

and similarly, if a worker of type x is unmatched, then this worker will get payoff equal to $\phi_{x0} + \beta(\mathbf{P}^\top u')_{x0} + \varepsilon_0$, and if a firm of type y is unmatched, then its payoff will be $\phi_{0y} + \beta(\mathbf{Q}^\top v')_{0y} + \eta_0$. These quantities are the sum of several terms:

- the systematic joint short-term payoff $(\Phi\lambda)_{xy}$ (for a matched pair) or the systematic reservation payoff $(\Phi\lambda)_{x0}$ or $(\Phi\lambda)_{0y}$ (for unmatched agents);
- the discounted expected future utility on the worker side $\beta(\mathbf{P}^\top u')_{xy}$, which is the average of the next-period utility associated to the

possible next-period type, weighted by the probability of the transitions, and similarly, the discounted expected future utility on the firm side, $\beta(\mathbf{Q}^\top v')_{xy}$. Although these terms originate without ambiguity on the worker's and the firm's side, respectively, they are pooled in the event of a match, and participate in the joint utility like the rest of the terms;

- the random preference shock (ε_y) that workers associate with a firm of type y; and, similarly, the random production shock (η_x) generated by firms with various worker types. These random utility shocks are drawn anew in each period. They are also pooled; but, just as in the static Choo and Siow matching model described in section 5.4, it will be a property of the equilibrium that, at equilibrium, the worker retains their utility shock ε_y, and the firm retains its utility shock η_x.

In the current period, agents solve a Choo and Siow matching model where the intertemporal payoff $(\Phi\lambda)_{xy} + \beta(\mathbf{P}^\top u')_{xy} + \beta(\mathbf{Q}^\top v')_{xy}$ replaces the single-period payoff $(\Phi\lambda)_{xy}$. Given the current numbers of agents of each type, indicated by n_x and m_y, and given the next-period utilities u'_x and v'_y that agents of each type anticipate receiving in the next period, one seeks to determine:

- u_x and v_y, the current-period average payoff associated with each type, similarly to what happens in the single-agent dynamic with the value iteration (backward induction), where the current-period utility is determined from the next-period utility; and
- n'_x and m'_y, the next-period numbers of agents in each type: these numbers are the result of the choices that agents make when they match, which determines the number of matches xy of each type in the current period, which then determines the number of agents in each state in the next period through the Markov transitions induced by the transition probabilities $\mathbf{P}$ and $\mathbf{Q}$, similarly to what happens in the single-agent case with policy iteration (forward induction).

We consider a sample of size I that consists of observations $x_i y_i$ that are either a pair of matched individuals, in which case $x_i \in [X]$ and $y_i \in [Y]$, or unmatched individuals, in which case either $x_i = 0$ or $y_i = 0$. We define $\hat{\mu}_{xy} = \sum_i \mathbf{1}_{x_i = x, y_i = y}$ to be the number of observations of each type, and $\hat{f}_{xy} = \hat{\mu}_{xy}/I$ to be the empirical frequencies associated with each observation. Our goal is to estimate the parameter λ of the matching surplus $\Phi\lambda$ based on our data.

Recall that the expression of μ is given by Choo and Siow's equations (5.4.7), which in the present context become

$$\begin{cases} \mu_{xy} = \sqrt{n_x m_y} \exp\left(\frac{(\Phi\lambda)_{xy} + \beta(\mathbf{P}^\top u')_{xy} + \beta(\mathbf{Q}^\top v')_{xy} - u_x - v_y}{2} \right), \\ \mu_{x0} = n_x \exp\left((\Phi\lambda)_{x0} + \beta(\mathbf{P}^\top u')_{x0} - u_x \right), \\ \mu_{0y} = m_y \exp\left((\Phi\lambda)_{0y} + \beta(\mathbf{Q}^\top v')_{0y} - v_y \right), \end{cases} \tag{6.9.1}$$

which suggests introducing the function Z as

$$Z = 2 \sum_{xy \in [X \times Y]} \mu_{xy} + \sum_{x \in [X]} \mu_{x0} + \sum_{y \in [Y]} \mu_{0y} - \sum_{x \in [X]} n_x - \sum_{y} m_y, \tag{6.9.2}$$

where μ is a function of the parameters λ, u, v, u', v', n, and m given by (6.9.1) above, so that the partial derivatives of Z with respect to its arguments are

$$\frac{\partial Z}{\partial u_x} = -\left(\sum_{y \in [Y]} \mu_{xy} + \mu_{x0} \right) \quad \text{and} \quad \frac{\partial Z}{\partial v_y} = -\left(\sum_{x \in [X]} \mu_{xy} + \mu_{0y} \right),$$

$$\frac{\partial Z}{\partial u'_x} = \beta \sum_{x'y \in [X \times 0:Y]} P_{x,x'y} \mu_{x'y} \quad \text{and} \quad \frac{\partial Z}{\partial v'_y} = \beta \sum_{xy' \in [0:X \times Y]} Q_{y,xy'} \mu_{xy'},$$

$$\frac{\partial Z}{\partial n_x} = \sum_{y \in [Y]} \frac{\mu_{xy}}{n_x} + \frac{\mu_{x0}}{n_x} - 1 \quad \text{and} \quad \frac{\partial Z}{\partial m_y} = \sum_{x \in [X]} \frac{\mu_{xy}}{m_y} + \frac{\mu_{0y}}{m_y} - 1,$$

$$\text{and} \quad \frac{\partial Z}{\partial \lambda_k} = \sum_{xy \in [X \times Y]} \mu_{xy} \phi_{xyk} + \sum_{x \in [X]} \mu_{x0} \phi_{x0k} + \sum_{y \in [Y]} \mu_{0y} \phi_{0yk}.$$

For each value of λ, we can therefore write the equilibrium equations of the model that express stationarity as $u = u'$, $v = v'$, $n = n'$, and $m = m'$, and we have the following equations on (u, v, n, m):

$$\sum_{y \in [Y]} \mu_{xy} + \mu_{x0} = n_x \text{ and } \sum_{x \in [X]} \mu_{xy} + \mu_{0y} = m_y, \tag{6.9.3}$$

$$\sum_{x'y \in [X \times 0:Y]} P_{x,x'y} \mu_{x'y} = n_x \text{ and } \sum_{xy' \in [0:X \times Y]} Q_{y,xy'} \mu_{xy'} = m_y, \tag{6.9.4}$$

where the expression of μ as a function of (u, v, n, m) is defined from (6.9.1), setting $u' = u$ and $v' = v$. We vectorize these equations in order to permit tractable computations. Set $p^\top = (u^\top, v^\top)$ and

$$\mathbf{\Sigma}^\top = \begin{pmatrix} \mathbf{I}_X \otimes \mathbf{1}_Y & \mathbf{1}_X \otimes \mathbf{I}_Y \\ \mathbf{I}_X & \mathbf{0}_{X \times Y} \\ \mathbf{0}_{Y \times X} & \mathbf{I}_Y \end{pmatrix} \text{ and } \mathbf{R}^\top = \begin{pmatrix} (\mathbf{P}^\top)_{xy,x'} & (\mathbf{Q}^\top)_{xy,y'} \\ (\mathbf{P}^\top)_{x0,x'} & \mathbf{0}_{X \times Y} \\ \mathbf{0}_{Y \times X} & (\mathbf{Q}^\top)_{0y,y'} \end{pmatrix}$$

and $\nu^\top = (\log n^\top, \log m^\top)$. Let $B = XY + X + Y$ be the number of entries of $\mu = \left(\mu_{xy}, \mu_{x0}, \mu_{0y}\right)$. For $b \in [B]$, let s_b be the size of match b, that is, $s_b = 2$ if $b \leq XY$, in which case b is one of the xy for $x \in [X]$ and $y \in [Y]$, and $s_b = 1$ if $b > XY$, in which case b stands for either $x0$ for $x \in [X]$ or $0y$ for $y \in [Y]$. Let $\mathbf{\Delta}_s$ be the diagonal matrix of size B whose diagonal is s. Let $\mathbf{\Psi} = \beta \mathbf{R}^\top + \mathbf{\Sigma}^\top$. Define $\theta^\top = (p^\top, \nu^\top)$, $D_1 = \mathbf{\Delta}_s^{-1}\Phi$, $D_2 = \left(\beta \mathbf{\Delta}_s^{-1}\mathbf{R}^\top - \mathbf{\Delta}_s^{-1}\mathbf{\Sigma}^\top, \quad \mathbf{\Delta}_s^{-1}\mathbf{\Sigma}^\top\right)$, $E_1 = \left(\mathbf{\Sigma}^\top, \quad \mathbf{R}^\top\right)$, $E_2 = \left(\mathbf{I}_{X+Y}, \quad \mathbf{I}_{X+Y}\right)$, and $D_3 = \left(\mathbf{0}_{(X+Y)\times(X+Y)}, \quad \mathbf{I}_{(X+Y)}\right)$. The probability of observing $b \in [B]$ is equal to f_b, where f has the following vectorized equation:

$$f := \frac{\mu}{\mathbf{1}_B^\top \mu} \quad \text{where } \mu = \exp\left(D_1 \lambda + D_2 \theta^\lambda\right), \tag{6.9.5}$$

where given the value of λ, the vector θ is implicitly determined by equations (6.9.3) and (6.9.4), which can be rewritten as

$$E_1^\top \exp\left(D_1 \lambda + D_2 \theta\right) = E_2^\top \exp\left(D_3 \theta\right). \tag{6.9.6}$$

By differentiation of equation (6.9.6), denoting $\mathbf{\Delta}_\mu = diag(\mu)$, and $\mathbf{\Delta}_{\exp(\nu)} = diag(\exp(\nu))$ where $\nu = D_3\theta$, we have that the differential of θ with respect to λ satisfies

$$E_1^\top \mathbf{\Delta}_\mu \left(D_1 + D_2\, d\theta\right) = E_2^\top \mathbf{\Delta}_{\exp(\nu)} D_3\, d\theta.$$

Letting $K := E_2^\top \mathbf{\Delta}_{\exp(\nu)} D_3 - E_1^\top \mathbf{\Delta}_\mu D_2$, a calculation similar to the one carried out in section 6.6 shows that the first and second differentials of θ with respect to λ are given by

$$\begin{cases} d\theta = K^{-1} E_1^\top \mathbf{\Delta}_\mu D_1, \\ d^2\theta = K^{-1}\left(E_2^\top \mathbf{\Delta}_{\exp(\nu)} \left(D_3\, d\theta\right)^{\circledast 2} - E_1^\top \mathbf{\Delta}_\mu \left(D_1 + D_2\, d\theta\right)^{\circledast 2}\right), \end{cases} \tag{6.9.7}$$

where we recall that the row-wise Kronecker product $\circledast$ is defined in expression (A.3.6), and that the notation $M^{\circledast 2}$ stands for $M \circledast M$.

Recall that I is the sample size, and that $\hat{f} := \hat{\mu}/I$ is the vector of the empirical frequencies of the observations. Letting $M(\lambda) := \log f = D_1\lambda + D_2\theta - \log(\mathbf{1}_B^\top \exp(D_1\lambda + D_2\theta))$, where θ is implicitly a function of λ, the expression for the log-likelihood of the sample (rescaled by I) is $\ell_I(\lambda) := \hat{f}^\top M(\lambda)$, and the maximum likelihood estimator $\hat{\lambda}$ then maximizes $\ell_I(\lambda)$. Recalling that the expression of f was given by (6.9.5), we can deduce from expressions (6.9.7) that

$$\begin{cases} d\ell_I(\lambda) = \hat{f}^\top \left(D_1 + D_2\, d\theta\right) - f^\top \left(D_1 + D_2\, d\theta\right), \\ d^2\ell_I(\lambda) = \hat{f}^\top D_2\, d^2\theta - f^\top D_2\, d^2\theta + \left(D_1 + D_2\, d\theta\right)^\top df, \end{cases} \tag{6.9.8}$$

and we can deduce the Hessian of ℓ_I by reshaping $d^2\ell$ into a $K \times K$ matrix. It follows that:

Theorem 6.9.1 (Corblet, Fox, and Galichon). *Assume that the observations $x_i y_i \in B$ are drawn from the population in an i.i.d. fashion from a data-generating process consistent with the model described in this section, and are collected in a sample of size I. Letting $\hat{\mu}_{xy}$ be the number of observations of type xy in the sample, the following convergence holds in distribution:*

$$\sqrt{I}(\hat{\lambda} - \lambda) \xrightarrow{d} \mathcal{N}(0, H^{-1}CH^{-1}),$$

where $H = D^2\left(f^\top M\right)(\lambda)$ and $C = dM(\lambda)^\top \operatorname{diag}\left(f\right) dM(\lambda) - dM(\lambda)^\top f^\top f dM(\lambda)$.

This model is coded in problem 6.2.

6.10 Exercises

Exercise 6.1. *What happens to the model of section 6.9 when we allow for coalitions of possibly more than two individuals, as it is done in section 5.5 in the static case?*

Exercise 6.2. *Consider the conditions associated with the stationary solutions as in definition 6.5.2. It follows from remark 6.5.2 that when $\beta = 1$, these conditions are the first-order conditions associated with a convex optimization/saddle point problem.*

(i) Defining Z as in (6.5.7), show that the Jacobian of

$$u \mapsto \left(\frac{\partial Z}{\partial u'_x}(u, u, n) + \beta \frac{\partial Z}{\partial u_x}(u, u, n) \right)_{x \in [X]} \tag{6.10.1}$$

is $(\mathbf{P} - \boldsymbol{\Sigma}_Y)\boldsymbol{\Delta}_\mu(\mathbf{P} - \beta\boldsymbol{\Sigma}_Y)$, where $\boldsymbol{\Sigma}_Y = \mathbf{I}_X \otimes \mathbf{1}_Y^\top$, and $\boldsymbol{\Delta}_\mu = \operatorname{diag}(\mu)$.

(ii) Deduce that, in general, stationarity equations cannot be expressed as the saddle point of an optimization problem for $\beta \in (0, 1]$ unless $\beta = 1$.

(iii) When $\beta = 1$, suggest a computational procedure to compute (u, n) when $\beta = 1$.

Exercise 6.3. *Let $T : \mathbb{R}^d \to \mathbb{R}^d$ be a contraction mapping with Lipschitz constant $K \in (0, 1)$, and call $x^\star$ its fixed point. Let x be such that $|x - T(x)| \leq \varepsilon$.*

Show that $|x - x^{\star}| \leq \varepsilon/(1-K)$. *Conversely, assume that* $|x - x^{\star}| \leq \varepsilon$. *Show that* $|x - T(x)| \leq \varepsilon(1+K)$.

Exercise 6.4. *Write down the equations that govern the evolution of* (n_{tx}, m_{ty}) *in the model of section 6.9. Show that the quantities* $\sum_x n_{tx}$ *and* $\sum_y m_{ty}$ *are independent of* t.

Exercise 6.5 (A dynamic hedonic model). *Write down a dynamic version of the hedonic model of section 5.6 in finite and infinite horizon. Simulate the finite-horizon model with probit heterogeneity.*

Exercise 6.6 (Convergence of the finite-horizon model). *Take* $X = 10$, $Y = 2$, *and* $\beta = 0.9$, *and use the code from appendix B.6.5 to generate* $\mathbf{P}$ *and* ϕ_{kxy}. *Set* $\phi = \boldsymbol{\phi}\lambda$ *with* $\lambda = (1,1,1)^{\top}$. *Assume that, at the initial time,* $n_x = 1$ *for all* $x \in [X]$. *Draw a sample of* $I = 1{,}000$ *vectors* $(\varepsilon_{ixy})_{xy}$ *from an i.i.d. Gumbel distribution, and solve the finite-horizon problem for* $T = 10$, $T = 50$, *and* $T = 100$. *Comment on the behavior of* n^T *and* u^1 *as* T *becomes large.*

6.11 Problems

Problem 6.1 (Estimation of dynamic discrete choice based on Rust's data). *This problem is based on Rust's (1987) original data from his landmark paper. The code in appendix B.6.8 loads that dataset, which consists of a three-column table organized as follows:*

- *In the first column: the mileage class of a bus engine at the start of the period, an integer* x *from 0 (new) to 89 (oldest), which corresponds to 5,000-mile brackets, so that engines in class* x *are on average* $5{,}000x$ *miles old;*
- *In the second column: whether the engine was replaced during the period—0 if not, 1 if yes;*
- *In the third column: by how many mileage classes the bus aged over the period—0 if it remained in the current class, 1 if it moved to the next one, 2 if it moved up two classes, and so on.*

Rust considers an infinite-horizon model of the decision-making process and assumes that the cost of operating is $0.001\lambda_0 x$, *and the cost of repairing is* λ_1.

(i) *Estimate the transition matrix* $P_{x'|xy}$ *representing the probability of transitioning to state* x' *conditional on being in state* x *in the previous period and making decision* y. *Assume that the increment* $x(t+1) - x(t)$, *representing the change in the bus's mileage class at time* t, *is* *independent of $x(t)$, and that only transitions from x to x, $x+1$, or $x+2$ are possible. In the case of engine replacement ($y=1$), assume that the bus resumes operation immediately during the same period, so transitions to $x=0$, 1, or 2 are possible. Set the discount factor to $\beta=0.9999$.*

(ii) *Compute the maximum likelihood estimator for (λ_0, λ_1) using Rust's dataset.*

Problem 6.2 (Dynamic matching). *The code in appendix B.6.9 generates dynamic matching data. Estimate the dynamic matching model of section 6.9 based on that dataset.*

Problem 6.3 (Multi-armed bandits, part 1: Basics). *Here we describe a "multi-armed bandit." A multi-armed bandit is so called because it is exemplified by a coin machine with a finite number of arms such that each can be pulled one at a time. Each arm has its own state to which it is associated. If an arm is pulled, then the agent gets a reward that depends both on which arm is pulled, and on the state that it is in.*

Formally, let Y be the number of arms, and let Z_y be the number of states of arm y. For each $y \in [Y]$, Q^y is a $Z_y \times Z_y$ matrix whose (z', z)th term $Q^y_{z'z}$ is the probability that arm y should transition from state z to z' if pulled. If it is not pulled, the arm remains in the same state. Consequently, we define $[X] = [Z_1 \times Z_2 \times \ldots \times Z_Y]$ as the state of the multi-armed bandit as a whole. We call $[X]$ the "set of bandit states," and $[Z_y]$ the "set of arm y states." Given $x = (z_1, \ldots, z_Y)$ and $x' = (z'_1, \ldots, z'_Y)$, the probability $P_{x'xy}$ of a transition from bandit state x to bandit state x' if arm y is pulled is equal to $Q^y_{x'_y|x_y}$ if all $x'_{-y} = x_{-y}$, and zero otherwise. We let P be the $X \times XY$ matrix of term $P_{x'xy}$. If arm y is pulled and is in state $z \in Z_y$, then the agent gets a reward equal to $\phi_{zy} > 0$. The discounting rate is β. The time horizon is infinite.

(i) *(This question about vectorization is slightly peripheral and may be skipped.) We represent $x \in [X]$ by identifying x with $z_1 z_2 \ldots z_Y$ in the row-major order. For example, if $Y=2$, $Z_1=2$, and $Z_2=3$, the pairs of indices $z_1 z_2$ are listed in the following order: 11, 12, 13, 21, 22, 23. Show that, in general,*

$$P = \sum_{y \in [Y]} \mathbf{I}_{Z_1 \times \ldots Z_{y-1}} \otimes Q^y \otimes \mathbf{I}_{Z_{y+1} \times \ldots \times Z_Y} \otimes \left(\mathbf{e}^y_Y\right)^\top,$$

where $\mathbf{e}^y_Y$ is the column vector of size Y whose yth entry is equal to 1, the others being 0.

(ii) *For each arm $y \in [Y]$, we consider a very simple binary bandit that consists of two arms. The first arm generates a payoff g and has only one state. The second arm has Z_y states, transition matrix Q^y, and reward ϕ_{zy} in state $z \in Z_y$. In that case, show that the value associated with state $z \in [Z_y]$ is*

$$u_z = \max\{g + \beta u_z, \phi_{zy} + \beta \sum_{z' \in [Z_y]} Q^y_{z'z} u_{z'}\}, \ \forall z \in [Z_y].$$

(iii) *The Gittins index $g(z, y)$ associated with arm $y \in [Y]$ in state $z \in [Z_y]$ is the value γ that makes the player indifferent between the two arms of the binary bandit described above. Show that $g(z, y)$ is the value of γ such that*

$$\begin{cases} u_w = \max\{\frac{\gamma}{1-\beta}, \phi_{wy} + \beta \sum_{z' \in [Z_y]} Q^y_{z'w} u_{z'}\}, \ \forall w \in [Z_y], \\ \frac{\gamma}{1-\beta} = \phi_{zy} + \beta \sum_{z' \in [Z_y]} Q^y_{z'z} u_{z'}. \end{cases} \tag{6.11.1}$$

(iv) *Going back to the multi-armed bandit problem, show that optimal strategy consists of pulling the arm whose current state has the largest Gittins index.*

Problem 6.4 (Multi-armed bandits, part 2: Computation of Gittins indices). *This problem is a sequel to problem 6.3, and uses some of the results there.*

(i) *Start from the characterization of the Gittins index $g(z, y)$ in (6.11.1) to show that $g(z, y)$ if only $g(z, y) = g$, where there exists a vector $u \in \mathbb{R}^Z$ such that the pair $(u, g) \in \mathbb{R}^Z \times \mathbb{R}$ is a fixed point of a mapping $T : \mathbb{R}^{Z+1} \to \mathbb{R}^{Z+1}$ that has the following properties:*
   - *T is order-preserving and continuous;*
   - *T is a strict contraction for the infinity norm;*
   - *$(u, g) \in \operatorname{Im} T$ implies $g \leq |\phi|_\infty$ and $|u|_\infty \leq |\phi|_\infty / (1 - \beta)$;*
   - *$T(0) \geq 0$.*

(ii) *Deduce that the sequence $(u^t_z, g^t) \in \mathbb{R}^Z \times \mathbb{R}$ defined by*

$$\begin{cases} \left(u^t, g^t\right) = (\mathbf{0}_Z, 0), \\ (u^{t+1}, g^{t+1}) = T(u^t, g^t) \end{cases}$$

converges to the unique fixed point of T, which provides the value of the Gittins index $g(z, y)$.

(iii) *Implement the algorithm and run it on simulated data.*

Problem 6.5 (Multi-armed bandits, part 3: Applications). *This problem is a sequel to problems 6.3 and 6.4, and uses some of the results there.*

(1) *(Single-machine scheduling). A single machine can run Y different jobs. Job $y \in [Y]$ takes τ_y periods to process and some reward ρ_y is obtained after completion. The future is discounted with discount factor β.*

(i) *Set up the problem as a multi-armed bandit problem and use the algorithm described in problem 6.4 in order to numerically compute the Gittins indices $g(z, y)$ based on simulated data.*

(ii) *Show that the Gittins index associated with scheduling job y if the machine is vacant can be obtained in closed form as*

$$\rho_y \frac{\beta^{\tau_y}(1-\beta)}{1-\beta^{\tau_y}}$$

and verify the numerical results obtained in (i).

(2) *(Weitzman's Pandora's boxes problem). An agent faces Y boxes and each box $y \in [Y]$ contains an unknown number ρ_y. According to the agent's belief, the distribution of the number in box y is F_y. The cost associated with opening box y is c_y. The agent can open as many boxes as she wants, in any order, and stop when she wants. Once she stops, she receives the maximum of the numbers in the boxes that were opened. As a result, if $O \subseteq [Y]$ is the set of boxes that were opened, the agent's payoff is*

$$\max_{y \in O} \{\rho_y\} - \sum_{y \in O} c_y.$$

Show that the optimal policy of the agent consists of opening the box with the greater Gittins index, where the Gittins of box y is ρ_y if box y has been opened, and is the solution g to $g = \mathbb{E}_{F_y}[\max\{\rho_y, g\}] - c_y$ if the box has not been opened yet.

6.12 References and Notes

The seminal Rust (1987) paper started the econometric literature on dynamic discrete choice, and proposed the nested fixed-point maximum likelihood estimation procedure (6.6.4). The conditional choice probability estimators were originally suggested by Hotz and Miller (1993). Other important contributions include Keane and Wolpin (1997), who use the framework to model career decisions, as well as Arcidiacono and Miller (2011), who focus on estimation issues, Aguirregabiria and Mira (2007), who extend the framework to a competitive setting, and Su and Judd (2012), who suggest the MPEC formulation (6.6.5). The expository chapter by Rust (1994), the one by Keane, Todd, and Wolpin (2011), and the survey by Aguirregabiria and Mira (2010) are excellent references. The lecture notes by Shum (2016) provide a nice treatment on dynamic discrete choice, among other topics. The paper by Magnac and Thesmar (2002) covers identification issues. Although not directly focused on dynamic discrete choice models, the paper by Pakes (1984) on patents contains early insights on dynamic models in the field of industrial organization. Rosaia (2020) studies the limit of unit discounting in a dynamic discrete choice model. The reference for the dynamic matching model of section 6.9 is Corblet, Fox, and Galichon (2024). A good introduction to multi-armed bandits and Gittins indices, which are the focus of problems 6.3, 6.4, and 6.5, is the set of lecture notes by Weber (2016) at Cambridge University; see also Gittins, Glazebrook, and Weber's (2011) book. The "Pandora's box" problem comes from Weitzman (1979).

The model presented in section 6.1 is called the "Markov decision process" when there is no capacity constraint, which is the case we called the "single-agent dynamic." Standard references are the two-volume monograph by Bertsekas (2012) and Puterman's (2014) book. In the language of Markov decision processes, $[Y]$ is called the set of "actions," $P_{x',xy}$ is called the "transition function," and ϕ_{txy} is called the "reward function." A number of researchers have replicated and expanded upon Rust's (1987) original study, as we invite the reader to do in problem 6.1; Lin's (2025b) GitHub repository is an example of such an exercise, carried out in Python.

6.13 Summary

SUMMARY OF CHAPTER 6

This chapter expands upon the concept of models by incorporating a dynamic perspective, where decision-makers face sequential choices that influence both their current options and future opportunities. Such decisions include savings, career paths, maintenance actions, and retirement planning, highlighting the difference between short-term, myopic decisions and those optimized for long-term benefits. The inclusion of dynamic programming elements allows for the valuation of states within a Markovian framework, considering both immediate payoffs and the expected future state values, thus capturing the essence of strategic decision-making over time.

The chapter starts with the analysis in section 6.1 of the model with no heterogeneity, and a finite horizon. The model is framed as a linear programming problem, introducing a methodological foundation for understanding the dynamic choice process.

Section 6.2, via i.i.d. Gumbel components, transforms the model to exhibit a logit structure, facilitating its estimation and analysis. The model is fully vectorized to facilitate the computational and statistical analysis, and the maximum likelihood estimator is computed in section 6.3, as well as its asymptotic distribution.

The role of a certain potential function called the Z function is then discussed in section 6.4, showcasing how the equations of the dynamic model emerge from the partial derivatives of Z. This exploration reveals the possibility of estimating the model using generalized linear model techniques, even as the discount factor approaches unity, shedding light on the model's adaptability to various temporal frameworks.

Next, the infinite horizon model is discussed in section 6.5 for equilibrium aspects, as well as inference issues, in section 6.6. A discussion of the respective merits of the nested fixed-point and MPEC approaches, as well as the asymptotic distribution of the maximum likelihood estimator, is provided.

Identification issues are discussed in section 6.8, where an alternative estimation method relying on the inversion of conditional choice probabilities (the "conditional choice probability approach") is discussed.

Finally, dynamic matching problems are considered in section 6.9, demonstrating the model's flexibility in extending to contexts where individuals or agents form pairs or groups, further broadening the application of the basic theory.

CHAPTER 7

Constrained Choice

In this chapter, we consider constrained choice problems, where the limited availability of certain options leads to the introduction of rationing. Many familiar markets fall under this setting: concert tickets, day care, walk-in-only restaurants. In these models, prices do not adjust supply and demand, but other quantities often do: this is the case of waiting lines, which decrease the utility associated with overdemanded options, thus regulating supply and demand. The main difference between the setting of this chapter and the setting of chapter 5, in particular, sections 5.4 and 5.6, is that those models are equilibrium models with money (or, more generally, with utility transfers), while the markets we study in this chapter are markets without money (or, more generally, with nontransferable utility).

7.1 Basics

Consider the demand for a certain service, for example, medical care at various hospitals $y \in [Y]$. We assume that there is also a zero option ($y=0$) associated with not receiving medical care. We assume that each hospital is subject to a capacity constraint, namely, the demand π_y for hospital $y \in [Y]$ should not exceed the capacity $\bar{\pi}_y$ of hospital y. However, in this economy, there are no prices to regulate demand and to ensure that it stays within capacity. Assume that agents pick their hospitals according to a discrete choice problem, picking the hospital y that maximizes $U_y + \varepsilon_y$ if the corresponding maximum is above the reservation utility ε_0, and picking the outside option $y=0$ otherwise. This will lead to demand $\boldsymbol{\pi}_y(U)$ for hospital y. There is no guarantee that the constraints $\boldsymbol{\pi}_y(U) \leq \bar{\pi}_y$ are satisfied, and any hospital for which this constraint is violated will face excess demand.

If hospital y is overdemanded, one may predict that there will be congestion. Waiting times will increase, quality of service may deteriorate, and so on. We shall assume that this will result in a decrease of systematic utility by $\tau_y \geq 0$, and therefore the systematic utility associated with hospital y will become $U_y - \tau_y$ instead of U_y. The quantity τ_y is interpreted as the "shadow price" of the capacity constraint. One associates such a nonnegative shadow price to each hospital $y \in [Y]$, and we denote by τ the Y-dimensional vector of shadow prices. The market share of hospital y becomes $\boldsymbol{\pi}_y(U - \tau)$, and the presence of τ ensures that the constraints $\boldsymbol{\pi}_y(U-\tau) \leq \bar{\pi}_y$ are satisfied for each hospital $y \in [Y]$. Of course, hospitals that are not overdemanded have no congestion, so $\tau_y > 0$ holds only if the capacity constraint is saturated.

These considerations lead us to investigate the solution $\tau \in \mathbb{R}^Y$ of the system

$$\begin{cases} \tau_y \geq 0 & \forall y \in [Y], \\ \bar{\pi}_y - \boldsymbol{\pi}_y(U-\tau) \geq 0 & \forall y \in [Y], \\ \tau_y \left(\bar{\pi}_y - \boldsymbol{\pi}_y(U-\tau)\right) = 0 & \forall y \in [Y]. \end{cases} \tag{7.1.1}$$

We have the following result:

Theorem 7.1.1. *Under assumptions 1.1 and 1.2, the system of equations* (7.1.1) *has a unique solution.*

Before we give the proof of this result, we provide a bunch of intermediate claims. First, we give a useful reformulation of system (7.1.1) as a set of nonlinear equations. The idea is to notice that for all $y \in [Y]$, the shadow price τ_y and the overcapacity in option y, defined by

$$\rho_y := \bar{\pi}_y - \boldsymbol{\pi}_y (U - \tau),$$

(i) are both greater than or equal to zero, and (ii) cannot be both strictly positive at the same time. We thus introduce the Y-dimensional vector θ whose element θ_y has positive part equal to τ_y, and negative part equal to ρ_y. Formally, the relationship between θ and (τ, ρ) is given by

$$\theta = \tau - \rho \quad \text{and} \quad \tau = \theta^+, \rho = \theta^-,$$

where we recall that, for $a \in \mathbb{R}^Y$, a^+ stands for the vector of the same size whose yth entry is $a_y^+ = \max(a_y, 0)$, and a^- for the vector whose yth entry is $a_y^- = \max(-a_y, 0)$. Hence if τ is the solution to (7.1.1), we have $\boldsymbol{\pi}\left(U - \theta^+\right) + \theta^- = \bar{\pi}$. Conversely, if θ satisfies that condition, then $\tau = \theta^+$ solves (7.1.1). We have thus proven:

Proposition 7.1.1. *Assume the distribution of ε satisfies assumptions 1.1 and 1.2. Let $\bar{\pi} \in \mathbb{R}^Y_+$ such that $\bar{\pi}_y > 0$ for all $y \in [Y]$. Then the solution τ to* *system* (7.1.1) *is given by* $\tau = \theta^+$, *where* $\theta \in \mathbb{R}^Y$ *is the solution to the set of equations*

$$\boldsymbol{\pi}\left(U - \theta^+\right) + \theta^- = \bar{\pi}. \tag{7.1.2}$$

Proposition 7.1.2. *Under assumptions 1.1 and 1.2, the map* $\mathbf{q}$ *defined by*

$$\mathbf{q}(\theta) = \bar{\pi} - \theta^- - \boldsymbol{\pi}(U - \theta^+) \tag{7.1.3}$$

is an M-function.

Proof. As $\boldsymbol{\pi} = \nabla G$ and G is submodular as a result of proposition 1.2.1, it follows that $\boldsymbol{\pi}$ is a Z-function, and, subsequently, that $\mathbf{q}$ is one as well. Let us show that it is strongly nonreversing (cf. definition A.3.9). Assume $\theta \leq \theta'$ and $\mathbf{q}(\theta) \geq \mathbf{q}\left(\theta'\right)$. Then $\theta^- \geq \theta'^-$, and $\bar{\pi} - \theta^- - \boldsymbol{\pi}(U - \theta^+) \geq \bar{\pi} - \theta'^- - \boldsymbol{\pi}(U - \theta'^+)$. By summation, it follows that $\boldsymbol{\pi}(U - \theta^+) \leq \boldsymbol{\pi}(U - \theta'^+)$, but as $\boldsymbol{\pi}$ is an M-function, this implies $U - \theta^+ \leq U - \theta'^+$. As a result, we have $\theta'^+ \leq \theta^+$, but since $\theta^+ \leq \theta'^+$ by assumption, we have $\theta^+ = \theta'^+$. The inequality $\mathbf{q}(\theta) \geq \mathbf{q}\left(\theta'\right)$ thus simplifies into $\theta^- \leq \theta'^-$, hence $\theta^- = \theta'^-$ as well, and therefore $\theta = \theta'$. ■

We are now able to provide the following result that guarantees the existence and uniqueness of a solution to the system of equations (7.1.2).

Proposition 7.1.3. *Under assumptions 1.1 and 1.2, a solution* θ *to equations* (7.1.2) *exists and is unique.*

Proof. Existence. Let $\mathbf{q}$ be as defined by (7.1.3). For $c > 0$ large enough, the vector $\bar{\theta} = c\mathbf{1}_Y$ is such that $\boldsymbol{\pi}_y(U - c\mathbf{1}_Y) < \bar{\pi}_y$ for all $y \in [Y]$. We therefore take $\theta^0 = c\mathbf{1}_Y$, and we have $\mathbf{q}\left(\theta^0\right) > 0$. As θ^0 has been constructed, we show recursively that we can construct θ^s such that $-\bar{\pi} \leq \theta^s \leq \theta^{s-1}$ and $\mathbf{q}(\theta^s) \geq 0$. Let us assume we have constructed a sequence in that manner up to θ^{s-1}. We have $\mathbf{q}\left(\theta^{s-1}\right) \geq 0$, and as $\theta_y \mapsto \mathbf{q}_y(\theta_y, \theta^{s-1}_{-y})$ is increasing, continuous, nonnegative for $\theta_y = \theta^{s-1}_y$, and negative for $\theta_y = -\bar{\pi}_y$, it follows from the intermediate value theorem that there is a $\theta^s_y \in [-\bar{\pi}_y, \theta^{s-1}_y]$ such that $\mathbf{q}_y\left(\theta_y, \theta^{s-1}_{-y}\right) = 0$. Given that $\mathbf{q}_y(\theta)$ is nonincreasing with respect to θ_z for $z \neq y$, and given $\theta^s \leq \theta^{s-1}$, we have $\mathbf{q}_y\left(\theta^s_y, \theta^s_{-y}\right) \geq \mathbf{q}_y\left(\theta^s_y, \theta^{s-1}_{-y}\right) = 0$, which shows the induction hypothesis. As θ^s_y is decreasing with respect to s and remains greater than or equal to $-\bar{\pi}_y$, it follows that it converges; let $\theta^\star_y$ be its limit. We have $\mathbf{q}_y\left(\theta^s_y, \theta^{s-1}_{-y}\right) = 0$, and since $\mathbf{q}$ is continuous, it follows that $\mathbf{q}_y(\theta^\star) = 0$. We have therefore proven that system (7.1.2) has a solution.

Uniqueness follows directly from proposition 7.1.2, which implies that the equation $\mathbf{q}(\theta) = 0$ cannot have multiple solutions. ■

This brings us to the proof of theorem 7.1.1.

Proof of theorem 7.1.1. From proposition 7.1.2, τ is a solution to (7.1.1) if and only if $\tau = \theta^+$, where θ is the solution to (7.1.2). From proposition 7.1.3, the solution exists and is unique. Hence, a solution to (7.1.1) exists and it is unique. ■

This leads to a few important formal definitions:

Definition 7.1.1 (Shadow prices, constrained market shares, rejection map). *Under assumptions 1.1 and 1.2, let τ be the unique solution to* (7.1.1). *Based on that vector τ, we define the following notions:*

(i) *the* shadow price map $\boldsymbol{\tau}(U, \bar{\pi})$ *that is τ itself,*

$$\boldsymbol{\tau}(U, \bar{\pi}) := \tau;$$

(ii) *the* constrained market share map $\tilde{\boldsymbol{\pi}}(U, \bar{\pi})$, *defined as*

$$\tilde{\boldsymbol{\pi}}(U, \bar{\pi}) := \boldsymbol{\pi}\,(U - \tau)\,;$$

(iii) *the* rejection map, *or* excess capacity map $\boldsymbol{\rho}(U, \bar{\pi})$, *defined as the remaining capacity,*

$$\boldsymbol{\rho}(U, \bar{\pi}) = \bar{\pi} - \boldsymbol{\pi}\,(U - \tau)\,.$$

The previous considerations allow us to provide a characterization of the constrained market share map and the shadow price map in the case of the logit model.

Example 1.2 (Logit model, continued). Consider the multinomial logit model with one default option 0 and Y alternatives. As before, we normalize the utility of the default option to $U_0 = 0$, and we assume that there is no constraint on it, so $\bar{\pi}_0 = 1$. We have $\exp(U_y - \tau_y - u) \leq \bar{\pi}_y$, and $\tau_y > 0$ implies $\exp(U_y - \tau_y - u) = \bar{\pi}_y$, that is, $\tau_y = U_y - u - \log \bar{\pi}_y$. Now $\tau_y = 0$ implies $\exp(U_y - u) \leq \bar{\pi}_y$, that is, $U_y - u - \log \bar{\pi}_y \leq 0$. Therefore $\tau_y = \max(U_y - u - \log \bar{\pi}_y, 0)$, and $\pi_y = \exp(U_y - \tau_y - u) = \min\left(\exp(U_y - u), \bar{\pi}_y\right)$. We have

$$\tilde{\boldsymbol{\pi}}_y(U, \bar{\pi}) = \min\left(e^{U_y - u}, \bar{\pi}_y\right), \tag{7.1.4}$$

where $u \in \mathbb{R}$ is obtained by the scalar equation

$$e^{-u} + \sum_{y \in [Y]} \min\left(e^{U_y - u}, \bar{\pi}_y\right) = 1.$$

Python code is provided in appendix B.7.1.

7.2 Constrained Welfare and Entropy

Remember from chapter 1 that in the case where the capacity constraints are absent, one can get the welfare function using the entropy of choice by

$$G(U) = \max_{\pi \in \mathbb{R}_+^Y} \left\{ \sum_{y \in [Y]} \pi_y U_y - G^*(\pi) \right\}, \tag{7.2.1}$$

and the market share map $\boldsymbol{\pi}(U)$ arises as the solution to the problem. It therefore makes sense to consider a version of (7.2.1) where we impose the constraint $\pi \leq \bar{\pi}$. We introduce $\tilde{G}(U, \bar{\pi})$, the constrained welfare function, which is given by

$$\tilde{G}(U, \bar{\pi}) = \max_{\pi \in \mathbb{R}_+^Y} \left\{ \sum_{y \in [Y]} \pi_y U_y - G^*(\pi) \right\} \tag{7.2.2}$$

$$s.t.\ \pi \leq \bar{\pi}.$$

Using theorem A.1.4, it is easy to see that the constrained market share map $\tilde{\boldsymbol{\pi}}(U, \bar{\pi})$ is the solution to (7.2.2), and that $\boldsymbol{\tau}(U, \bar{\pi})$ arises as the vector of Lagrange multipliers of the scarcity constraints $\pi \leq \bar{\pi}$ in that problem. The Lagrangian associated with problem (7.2.2) is $\mathcal{L}(\tau, \pi) = \sum_y \bar{\pi}_y \tau_y + \sum_y \pi_y (U_y - \tau_y) - G^*(\pi)$, and an easy calculation shows that

$$\tilde{G}(U, \bar{\pi}) = \min_{\tau \geq 0} \left\{ \sum_{y \in [Y]} \bar{\pi}_y \tau_y + G(U - \tau) \right\}, \tag{7.2.3}$$

from which it follows that:

Theorem 7.2.1. *Under assumptions 1.1 and 1.2, the constrained market share map $\tilde{\boldsymbol{\pi}}(U, \bar{\pi})$ is the solution to (7.2.2), and $\boldsymbol{\tau}(U, \bar{\pi})$ is the solution to (7.2.3), and we have*

$$\tilde{\boldsymbol{\pi}}(U, \bar{\pi}) = \nabla_U \tilde{G}(U, \bar{\pi}) \quad \text{and} \quad \boldsymbol{\tau}(U, \bar{\pi}) = \nabla_{\bar{\pi}} \tilde{G}(U, \bar{\pi}).$$

As a result, one may provide results for the simulated constrained market share map.

Example 7.1. If we simulate a sample ε_{iy} such that for each $i \in [I]$, $(\varepsilon_{iy})_y$ is a random sample drawn from distribution $\mathcal{P}$, we recall from (1.4.2) that the opposite of the entropy of choice of the sample $-G_I^*(\pi)$ is $\max_{\lambda_{iy} \geq 0} \sum_{iy} \lambda_{iy}\varepsilon_{iy}$ subject to constraints $\sum_y \lambda_{iy} = 1/I$ and $\sum_i \lambda_{iy} = \pi_y$, and plugging this into expression (7.2.2) yields

$$\begin{aligned}
\tilde{G}_I(U, \bar{\pi}) = \max_{\pi \in \mathbb{R}_+^Y, \lambda \in \mathbb{R}_+^{I \times Y}} & \sum_{y \in [Y]} \pi_y U_y + \sum_{iy \in [I \times Y]} \lambda_{iy}\varepsilon_{iy} \\
s.t. & \sum_{y \in [Y]} \lambda_{iy} = 1/I, \\
& \sum_{i \in [I]} \lambda_{iy} = \pi_y, \\
& \pi \leq \bar{\pi},
\end{aligned}$$

which we can slightly simplify as

$$\begin{aligned}
\tilde{G}_I(U, \bar{\pi}) = \max_{\lambda \in \mathbb{R}_+^{I \times Y}} & \sum_{iy \in [I \times Y]} \lambda_{iy}\left(U_y + \varepsilon_{iy}\right) \\
s.t. & \sum_{y \in [Y]} \lambda_{iy} = 1/I \qquad [u_i], \\
& \sum_{i \in [I]} \lambda_{iy} \leq \bar{\pi} \qquad [\tau_y \geq 0],
\end{aligned}$$

whose dual formulation is

$$\begin{aligned}
\tilde{G}_I(U, \bar{\pi}) = \min_{u \in \mathbb{R}^I, \tau \in \mathbb{R}_+^Y} & \sum_{i \in [I]} \frac{1}{I} u_i + \sum_{y \in [Y]} \bar{\pi}_y \tau_y \\
s.t. \ & u_i \geq U_y - \tau_y + \varepsilon_{iy} \qquad [\lambda_{iy} \geq 0].
\end{aligned}$$

Python code based on Gurobi is provided in appendix B.7.2.

7.3 Comparative Statics

In this section, we continue the discussion on monotone comparative statics started in section 1.6, and we focus on how the vector of shadow prices τ

responds with respect to the constraints $\bar{\pi}$. Intuitively, we expect that when $\bar{\pi}$ decreases, the constraints are tighter, and therefore the shadow prices increase. As we shall see, this can be proven rigorously. Recall that it follows from propositions 7.1.1 and 7.1.3 that the vector of shadow prices τ is given by $\tau = \theta^+$, where $\theta \in \mathbb{R}^Y$ is the unique solution to

$$\boldsymbol{\pi}(U - \theta^+) + \theta^- = \bar{\pi}. \tag{7.3.1}$$

Assume that $\bar{\pi} \leq \bar{\pi}'$ are such that $\boldsymbol{\pi}(U - \theta^+) + \theta^- = \bar{\pi}$ and $\boldsymbol{\pi}(U - \theta'^+) + \theta'^- = \bar{\pi}'$. By proposition 7.1.2, the map $\theta \mapsto \bar{\pi} - \boldsymbol{\pi}(U - \theta^+) - \theta^-$ is an M-function, and therefore it follows that $\theta \geq \theta'$. As a result, $\boldsymbol{\tau}(U, \bar{\pi}) \geq \boldsymbol{\tau}(U, \bar{\pi}')$ and $\boldsymbol{\rho}(U, \bar{\pi}) \leq \boldsymbol{\rho}(U, \bar{\pi}')$, and we have proven (recalling that *isotone* means order-preserving, and *antitone* means order-reversing):

Theorem 7.3.1. *Under assumptions 1.1 and 1.2:*

(i) *the map* $\bar{\pi} \mapsto \boldsymbol{\tau}(U, \bar{\pi})$ *is antitone, and*
(ii) *the map* $\bar{\pi} \mapsto \boldsymbol{\rho}(U, \bar{\pi})$ *is isotone.*

We discuss how theorem 7.3.1 is a manifestation of substitutability, a property that is inherent to discrete choice models. Indeed, in discrete choice models, agents form demand for only one option, and therefore, they must view options as substitutable. A classic example of goods that are complements, as opposed to substitutes, are gloves: right gloves and left gloves are complements, not substitutes; this is due to the fact that most individuals want to purchase a right glove and a left glove together. But in the type of discrete choice models that we have seen so far, agents can purchase only one glove, which makes these models ill-suited to account for the complementarities between right and left gloves.

The first part of theorem 7.3.1 expresses the idea that tightening constraints leads to (weakly) higher shadow prices. Intuitively, this is plausible: tighter constraints require stronger incentives (i.e., higher prices) to be enforced. However, since the problem involves multiple options—each with its own constraint and corresponding shadow price—the conclusion is not immediately obvious.

The theorem asserts that if the constraint on option y becomes tighter (i.e., if $\bar{\pi}_y$ decreases), then not only does the shadow price τ_y for that option weakly increase, but so does $\tau_{y'}$ for every other option $y' \neq y$. This result relies on the *substitutability* of options: reducing the availability of option y causes some of its demand to shift to substitutes y', thereby pushing these alternatives closer to their own capacity limits. As a result, their corresponding shadow prices $\tau_{y'}$ may also rise, even though their constraints have not changed.

In contrast, if options are *complements*, the result no longer holds. Consider the case where y represents right gloves and y' represents left gloves, and start from a situation where the capacity constraint for left gloves is binding, so $\tau_{y'}$ is initially positive. If the capacity $\bar{\pi}_y$ for right gloves is reduced to zero, then right gloves are no longer available and τ_y becomes strictly positive. However, since left gloves are useless without right gloves, the demand for left gloves collapses to zero as well. The capacity constraint for left gloves is no longer binding, and $\tau_{y'}$ drops to zero.

This example highlights that the antitonicity of the vector of shadow prices τ with respect to the vector of capacities $\bar{\pi}$ is a structural feature that reflects the substitutability of the options under consideration.

The second statement of theorem 7.3.1 states that if some options are rejected (i.e., not chosen) under a tighter constraint, they are dominated by the options that are chosen; when the constraints are loosened by increasing availability, the previously chosen offers remain feasible and continue to dominate the previously rejected ones, which therefore remain rejected. This monotonicity property is again a consequence of *substitutability*: under substitutability, loosening constraints cannot cause an option that was previously chosen to be displaced in favor of one that was previously rejected.

This implication would fail in the presence of *complementarities*. Reconsider the glove example: suppose that a pair of gloves (left and right) is highly demanded, but initially only right gloves are available, while no left gloves are. In that case, right gloves are rejected due to the absence of their complementary counterparts. Now suppose that left gloves become available. The pair becomes usable and thus demanded, so right gloves, which were previously rejected, now become chosen. In this situation, relaxing the capacity constraint on left gloves *increases* the demand for right gloves. Such a reversal is incompatible with substitutability.

Under substitutability, however, such effects cannot occur. This guarantees that the allocation mapping $\boldsymbol{\rho}(U, \bar{\pi})$ is isotone in $\bar{\pi}$.

7.4 A Dynamic Model of Waiting Lines

We assume that there is a flow $\bar{\pi}_y \geq 0$ of arrivals of options of type $y \in [Y]$ per period. The discrepancy between the supply and demand flows will induce waiting times for the options that are overdemanded, as customers will wait in line for these options. On the contrary, for options that are underdemanded, stocks will pile up and go to infinity. We assume that individuals arriving at time t anticipate waiting in line for option y an amount of time $\tau_y^t \geq 0$, and we postulate that the utility associated with choosing option y is $U_y - \tau_y^t$, which is the systematic utility associated with option y minus the waiting time. The market share of option y at time t will then be $\boldsymbol{\pi}_y\left(U - \tau^t\right)$. Letting N_y^t be the volume of option y available at the start of period t, we have

$$N_y^{t+1} = N_y^t + \bar{\pi}_y - \boldsymbol{\pi}_y\left(U - \tau^t\right).$$

A stationary solution has $N_y^t = N_y \in \mathbb{R} \cup \{\infty\}$ and $\tau_y^t = \tau_y$ for all t, therefore

$$\begin{cases} N_y = +\infty \implies \boldsymbol{\pi}_y (U - \tau) \leq \bar{\pi}_y, \\ N_y < +\infty \implies \boldsymbol{\pi}_y (U - \tau) = \bar{\pi}_y. \end{cases}$$

Further, $\tau_y > 0$ implies that there is a waiting line for option y, and thus $\boldsymbol{\pi}_y (U - \tau) = \bar{\pi}_y$. One has to conclude therefore that τ is the solution of (7.1.1), and hence the stationary solution to the model is such that the stationary length of the waiting line τ_y for option y is given by

$$\tau = \boldsymbol{\tau} (U, \bar{\pi}).$$

7.5 An Empirical Model of Matching without Prices

Consider a matching market between firms and workers where wages are fixed and given. A firm i of type x has utility $\alpha_{xy} + \varepsilon_{iy}$ associated with hiring a worker of type y, and utility ε_{i0} associated with being unmatched, where α_{xy} is deterministic and exogenous, and $(\varepsilon_{iy})_y \sim \mathbf{P}_x$ is the vector of random utilities drawn by the individual firm i. The term α_{xy} includes the fixed wage paid to the worker. Similarly, a worker j of type y has utility $\gamma_{xy} + \eta_{jx}$ associated with working with a firm of type x, and utility η_{j0} associated with being unmatched, where, symmetrically, γ_{xy} is deterministic and exogenous, and $(\eta_{jx})_x \sim \mathbf{Q}_y$ is the vector of random utilities drawn by the individual worker j. Again, γ_{xy} includes the fixed wage that the worker receives from the firm. Our goal in this section will be to investigate a notion of equilibrium matching when prices are absent, following Galichon, Hsieh, and Sylvestre. Before we detail what we mean, it is useful to recall some concepts related to welfare and choices.

Aggregate welfare functions. Given that in this section we are dealing with multiple observable types of workers and firms, we need to slightly adapt our notations to be able to deal with aggregate quantities. If U_{xy} and V_{xy} are the systematic utilities of firms and workers, we can use these vectors as parameters to define the unconstrained welfare functions of firms and workers respectively as $G(U)$ and $H(V)$,

$$G(U) = \sum_{x \in [X]} n_x \mathbb{E}_{\mathcal{P}_x} \left[\max_{y \in [Y]} \left\{ U_{xy} + \varepsilon_y, \varepsilon_0 \right\} \right], \tag{7.5.1}$$

$$H(V) = \sum_{y \in [Y]} m_y \mathbb{E}_{\mathcal{Q}_y} \left[\max_{x \in [X]} \left\{ V_{xy} + \eta_x, \eta_0 \right\} \right], \tag{7.5.2}$$

and the corresponding functions of entropy of choice as their convex conjugates, whose expressions are given by

$$G^*(\mu) = \max_{U \in \mathbb{R}^{X \times Y}} \left\{ \sum_{xy \in [X \times Y]} \mu_{xy} U_{xy} - G(U) \right\}, \tag{7.5.3}$$

$$H^*(\mu) = \max_{V \in \mathbb{R}^{X \times Y}} \left\{ \sum_{xy \in [X \times Y]} \mu_{xy} V_{xy} - H(V) \right\}. \tag{7.5.4}$$

Finally, we define the constrained welfare functions of firms and workers $\tilde{G}(U, \bar{\mu})$ and $\tilde{H}(V, \bar{\mu})$ under capacity constraints $\mu_{xy} \leq \bar{\mu}_{xy}$ as

$$\tilde{G}(U, \bar{\mu}) = \max_{\mu \in \mathbb{R}_+^{X \times Y}} \left\{ \sum_{xy \in [X \times Y]} \mu_{xy} U_{xy} - G^*(\mu) \quad s.t.\ \mu_{xy} \leq \bar{\mu}_{xy}\ \forall xy \right\}, \tag{7.5.5}$$

$$\tilde{H}(V, \bar{\mu}) = \max_{\mu \in \mathbb{R}_+^{X \times Y}} \left\{ \sum_{xy \in [X \times Y]} \mu_{xy} V_{xy} - H^*(\mu) \quad s.t.\ \mu_{xy} \leq \bar{\mu}_{xy}\ \forall xy \right\}, \tag{7.5.6}$$

and we define $\tilde{\boldsymbol{\mu}}^{\alpha}(\bar{\mu})$ as the solution μ to the problem expressed by $\tilde{G}(U, \bar{\mu})$ with $U = \alpha$, and, similarly, $\tilde{\boldsymbol{\mu}}^{\gamma}(\bar{\mu})$ as the solution μ to the problem expressed by $\tilde{H}(V, \bar{\mu})$ with $V = \gamma$. Accordingly, we define the corresponding shadow price maps $\boldsymbol{\tau}^{\alpha}$ and $\boldsymbol{\tau}^{\gamma}$, as well as rejection maps of firms and workers $\boldsymbol{\rho}^{\alpha}$ and $\boldsymbol{\rho}^{\gamma}$. To summarize, we define:

$$\tilde{\boldsymbol{\mu}}^{\alpha}(\bar{\mu}) = \frac{\partial \tilde{G}}{\partial U}(\alpha, \bar{\mu}) \quad \text{and} \quad \tilde{\boldsymbol{\mu}}^{\gamma}(\bar{\mu}) = \frac{\partial \tilde{H}}{\partial V}(\gamma, \bar{\mu}), \tag{7.5.7}$$

$$\boldsymbol{\tau}^{\alpha}(\bar{\mu}) = \frac{\partial \tilde{G}}{\partial \bar{\mu}}(\alpha, \bar{\mu}) \quad \text{and} \quad \boldsymbol{\tau}^{\gamma}(\bar{\mu}) = \frac{\partial \tilde{H}}{\partial \bar{\mu}}(\gamma, \bar{\mu}), \tag{7.5.8}$$

$$\boldsymbol{\rho}^{\alpha}(\bar{\mu}) = \bar{\mu} - \tilde{\boldsymbol{\mu}}^{\alpha}(\bar{\mu}) \quad \text{and} \quad \boldsymbol{\rho}^{\gamma}(\bar{\mu}) = \bar{\mu} - \tilde{\boldsymbol{\mu}}^{\gamma}(\bar{\mu}). \tag{7.5.9}$$

Equilibrium matching without prices. We go back to defining an equilibrium without prices. In contrast with the setting of section 5.4, we do not assume that there are market wages—or other prices, or any other form of utility transfers—that adjust to clear the market. For that reason, there will be congestion: in the absence of market-clearing prices, some categories x or y of firms or workers will be overdemanded, leading to waiting times, crowding, or other negative effects of excessive demand. If they compete to attract certain types of highly demanded employees, firms may run costly targeted advertising campaigns, which will decrease their utility associated with hiring these types of employees, but the resulting utility loss will not be transferred to employees. We shall therefore assume that there is a shadow price $\tau_{xy}^{\alpha} \geq 0$

that is induced by the limited availability of employees of type y for firms of type x, and the utility of a firm i hiring an employee of type y will become $\alpha_{xy} - \tau^{\alpha}_{xy} + \varepsilon_{iy}$; similarly, there is a shadow price $\tau^{\gamma}_{xy} \geq 0$ on the other side of the market that is such that the utility of a worker j working for a firm of type x is $\gamma_{xy} - \tau^{\gamma}_{xy} + \eta_{jx}$. Clearly, the inequality $\tau^{\alpha}_{xy} > 0$ means that some firms of type x compete to attract workers of type y; this means that x is the "long side" in the xy relation. This implies that y is the "short side," and therefore that $\tau^{\gamma}_{xy} = 0$. As a result, we have $\min\{\tau^{\alpha}_{xy}, \tau^{\gamma}_{xy}\} = 0$. An equilibrium matching without prices (also called stable matching without transfers) is a vector (μ_{xy}) of matching patterns that are demanded by both sides, firms and workers, which means that $\mu = \nabla G(\alpha - \tau^{\alpha}) = \nabla H(\gamma - \tau^{\gamma})$. This leads to the following definition, which requires that the distribution of the random utility shocks ε and η satisfy assumptions 1.1 so that the demand maps ∇G and ∇H exist:

Definition 7.5.1 (Equilibrium matching without prices). *A vector $\mu \in \mathbb{R}^{X \times Y}_{+}$ is an equilibrium matching without prices if there exist two vectors τ^{α} and τ^{γ} in $\mathbb{R}^{X \times Y}_{+}$ such that*

$$\begin{cases} \min\left\{\tau^{\alpha}_{xy}, \tau^{\gamma}_{xy}\right\} = 0 \quad \forall xy \in [X \times Y], \\ \mu = \nabla G\left(\alpha - \tau^{\alpha}\right) = \nabla H\left(\gamma - \tau^{\gamma}\right). \end{cases} \tag{7.5.10}$$

While it may not look obvious at first glance, this definition is a strict generalization of Gale and Shapley's definition of a stable matching with nontransferable utility, which allows for random utility, nonstrict preferences, and more than one individual per type. The equivalence in the case when these complications are absent is worked out in problem 7.1.

Note that by replacing the condition $\min\{\tau^{\alpha}_{xy}, \tau^{\gamma}_{xy}\} = 0$ by $\tau^{\alpha}_{xy} + \tau^{\gamma}_{xy} = 0$, one would obtain the definition of an equilibrium matching with prices, such as the one discussed in section 5.4. In fact, the model discussed there was presented in the particular case of logit random utility, and in order to zoom in on the connection between equilibrium with and without prices, it is useful to investigate what becomes of definition 7.5.1 in the logit case.

We have the following result:

Theorem 7.5.1. *Assume that the distribution of the random utility shocks ε and η satisfies assumptions 1.1 and 1.2. Then an equilibrium matching without prices exists and is unique.*

Sketch of proof. The proof of uniqueness relies on the fact that the map from $\mathbb{R}^{X \times Y}$ to itself defined by $\tau \mapsto \nabla G(\alpha - \tau^{-}) - \nabla H(\gamma - \tau^{+})$ is an M-function. Existence is constructive, and one way to establish it is to show that the deferred acceptance algorithm, described in section 7.6 below, converges to an equilibrium matching without prices. ■

Example 7.2 (Equilibrium matching without prices, logit case). In the logit case, one can write the usual relations between the choice ratios and the systematic utilities on both sides of the market,

$$\frac{\mu_{xy}}{\mu_{x0}} = \exp\left(\alpha_{xy} - \tau^{\alpha}_{xy}\right) \quad \text{and} \quad \frac{\mu_{xy}}{\mu_{0y}} = \exp\left(\alpha_{xy} - \tau^{\gamma}_{xy}\right),$$

where $\min\left\{\tau^{\alpha}_{xy}, \tau^{\gamma}_{xy}\right\} = 0$, which yields

$$\mu_{xy} = \min\left\{\mu_{x0} \exp \alpha_{xy}, \mu_{0y} \exp \gamma_{xy}\right\}, \tag{7.5.11}$$

and therefore μ_{x0} and μ_{0y} are solutions to

$$\begin{cases} \mu_{x0} + \sum_{y\in[Y]} \min\left\{\mu_{x0} \exp \alpha_{xy}, \mu_{0y} \exp \gamma_{xy}\right\} = n_x, \\ \mu_{0y} + \sum_{x\in[X]} \min\left\{\mu_{x0} \exp \alpha_{xy}, \mu_{0y} \exp \gamma_{xy}\right\} = m_y. \end{cases} \tag{7.5.12}$$

One can appreciate the contrast between this model of equilibrium matching without prices and the model of equilibrium matching with prices described in section 5.4, where the equilibrium relation (7.5.11) is replaced by Choo and Siow's formula (5.4.7), which, as we recall, is

$$\mu_{xy} = \sqrt{\mu_{x0}\mu_{0y} \exp\left(\alpha_{xy} + \gamma_{xy}\right)}. \tag{7.5.13}$$

In other words, the min from expression (7.5.11) for the case without prices is replaced by the geometric mean in Choo and Siow's formula (7.5.11) in the case with prices.

7.6 Deferred Acceptance

In this section, we study the celebrated deferred acceptance algorithm by Gale and Shapley. Our presentation is not standard, as the Gale and Shapley algorithm is usually not presented in a discrete choice framework; we follow here the approach of Galichon, Hsieh, and Sylvestre, who proposed a notion of aggregate stable matching with nontransferable utility (which we introduced in the previous section), and who recast deferred acceptance in the setting of discrete choice analysis. Throughout the section, we assume the distribution of the random utility shocks ε and η should satisfy assumptions 1.1 and 1.2.

We consider offers that firms of type $x \in [X]$ may make to workers of type $y \in [Y]$. As there are n_x individuals of type x and m_y individuals of type y, the number of offers initially available $\mu_{xy}^{A,0}$ between firms of type x and workers of type y is equal to the maximum possible number of matches between these types, that is, $M_{xy} = \min(n_x, m_y)$. If firms were to pick workers in an unconstrained manner, firm i would pick the worker y maximizing the firm's utility, $\max_y \{\alpha_{xy} + \varepsilon_{iy}, \varepsilon_{i0}\}$, which would yield the maximum possible unconstrained social welfare $G(\alpha)$ to firms, and lead to a number of initial proposals $\mu_{xy}^{P,1}$ from firms of type x to workers of type y equal to $\mu^{P,1} = \nabla G(\alpha)$, which expresses that each firm has picked its most preferred worker.

However, $\mu_{xy}^{P,1}$ is the number of offers that are initially proposed by firms of type x to workers of type y, but it is not the final number of matches of type xy. Indeed, some categories of workers may receive too many offers to respond to, especially if they are popular. In that regard, if $\sum_x \mu_{xy}^{P,1} > m_y$, some offers will need to be turned down by worker category y. In that case, the workers of type y can pick from among the offers that have been proposed to them the ones they like best. They will solve a discrete choice problem under availability constraint, where the constraint is that the maximum number of offers they can keep is the number of offers that have been proposed to them. At this stage, the welfare of workers is therefore $\tilde{H}(\gamma, \mu^{P,1})$, and the number of offers from firms of type x kept by workers of type y is

$$\mu_{xy}^{K,1} = \tilde{\boldsymbol{\mu}}^{\gamma}(\mu^{P,1}),$$

and the number of rejected offers of type xy is $\mu_{xy}^{P,1} - \mu_{xy}^{K,1}$. As these offers have been rejected, they become unavailable; and we therefore update the number of available offers μ_{xy}^{A} at the end of the first period by

$$\mu_{xy}^{A,1} = \mu_{xy}^{A,0} - (\mu_{xy}^{P,1} - \mu_{xy}^{K,1}).$$

The second period of the algorithm resembles the first one, with the difference that the offers that were rejected in the first period can no longer be made in the second one. Therefore, while in the first period the firms solved an unconstrained discrete choice problem, in the second one the number of offers $\mu_{xy}^{P,2}$ that can be made by firms of type x to workers of type y is constrained by the number of available offers $\mu_{xy}^{A,1}$, offers that have not been rejected. The provisional welfare of workers in the second period is therefore $\tilde{G}(\alpha, \mu_{xy}^{A,1})$, and the number of offers made by type x firms to type y workers is

$$\mu_{xy}^{P,2} = \tilde{\boldsymbol{\mu}}^{\alpha}(\mu^{A,1}),$$

to which workers will again respond by picking their favorite offers from among those offered, as in the previous step. We can iterate in that

Algorithm 7.6.1 Deferred acceptance algorithm

Input : Systematic utilities α_{xy} and γ_{xy}, tolerance threshold $\epsilon > 0$
Output : Matching patterns (μ^K_{xy})
Initialize : $t = 0$ and $\mu^{A,0}_{xy} = \min(n_x, m_y)$
repeat
  Proposal phase: $\mu^{P,t+1}_{xy} = \tilde{\boldsymbol{\mu}}^{\alpha}(\mu^{A,t})$
  Disposal phase: $\mu^{K,t+1}_{xy} = \tilde{\boldsymbol{\mu}}^{\gamma}(\mu^{P,t+1})$
  Update phase: $\mu^{A,t+1}_{xy} = \mu^{A,t}_{xy} - (\mu^{P,t+1}_{xy} - \mu^{K,t+1}_{xy})$ and $t = t+1$
until $\max_{x,y}\{\mu^{P,t+1}_{xy} - \mu^{K,t+1}_{xy}\} < \epsilon$;
return $(\mu^{K,t+1}_{xy})$

fashion, and this leads us to the formal description of the procedure in algorithm 7.6.1 below. A Python implementation of the algorithm is provided in appendix B.7.4, which is generic and requires us to specify the constrained choice functions $\tilde{\boldsymbol{\mu}}^{\alpha}$ and $\tilde{\boldsymbol{\mu}}^{\gamma}$.

We have:

Theorem 7.6.1. *Under assumptions 1.1 and 1.2, the quantities $\mu^{P,t}$ and $\mu^{K,t}$ defined in the deferred acceptance algorithm 7.6.1 both converge to an equilibrium matching without prices.*

Proof. The proof of theorem 7.6.1 is decomposed in several claims. We let $\tau^{\alpha,t}_{xy} = \boldsymbol{\tau}^{\alpha}_{xy}\left(\mu^{A,t}\right)$ and $\tau^{\gamma,t}_{xy} = \boldsymbol{\tau}^{\gamma}_{xy}\left(\mu^{P,t}\right)$.

Claim 1. *There cannot be fewer offers proposed at time $t+1$ than offers proposed and kept at time t, i.e., $\mu^{P,t+1} \geq \mu^{K,t}$.* Indeed, $\mu^{A,t}_{xy} \leq \mu^{A,t-1}_{xy}$ and the fact that $\boldsymbol{\rho}^{\alpha}$ is isotone imply $\mu^{A,t}_{xy} - \mu^{P,t+1}_{xy} \leq \mu^{A,t-1}_{xy} - \mu^{P,t}_{xy}$, but as $\mu^{A,t}_{xy} = \mu^{A,t-1}_{xy} - (\mu^{P,t}_{xy} - \mu^{K,t}_{xy})$, we have $\mu^{K,t}_{xy} \leq \mu^{P,t+1}_{xy}$.

Claim 2. *$\tau^{\alpha,t}_{xy}$ is nondecreasing in t and $\tau^{\gamma,t}_{xy}$ is nonincreasing in t.* Indeed, $\tau^{\alpha,t} \leq \tau^{\alpha,t+1}$ comes from the fact that $\mu^{A,t} \leq \mu^{A,t-1}$ and that $\boldsymbol{\tau}^{\alpha}$ is antitone. Next, we have $\tau^{\gamma,t}_{xy} = \boldsymbol{\tau}^{\gamma}_{xy}\left(\mu^{P,t}\right)$ and $\tau^{\gamma,t+1}_{xy} = \boldsymbol{\tau}^{\gamma}_{xy}\left(\mu^{P,t+1}\right)$; but as $\boldsymbol{\tau}^{\gamma}_{xy}\left(\mu^{P,t}\right) = \boldsymbol{\tau}^{\gamma}_{xy}\left(\mu^{K,t}\right)$, we have also $\tau^{\gamma,t}_{xy} = \boldsymbol{\tau}^{\gamma}_{xy}\left(\mu^{K,t}\right)$; and as $\mu^{P,t+1} \geq \mu^{K,t}$, we have by the fact that $\tau^{\gamma,t}$ is antitone that $\tau^{\gamma,t+1}_{xy} \leq \tau^{\gamma,t}_{xy}$.

Claim 3. *One has $\min\{\tau^{\alpha,t}_{xy}, \tau^{\gamma,t}_{xy}\} = 0$ for all txy.* Assume by contradiction that $\min\{\tau^{\alpha,t}_{xy}, \tau^{\gamma,t}_{xy}\} > 0$ for some t and xy. Then $\tau^{\gamma,s}_{xy} > 0$ for all $s \leq t$, which implies $\mu^{K,s}_{xy} = \mu^{P,s}_{xy}$ for $s \leq t$, and, as a result, $\mu^{A,s}_{xy} = \mu^{A,s-1}_{xy}$ for all $s \leq t$, which in turn implies $\mu_{xy}^{A,t} = N_{xy}$. But $\min\{\tau_{xy}^{\alpha,t}, \tau_{xy}^{\gamma,t}\} > 0$ implies $\tau_{xy}^{\alpha,t} > 0$ and therefore $\mu_{xy}^{P,t} = \mu_{xy}^{A,t-1} = N_{xy}$, which cannot hold under assumption 1.1.

Claim 4. *Each of the sequences $\mu_{xy}^{A,t}$, $\mu_{xy}^{P,t}$, $\mu_{xy}^{K,t}$, $\tau_{xy}^{\alpha,t}$, and $\tau_{xy}^{\gamma,t}$ converges as $t \to +\infty$ for each xy.* The convergence of $\mu_{xy}^{A,t}$ comes from the fact that this sequence is nonincreasing and bounded below by 0. The convergence of $\mu_{xy}^{P,t}$ comes from the fact that $\mu^{P,t} = \tilde{\boldsymbol{\mu}}^{\alpha}\left(\mu^{A,t-1}\right)$ and the fact that $\tilde{\boldsymbol{\mu}}^{\alpha}$ is continuous. The convergence of $\mu_{xy}^{K,t}$ comes from $\mu^{K,t} = \tilde{\boldsymbol{\mu}}^{\gamma}\left(\mu^{P,t}\right)$ and the fact that $\tilde{\boldsymbol{\mu}}^{\gamma}$ is continuous. The convergence of $\tau_{xy}^{\alpha,t}$ and $\tau_{xy}^{\gamma,t}$ follows in a similar fashion from the continuity of $\boldsymbol{\tau}^{\alpha}$ and $\boldsymbol{\tau}^{\gamma}$.

Claim 5. *Both $\mu^{P,t}$ and $\mu^{K,t}$ converge as $t \to +\infty$ to an equilibrium matching without prices.* Indeed, one has $\mu^{A,t} = \mu^{A,t-1} - (\mu^{P,t} - \mu^{K,t})$, so as all the quantities involved converge, it follows that $\mu^{P,t}$ and $\mu^{K,t}$ have the same limit, denoted by $\mu^{\star}$. Let τ^{α} and τ^{γ} be the limits as $t \to +\infty$ of $\tau^{\alpha,t}, \tau^{\gamma,t}$. By taking the limits in $\mu^{P,t} = \nabla G\left(\alpha - \tau^{\alpha,t}\right)$ and $\mu^{K,t} = \nabla H\left(\gamma - \tau^{\gamma,t}\right)$, we have $\mu^{\star} = \nabla G(\alpha - \tau^{\alpha})$ and $\mu^{K} = \nabla H(\gamma - \tau^{\gamma})$, and by taking the limit in $\min\{\tau_{xy}^{\alpha,t}, \tau_{xy}^{\gamma,t}\} = 0$, we have $\min\{\tau_{xy}^{\alpha}, \tau_{xy}^{\gamma}\} = 0$. As a result, $\mu^{\star}$ is an equilibrium matching without prices. ∎

In problem 7.2, the reader is asked to verify that when unobserved heterogeneity is absent, then this algorithm coincides with the classical deferred acceptance by Gale and Shapley.

7.7 Reinterpretation à la Hatfield–Milgrom

Here we discuss an important reinterpretation of deferred acceptance algorithms that is due to Hatfield and Milgrom, but which Galichon, Hsieh, and Sylvestre have adapted in the discrete choice setting. Recall that ∇G is the unconstrained demand map associated with the choice problem of the firms, and ∇H is the unconstrained demand map associated with the choice problem of the workers. We go through the steps of algorithm 7.6.1 and revisit them under the light of characterization provided by proposition 7.1.1.

Period 1, proposal: All the matches are available to firms, so the number of matches of type xy that are initially available to the proposing side is $\mu_{xy}^{A,0} = M_{xy} := \min\{n_x, m_y\}$. In that case, we look for the vector of shadow prices $\tau^{\alpha,1}$ and the vector $R^{\alpha,1}$ of offers not made such that $\nabla G\left(\alpha - \tau^{\alpha,1}\right) + R^{\alpha,1} = M$; as $\nabla G(\alpha) \leq M$, the solution is $\tau^{\alpha,1} = 0$, and

$$\nabla G(\alpha) + R^{\alpha,1} = M. \tag{7.7.1}$$

Period 1, disposal: The vector of the number of proposals is $\nabla G(\alpha)$, and therefore we seek $\tau^{\gamma,1}$ and $\rho^{\gamma,1}$ such that $\min\{\tau^{\gamma,1},\rho^{\gamma,1}\}=0$ and $\nabla H\left(\gamma-\tau^{\gamma,1}\right)+\rho^{\gamma,1}=\nabla G(\alpha)$; but as we just saw, $\nabla G(\alpha)=M-R^{\alpha,1}$, and so we are seeking to solve

$$\nabla H\left(\gamma-\tau^{\gamma,1}\right)+\rho^{\gamma,1}=M-R^{\alpha,1}. \tag{7.7.2}$$

Period 1, update: We decrease the number of available offers by the number of those that have just been rejected, that is, we set $\mu^{A,1}=M-\rho^{\gamma,1}$.

We gave a complete description of the first iteration of the algorithm in period 1; however, that period is a bit particular because initially all matches are available. To get a better picture, it is useful to carry on the exploration of a second iteration by going to period 2.

Period 2, proposal: We solve a choice problem under capacity constraint determined by $\mu^{A,1}=M-\rho^{\gamma,1}$, and therefore we seek a pair $(\tau^{\alpha,2},R^{\alpha,2})$ such that $\min\{\tau^{\alpha,2},R^{\alpha,2}\}=0$ and which is a solution to

$$\nabla G\left(\alpha-\tau^{\alpha,2}\right)+R^{\alpha,2}=M-\rho^{\gamma,1}. \tag{7.7.3}$$

Period 2, disposal: We now solve for $(\tau^{\gamma,2},\rho^{\gamma,2})$ such that $\min\{\tau^{\gamma,2},\rho^{\gamma,2}\}=0$ and $\nabla H\left(\gamma-\tau^{\gamma,2}\right)+\rho^{\gamma,2}=\nabla G\left(\alpha-\tau^{\alpha,2}\right)$, but as $\nabla G\left(\alpha-\tau^{\alpha,2}\right)=M-\rho^{\gamma,1}-\rho^{\alpha,2}$, we are solving for

$$\nabla H\left(\gamma-\tau^{\gamma,2}\right)+\rho^{\gamma,2}=M-\rho^{\gamma,1}-R^{\alpha,2}. \tag{7.7.4}$$

At the disposal phase of period 2, it is worth noting that if one adds to the proposals the offers that were previously rejected, one can show that the offer will be rejected again and this will lead to the same kept offers. Indeed, one can introduce $R^{\gamma,2}:=\rho^{\gamma,1}+\rho^{\gamma,2}$, and, more generally,

$$R^{\gamma,t}:=\rho^{\gamma,1}+\cdots+\rho^{\gamma,t}, \tag{7.7.5}$$

which is the *cumulative number of rejected offers* up until period t. Remember from claim 2 in the proof of theorem 7.6.1 that $\tau^{\gamma,t}$ is decreasing with respect to t. As a result, $\tau_{xy}^{\gamma,t}>0$ implies $\tau_{xy}^{\gamma,s}>0$ for all $s\leq t$ and therefore $\rho^{\gamma,s}=0$ for all $s\leq t$, which implies $R^{\gamma,t}=0$. Thus we have

$$\min\left\{\tau^{\gamma,t},R^{\gamma,t}\right\}=0, \tag{7.7.6}$$

and therefore the disposal phase of period 2 can be rewritten:

Period 2, disposal (bis): We solve for $(\tau^{\gamma,2}, R^{\gamma,2})$ such that $\min\{\tau^{\gamma,2}, R^{\gamma,2}\}=0$ and

$$\nabla H\left(\gamma-\tau^{\gamma,2}\right)+R^{\gamma,2}=M-R^{\alpha,2}. \tag{7.7.7}$$

The update phase in period 2 can be rewritten in a straightforward manner using the cumulative number of rejected offers, as the number of available offers $\mu^{A,t}$ in period t equals the maximum number of matches $M_{xy}=\min\{n_x, m_y\}$ minus the cumulative number of rejected offers, that is,

$$\mu^{A,t}=M-R^{\gamma,t}, \tag{7.7.8}$$

which allows us to rewrite the proposal phase in the next period, period 3, as follows:

Period 3, proposal: We look for $(\tau^{\alpha,3}, R^{\alpha,3})$ such that $\min\{\tau^{\alpha,3}, R^{\alpha,3}\}=0$ and which is a solution to $\nabla G\left(\alpha-\tau^{\alpha,3}\right)+R^{\alpha,3}=\mu^{A,2}$. Making use of the relation between μ^{A} and R^{γ}, one has

$$\nabla G\left(\alpha-\tau^{\alpha,3}\right)+R^{\alpha,3}=M-R^{\gamma,2}. \tag{7.7.9}$$

The general evolution of the algorithm is now clear. We initially set $R^{\alpha}=M$ and $R^{\gamma}=0$. The proposal phase in period t is written:

Period t, proposal: We look for $(\tau^{\alpha,t}, R^{\alpha,t})$ such that $\min\{\tau^{\alpha,t}, R^{\alpha,t}\}=0$ and which is a solution to

$$\nabla G\left(\alpha-\tau^{\alpha,t}\right)=M-R^{\alpha,t}-R^{\gamma,t-1}, \tag{7.7.10}$$

while the disposal phase is written:

Period t, disposal: We look for $(\tau^{\gamma,t}, R^{\gamma,t})$ such that $\min\{\tau^{\gamma,t}, R^{\gamma,t}\}=0$ and which is a solution to

$$\nabla H\left(\gamma-\tau^{\gamma,t}\right)=M-R^{\alpha,t}-R^{\gamma,t}. \tag{7.7.11}$$

There is a further simplification of equations (7.7.10) and (7.7.11). Note that these equations can be rewritten as

$$\begin{cases} R^{\alpha,t}=\rho^{\alpha}\left(M-R^{\gamma,t-1}\right), \\ R^{\gamma,t}=\rho^{\gamma}\left(M-R^{\alpha,t}\right), \end{cases} \tag{7.7.12}$$

where we recall that we have defined the rejection maps of firms ρ^{α} and of workers ρ^{γ} in (7.5.9). Following Hatfield and Milgrom, we are able to provide an elegant reformulation of algorithm 7.6.1 as algorithm 7.7.2.

Algorithm 7.7.2 Deferred acceptance algorithm (bis)

Input : Systematic utilities α_{xy} and γ_{xy}, tolerance threshold $\epsilon > 0$
Output : Matching patterns (μ^K_{xy})
Initialize : $t = 0$ and $R^\alpha = M$ and $R^\gamma = 0$
repeat
  Proposal phase: $R^{\alpha,t+1} = \boldsymbol{\rho}^\alpha \left(M - R^{\gamma,t}\right)$
  Disposal phase: $R^{\gamma,t+1} = \boldsymbol{\rho}^\gamma \left(M - R^{\alpha,t+1}\right)$
until $\max_{x,y}\{\mu^{R,t+1}_{xy} - \mu^{R,t}_{xy}\} < \epsilon$;
return $\mu^{K,t} = \tilde{\boldsymbol{\mu}}^\gamma \left(M - R^{\alpha,t+1}\right)$

The previous discussion has allowed us to establish the equivalence between algorithms 7.6.1 and 7.7.2, which we formally state as:

Theorem 7.7.1. *Under assumptions 1.1 and 1.2, the quantities $\tilde{\boldsymbol{\mu}}^\alpha(M - R^{\gamma,t})$ and $\tilde{\boldsymbol{\mu}}^\gamma(M - R^{\alpha,t+1})$, where $R^{\alpha,t}$ and $R^{\gamma,t+1}$ are defined in algorithm 7.7.2, respectively coincide with $\mu^{P,t+1}$ and $\mu^{K,t+1}$, the number of offers proposed and kept that are defined in algorithm 7.6.1.*

We make some final remarks. First, one sees that algorithm 7.7.2 consists of the repeated application of $F(R^\alpha, R^\gamma) = (F^\alpha(R^\alpha, R^\gamma), F^\gamma(R^\alpha, R^\gamma))$, where

$$F^\alpha(R^\alpha, R^\gamma) := \boldsymbol{\rho}^\alpha(M - R^\gamma) \quad \text{and} \quad F^\gamma(R^\alpha, R^\gamma) := \boldsymbol{\rho}^\gamma(M - R^\alpha). \tag{7.7.13}$$

Since $\boldsymbol{\rho}^\alpha$ and $\boldsymbol{\rho}^\gamma$ are isotone, F^α and F^γ are antitone, and therefore F is antitone. As a result, $F^2 = F \circ F$ is isotone, and one may view the quantity R^α produced by algorithm 7.7.2 as a fixed point of $R^\alpha \mapsto \boldsymbol{\rho}^\alpha(M - \boldsymbol{\rho}^\gamma(M - R^\alpha))$, which is an isotone map.

Second, the discussion above implies that the equilibrium matching μ given by $\mu = \tilde{\boldsymbol{\mu}}^\alpha\ (M - R^\gamma) = \tilde{\boldsymbol{\mu}}^\gamma\ (M - R^\alpha)$ satisfies

$$M_{xy} - \mu_{xy} = R^\alpha_{xy} + R^\gamma_{xy},$$

which can be interpreted as saying that the number $M_{xy} - \mu_{xy}$ of matches of type xy that are not demanded can be broken down into the sum of the number R^α_{xy} of matches that are not of interest to firms of type x and the number R^γ_{xy} of matches that are not of interest to workers of type y. Indeed, it is very clear from the deferred acceptance algorithm that an offer of type xy is either not made by x, or turned down at some point by y, or is kept.

7.8 Exercises

Exercise 7.1. *The goal of this exercise is to improve the computation of equilibrium in example 7.2 by improving on the algorithm demonstrated in appendix B.7.1.*

(i) *Describe an algorithm to more efficiently find the root of an increasing, piece-wise affine and concave function than what is done by Brent's method in appendix B.7.1.*

(ii) *Based on the previous answer, code an algorithm to improve on the one provided in appendix B.7.4.*

(iii) *Compare the execution time of both methods.*

Exercise 7.2 (Constrained logit model). *We consider a model of constrained choice when the distribution of random utility $\mathcal{P}$ is the nested logit specification given in* (2.3.5).

(i) *Show that $\pi = \tilde{\pi}(U, \bar{\pi})$ if and only if there is a vector $\theta \in \mathbb{R}^Y$ such that for all $y \in [Y]$, one has*

$$\ln \pi_y = \ln \bar{\pi}_y - \theta_y^- = \frac{U_y - \theta_y^+ - u_x}{\lambda_{x_y}} + u_{x_y} - u,$$

where

$$u_x = \lambda_x \ln \sum_{y \in \mathcal{Y}_x} e^{\frac{U_y - \theta_y^+}{\lambda_x}} \text{ and } u = \ln \sum_{x \in [X]} \left(\sum_{y \in \mathcal{Y}_x} e^{\frac{U_y - \theta_y^+}{\lambda_x}} \right)^{\lambda_x}.$$

(ii) *Deduce that*

$$\pi_y = \min \left(\bar{\pi}_y, \exp \left(\frac{U_y - u_x}{\lambda_x} \right) \exp \left(u_x - u \right) \right),$$

where (u_x) and u are the solutions to

$$\begin{cases} \sum_{y \in [Y]} \min \left(\bar{\pi}_y, \exp \left(\frac{U_y - u_x}{\lambda_x} + u_x - u \right) \right) = 1, \\ \sum_{y \in \mathcal{Y}_x} \min \left(\bar{\pi}_y \exp \left(u - u_x \right), \exp \left(\frac{U_y - u_x}{\lambda_x} \right) \right) = 1. \end{cases} \tag{7.8.1}$$

(iii) *Suggest a computational method to solve for (u_x, u) in* (7.8.1) *and implement it.*

Exercise 7.3 (Labor market equilibrium with wage caps). *We consider a matching model with transfers and logit heterogeneity similar to the one described in section 5.4, with the exception that wages are capped. Assume that the maximum wage that workers of type x are allowed to receive from firms of type y is $\bar{w}_{xy}$. Let U_{xy} and V_{xy} be the respective systematic utilities that a worker and a firm get in a match between a worker of type x and a firm of type y.*

(i) *Show that the equations of the model are*

$$U_{xy} = \alpha_{xy} + \min\left\{w_{xy}, \bar{w}_{xy}\right\} \text{ and } V_{xy} = \gamma_{xy} - w_{xy}$$

and interpret them. Why can V_{xy} go below $\gamma_{xy} - \bar{w}_{xy}$ on the side of firms?

(ii) *Replacing $U_{xy} = \ln \frac{\mu_{xy}}{\mu_{x0}}$ and $V_{xy} = \ln \frac{\mu_{xy}}{\mu_{0y}}$, argue that, at equilibrium,*

$$\mu_{xy} = \min\left\{\mu_{x0} e^{\alpha_{xy} + \bar{w}_{xy}}, \sqrt{\mu_{x0}\mu_{0y}} e^{\frac{\Phi_{xy}}{2}}\right\},$$

where μ_{x0} and μ_{0y} are determined by

$$\begin{cases} \mu_{x0} + \sum_{y \in [Y]} \min\left\{\mu_{x0} e^{\alpha_{xy} + \bar{w}_{xy}}, \sqrt{\mu_{x0}\mu_{0y}} e^{\frac{\Phi_{xy}}{2}}\right\} = n_x, \\ \mu_{0y} + \sum_{x \in [X]} \min\left\{\mu_{x0} e^{\alpha_{xy} + \bar{w}_{xy}}, \sqrt{\mu_{x0}\mu_{0y}} e^{\frac{\Phi_{xy}}{2}}\right\} = m_y. \end{cases} \tag{7.8.2}$$

(iii) *Describe an algorithm to solve* (7.8.2).

(iv) *Implement that algorithm.*

Exercise 7.4 (Simulations of a constrained logit model). *In the context of appendix B.7.1 and the data generated in the code there, simulate the constrained logit model by drawing i.i.d. Gumbel random utility shocks and retrieve the constrained market shares and the Lagrange multipliers by linear programming. Assess the convergence of the simulated solution to the exact solution provided in the code displayed in appendix B.7.1.*

Exercise 7.5 (A Newton algorithm for equilibrium computation in the NTU logit matching model). *Compute the Hessian with respect to (μ_{x0}, μ_{0y}) of the right-hand side of* (7.5.12). *Deduce a Newton algorithm to compute (μ_{x0}, μ_{0y}) based on (n_x, m_y). Compare numerically the performance of that method with the ones proposed in appendix B.7.4.*

Exercise 7.6 (The α-axiom). *This exercise is about an important axiom in choice theory often called Sen's α-axiom, although Sen gives the credit for it to Chernoff. If E is a finite set, a (discrete) choice function is a function c defined on 2^E such that for $B \subseteq E$, $c(B) \subseteq B$. Sen's α-axiom is formulated as follows:*

$$\text{if } A \subseteq B \text{ and } b \in A \text{ and } b \in c(B), \text{ then } b \in c(A).$$

Show that the natural translation of Sen's α-axiom in the setting of the present chapter is the isotonicity of $\bar{\mu} \mapsto \boldsymbol{\rho}(U, \bar{\mu})$ as defined in definition 7.1.1.

Exercise 7.7 (Consistency and persistence). *Alkan and Gale study the properties of continuous choice functions, which are functions $c : \mathbb{R}_+^Y \to \mathbb{R}_+^Y$ such that $c(\bar{\mu}) \leq \bar{\mu}$ for all $\bar{\mu} \in \mathbb{R}_+^Y$. They define consistency of c as*

$$c(\bar{\mu}) \leq \bar{\mu}' \leq \bar{\mu} \text{ implies } c(\bar{\mu}) = c\left(\bar{\mu}'\right) \tag{7.8.3}$$

and persistence of c as

$$\bar{\mu}' \leq \bar{\mu} \text{ implies } c\left(\bar{\mu}'\right) \geq c(\bar{\mu}) \wedge \bar{\mu}'.$$

(i) *Show that persistence of c implies that $\bar{\mu} \mapsto \bar{\mu} - c(\bar{\mu})$ is isotone.*

(ii) *Show that the function $\bar{\mu} \mapsto \tilde{\boldsymbol{\pi}}(U, \bar{\mu})$ from definition 7.1.1 satisfies both consistency and persistence.*

7.9 Problems

Problem 7.1 (Equivalence with the classical notion of stability). *This exercise aims at showing that the notion of equilibrium matching without prices defined in definition 7.5.1, when $(\varepsilon_y) = 0$ and $(\eta_x) = 0$ and $n_x = 1$ and $m_y = 1$ for all $x \in [X]$ and $y \in [Y]$, is a direct generalization of the notion of stable matching with nontransferable utility defined in Gale and Shapley's 1962 paper. Assume that $\alpha_{x0} = 0$ and $\gamma_{0y} = 0$, and that $\alpha_{xy} = \alpha_{xy'}$ implies $y = y'$ and $\gamma_{xy} = \gamma_{x'y}$ implies $x = x'$. Let μ_{xy}, $xy \in [X \times Y] \cup [X] \times \{0\} \cup \{0\} \times [Y]$ be a vector with positive entries. One says that μ is a stable matching in the sense of Gale and Shapley if (i) $\mu_{xy} \in \{0,1\}$, $\sum_{y \in [Y]} \mu_{xy} \leq 1$, $\sum_{x \in [X]} \mu_{xy} \leq 1$ for all $x \in [X]$ and $y \in [Y]$; (ii) there is no blocking pair, that is, defining $u_x^\mu = \sum_y \mu_{xy}\alpha_{xy}$ and $v_y^\mu = \sum_x \mu_{xy}\gamma_{xy}$ as the utilities that x and y obtain under matching μ, we have*

$$u_x^\mu \leq \alpha_{xy} \text{ and } v_y^\mu \leq \gamma_{xy} \text{ jointly imply } \mu_{xy} = 1,$$

and $u_x^\mu \leq 0$ implies $\mu_{x0} = 0$, while $v_y^\mu \leq 0$ implies $\mu_{0y} = 0$.

(i) *Argue that when* $\varepsilon_y = 0$ *and* $\eta_x = 0$ *(and therefore when assumptions 1.1 and 1.2 are not satisfied), the natural generalization of an equilibrium matching without prices in definition 7.5.1 consists of the determination of vectors* τ^{α}_{xy} *and* τ^{γ}_{xy} *such that* $\min\{\tau^{\alpha}_{xy}, \tau^{\gamma}_{xy}\} = 0$ *for all* $xy \in [X \times Y]$ *and such that*

$$\begin{cases} \mu_{xy} > 0 \implies y \in \arg\max_{y' \in [0:Y]} \left\{\alpha_{xy'} - \tau^{\alpha}_{xy'}\right\}, & xy \in [X \times (0:Y)], \\ \mu_{xy} > 0 \implies x \in \arg\max_{x' \in [0:X]} \left\{\gamma_{x'y} - \tau^{\gamma}_{x'y}\right\}, & xy \in [(0:X) \times Y] \end{cases} \tag{7.9.1}$$

(for convenience one has set $\tau^{\alpha}_{x0} = 0$ *and* $\tau^{\gamma}_{0y} = 0$*).*

(ii) *Let* μ *be a stable matching in the sense of Gale and Shapley. Show that one can find* τ^{α}_{xy} *and* τ^{γ}_{xy} *such that conditions* (7.9.1) *hold.*

(iii) *Conversely, suppose conditions* (7.9.1) *hold and* $\mu_{xy} \in \{0,1\}$, $\sum_{y \in [Y]} \mu_{xy} \leq 1$, $\sum_{x \in [X]} \mu_{xy} \leq 1$ *for all* $x \in [X]$ *and* $y \in [Y]$*. Show that* μ *is a stable matching in the sense of Gale and Shapley.*

Problem 7.2 (Equivalence with Gale and Shapley). *This problem assumes some basic knowledge about the Gale and Shapley algorithm; see the original 1962 paper of these authors. Consider the case when the unobserved heterogeneities* ε *and* η *are zero.*

(i) *When the entries of* $\bar{\mu}$ *are in* $\{0,1\}$*, how are the quantities* $\tilde{\boldsymbol{\mu}}^{\alpha}(\bar{\mu})$ *and* $\tilde{\boldsymbol{\tau}}^{\alpha}(\bar{\mu})$ *determined?*

(ii) *Deduce that in the case when* $n_x = 1$ *for all* $x \in [X]$ *and* $m_y = 1$ *for all* $y \in [Y]$*, algorithm 7.6.1 coincides with the classical Gale and Shapley algorithm.*

(iii) *Same question as in (ii) when there may be multiple individuals per type, and we no longer restrict* $n_x = 1$ *for all* $x \in [X]$ *and* $m_y = 1$ *for all* $y \in [Y]$*.*

Problem 7.3 (Adachi's algorithm). *Consider a one-to-one model of matching between workers and firms. There is a finite set of workers indexed by* $x \in [X]$*, and a finite set of firms indexed by* $y \in [Y]$*. There is one firm as well as one worker per type, so* $n_x = 1$ *and* $m_y = 1$ *for all* x *and* y*. If worker* x *matches with firm* y*, then the worker gets* α_{xy} *and the firm gets* γ_{xy}*. If* x *or* y *remain unmatched, they get respectively* α_{x0} *and* γ_{0y}*. As in problem 7.1, we shall assume that* $\alpha_{x0} = 0$ *and* $\gamma_{0y} = 0$*, and that* $\alpha_{xy} = \alpha_{xy'}$ *implies* $y = y'$ *and* $\gamma_{xy} = \gamma_{x'y}$ *implies* $x = x'$*.*

(i) *Show that* (u_x) *and* (v_y) *are the payoffs of a stable matching in the sense of Gale and Shapley (as recalled in problem 7.1 above) if and only if*

$$\begin{cases} u_x = \max\left\{\alpha_{xy} : y \in [Y], \gamma_{xy} \geq v_y\right\} \cup \{\alpha_{x0}\}, \\ v_y = \max\left\{\gamma_{xy} : x \in [X], \alpha_{xy} \geq u_x\right\} \cup \left\{\gamma_{0y}\right\}. \end{cases} \tag{7.9.2}$$

(ii) *By introducing the vector* $p \in \mathbb{R}^{X+Y}$ *such that* $p_x = u_x$ *and* $p_{X+y} = -v_y$, *show that one can restate system* (7.9.2) *in terms of a fixed point problem for an order-preserving mapping* $F(p)$, $F : \mathbb{R}^{X+Y} \to \mathbb{R}^{X+Y}$.

(iii) *Letting* $p^0 \in \mathbb{R}^{X+Y}$ *be such that* $p_z^0 = -\infty$ *for all* $z \in [X+Y]$ *and defining* $p^{t+1} = F(p^t)$, *show that* p^t *converges in a nondecreasing fashion and in a finite number of steps toward a fixed point of* F.

(iv) *Implement the previous algorithm on simulated data.*

7.10 References and Notes

The main reference for the chapter is the paper by Galichon, Hsieh, and Sylvestre (2024). Sections 7.1 and 7.2 are directly taken from there. Section 7.3 also originates from there, but the isotonicity of inverse demand under substitutability was observed by a number of authors, in particular by Alkan and Gale (2003) and Hatfield and Milgrom (2005). Azevedo and Leshno (2016) generalize stable matching in a continuous setting.

Section 7.6 is a reinterpretation of Gale and Shapley's (1962) deferred acceptance algorithm in a discrete choice setting by Galichon, Hsieh, and Sylvestre (2024); that paper is more general than the presentation in section 7.6 because it dispenses with assumptions 1.1 and 1.2, and therefore applies to choice correspondences instead of choice functions.

Section 7.7 draws inspiration from the paper by Hatfield and Milgrom (2005). Actually that paper is more general as it also handles contracts and one-to-many matching, but in section 7.7 we restrict ourselves to the case of one-to-one matching with a unique contract per match.

References for the α-axiom in exercise 7.6 are the paper by Chernoff (1954) and the paper by Sen (1971). The notions of consistency and persistence introduced in exercise 7.7 are taken from the paper by Alkan and Gale (2003). Problems 7.1 and 7.2 connect definition 7.5.1 and algorithm 7.6.1 with the notion of stable matching with nontransferable utility, as well as with the original deferred acceptance algorithm, which were both introduced in Gale and Shapley (1962); they both draw on material from Galichon, Hsieh, and Sylvestre

(2024). Problem 7.3 is based on the paper by Adachi (2000). A reference for exercise 7.3 is the paper by Fahli and Jacquet (2024).

7.11 Summary

SUMMARY OF CHAPTER 7

Chapter 7 covers constrained choice models, which are discrete choice models with capacity constraints that limit the availability of some options. This is tyically the case when prices are rigid or capped and cannot fully operate to adjust demand and supply.

Section 7.1 introduces the main notions. Let $\bar{\pi}_y$ be the capacity of y, which is an upper bound on the market share of y. Under the standard assumptions on the distribution of the random utility vector ε, the classical discrete choice model associates utility $U_y + \varepsilon_y$ to option y, leading to the prediction of a market share $\boldsymbol{\pi}_y(U)$ for option y. If some of the predicted market shares exceed the capacities, a vector of shadow prices (τ_y) will regulate demand to ensure it meets the capacity constraints. More specifically, letting $\rho = \bar{\pi} - \boldsymbol{\pi}(U - \tau)$ be the vector of excess capacity, τ and ρ will be defined by $\tau \geq 0$, $\rho \geq 0$, and the complementarity relation $\tau^\top \rho = 0$. Theorem 7.1.1 ensures that both τ and ρ exist and are unique under the standard assumptions. It uses a reformulation of the conditions on τ on ρ as the system of nonlinear equations

$$\boldsymbol{\pi}\left(U - \theta^+\right) + \theta^- = \bar{\pi}, \tag{7.11.1}$$

where one has set $\theta = \tau - \rho$, and thus $\tau = \theta^+$ and $\rho = \theta^-$. Proposition 7.1.2 guarantees that the function on the left-hand side of (7.11.1) is an M-function, which is the key to the existence and the uniqueness of a solution θ. Consequently, one defines $\tau = \theta^+$ as $\boldsymbol{\tau}(U, \bar{\pi})$ (the shadow price map), $\boldsymbol{\pi}\left(U - \theta^+\right)$ as $\tilde{\boldsymbol{\pi}}(U, \bar{\pi})$ (the constrained market share map), and $\rho = \theta^-$ as $\boldsymbol{\rho}(U, \bar{\pi})$ (the rejection map, or excess capacity map).

The derivations of these quantities in the case of the logit model are provided.

Section 7.2 takes the point of view of welfare and defines the notion of constrained welfare, which is given by

$$\tilde{G}(U, \bar{\pi}) = \max_{\pi}\{\sum_y \pi_y U_y - G^*(\pi) : \pi \leq \bar{\pi}\}.$$

The Daly–Zachary–Williams theorem extends to this case and one has $\tilde{\boldsymbol{\pi}}(U, \bar{\pi}) = \nabla_U \tilde{G}(U, \bar{\pi})$; similarly, one can show that $\boldsymbol{\tau}(U, \bar{\pi}) = \nabla_{\bar{\pi}} \tilde{G}(U, \bar{\pi})$.

Outside of the logit case, it is necessary to resort to simulation, and it is shown in example 7.1 how these simulated quantities can be obtained by linear programming.

Section 7.3 offers two key monotone comparative statics results: the fact that the shadow price map $\tau(U, \bar{\pi})$ is antitone (i.e., order-reversing) with respect to the vector of capacities $\bar{\pi}$, and the fact that the rejection map $\rho(U, \bar{\pi})$ is isotone (i.e., order-preserving) with respect to $\bar{\pi}$. As discussed, this is a manifestation of the substitutability property that is intrinsic to discrete choice models.

Section 7.4 is an attempt to find a concrete manifestation of the shadow prices in a constrained optimization problem as the waiting lines in a stationary dynamic model of queuing.

Section 7.5 seeks to adapt the notion of equilibrium matching to contexts where monetary transfers are ruled out, extending the notion of stable matching to an equilibrium setting. Accordingly, definition 7.5.1 introduces the notion of equilibrium matching without prices.

Section 7.6 investigates an adaptation of the deferred acceptance algorithm by Gale and Shapley to the discrete choice setting, which is presented in algorithm 7.6.1. Theorem 7.6.1 ensures that the algorithm converges to an equilibrium matching without prices, relying in particular on the comparative statics results obtained in section 7.3.

Finally, section 7.7 reinterprets algorithm 7.6.1 as the iteration of an isotone fixed point operator, in the spirit of a paper by Hatfield and Milgrom.

CHAPTER 8

Models with General Heterogeneities

In chapter 1, we presented the general theory of random utility models, which did not necessitate making any assumption on the distribution of the random utility vector ε (aside from the continuous distribution or full support assumption), and we introduced important objects such as the welfare function G, the entropy of choice G^*, the market share map ∇G, and other related quantities. In chapters 2 and 4, we provided examples of random utility models where the computation of some of these objects could be simplified: either by obtaining them in closed form, as in the nested logit model for which all objects can be obtained explicitly, or by relating them to existing machinery, as in the pure characteristics model, which can be reformulated using power diagrams and semi-discrete optimal transport. In chapter 7, we discussed general considerations on constrained choice models, which also rested only on the assumptions of continuous distribution and full support, and which led to the introduction of new objects such as the constrained welfare function $\tilde{G}$ and the constrained market share map $\tilde{\boldsymbol{\pi}}$.

However, in the four chapters from 3 to 6, we relied almost only on Gumbel heterogeneity and the logistic framework. While the logistic regression introduced in chapter 3 is obviously deeply rooted in the logit model, it would nevertheless be interesting to understand what an extension to more general structures of random utilities would entail. Likewise, the allocation and equilibrium models studied in chapter 5 need not rely on the logit model; that particular structure of random utilities has been chosen purely for convenience and tractability. The same can be said about the dynamic discrete choice models of chapter 6, which in that chapter are presented under the logit framework, but could extend to a general heterogeneity structure.

The goal of the current chapter is therefore to extend the models of choice, allocation, and equilibrium, both in the static and dynamic cases, as seen in chapters 5 and 6, to general heterogeneities. We start by a reflection on how to extend logistic regressions to any structure of heterogeneity.

8.1 One-Sided Choice and Minimax Regret

There are Y options $y \in [Y]$, and a decision-maker of observable type $x \in [X]$ associates a random utility $U_{xy} + \varepsilon_y$ to option y, where ε is a random vector of $\mathbb{R}^Y$ drawn according to a probability distribution $\mathcal{P}_x$ that may depend on x, and that satisfies assumptions 1.1 and 1.2. We assume a parametric form for the systematic utility term U_{xy}, and we postulate that $U_{xy} = \sum_k \phi_{xyk} \lambda_k$, which we write in vector terms as $U = \Phi\lambda$, where Φ is the $XY \times K$ matrix whose (xy, k)th term is ϕ_{xyk}. For $U \in \mathbb{R}^Y$, the welfare function associated with distribution $\mathcal{P}_x$ is denoted by $G_x(U) = \mathbb{E}_{\mathcal{P}_x}[\max_y\{U_y + \varepsilon_y\}]$, and the corresponding entropy of choice is $G_x^*(\pi)$, which, as we recall from chapter 1, is defined by $G_x^*(\pi) = \max_U\{\pi^\top U - G_x(U)\}$. Assuming that we observe a sample where $\hat{\mu}_{xy}$ is the number of individuals of type x who choose option y, the natural generalization of the logistic regression (5.1.1) of chapter 5 is given by

$$\max_{\lambda \in \mathbb{R}^K} \{\hat{\mu}^\top \Phi\lambda - \sum_{x \in [X]} n_x G_x(\Phi\lambda)\}, \tag{8.1.1}$$

where we have written $G_x(\Phi\lambda)$ instead of $G_x((\Phi\lambda)_{y\in[Y]})$ in order to simplify notation, thereby implicitly viewing G_x as a function of $\mathbb{R}^{X\times Y}$ to $\mathbb{R}$. The first-order conditions associated with (8.1.1) are

$$\Phi^\top \hat{\mu} = \Phi^\top \mu^\lambda, \tag{8.1.2}$$

where $\mu^\lambda_{xy} = n_x \partial G_x / \partial U_y(\Phi\lambda)$ is the predicted number of individuals of type x who choose option y. Equation (8.1.2) expresses that the model generates the same K moments of ϕ as those that are observed in the sample. We therefore see that this holds in the general case, not solely in the logistic case. In the logistic case, we saw in chapter 3 that equation (8.1.1) could be interpreted as a maximum likelihood estimation procedure, but this interpretation holds only in the logistic case. It is possible to interpret problem (8.1.1) in the general case as a *multivariate generalized linear model*, although this concept is not widely used; we provide references in section 8.7.

Interpretation as a minimax regret procedure. Note that one may view problem (8.1.1) as a minimax regret procedure. Indeed, if one observes that a decision maker type x makes decision $\tilde{y}$, then the regret of the decision-maker with respect to alternative $y \in [Y]$ is given by

$$R^\lambda_{xy\tilde{y}}(\varepsilon) := (\Phi\lambda)_{xy} + \varepsilon_y - (\Phi\lambda)_{x\tilde{y}} - \varepsilon_{x\tilde{y}}. \tag{8.1.3}$$

Of course, we do not know the realization of the random vector ε, only its distribution; and we do not know the joint distribution of ε and $\tilde{y}$ either. The maximum regret with respect to an alternative is defined as $\max_y R^{\lambda}_{xy\tilde{y}}(\varepsilon) = \max_y\{(\Phi\lambda)_{xy} + \varepsilon_{xy}\} - (\Phi\lambda)_{x\tilde{y}} - \varepsilon_{x\tilde{y}}$, and the integrated maximum regret is given by

$$\sum_{x\in[X]} n_x \mathbb{E}_{\mathcal{P}_x}[\max_y R^{\lambda}_{xy\tilde{y}}(\varepsilon)] = \sum_{x\in[X]} n_x G_x(\Phi\lambda) - \hat{\mu}^\top \Phi\lambda - \sum_{x\in[X]} n_x \mathbb{E}_{\mathcal{P}_x}[\varepsilon_{x\tilde{y}}]. \tag{8.1.4}$$

The last term is a weighted sum of expectations $\mathbb{E}_{\mathcal{P}_x}[\varepsilon_{x\tilde{y}}]$, which depend on the joint distribution of ε and $\tilde{y}$, which is unknown; however, this joint distribution does not depend on the parameter λ, so the exact value of this term is irrelevant for our estimation problem, and we may, for instance, take the joint conditional distribution of $(\tilde{y}, \varepsilon)$ conditional on x that achieves the largest value of $\mathbb{E}_{\mathcal{P}_x}[\varepsilon_{x\tilde{y}}]$. By the virtues of theorem 1.3.3, this largest value is equal to $-G^*_x(\mu/n_x)$, and therefore we may define our *maximum choice regret* as:

Definition 8.1.1 (Maximum choice regret). *The maximum choice regret associated with the parameter vector $\lambda \in \mathbb{R}^K$ is*

$$R(\lambda) = \sum_{x\in[X]} n_x G_x(\Phi\lambda) - \hat{\mu}^\top \Phi\lambda + \sum_{x\in[X]} n_x G^*_x\left(\frac{\hat{\mu}}{n}\right), \tag{8.1.5}$$

and the minimax regret estimator $\hat{\lambda}$ is the parameter vector λ that achieves the minimum of $R(\lambda)$.

As the last term in the expression of $R(\lambda)$ in (8.1.5) does not depend on λ, one sees immediately the minimax regret estimator coincides with the solution to (8.1.1).

A reformulation. It is possible to rewrite problem (8.1.1) in an equivalent, yet slightly different way. Introducing the margining-out matrix $\mathbf{\Sigma}_Y = \mathbf{I}_X \otimes \mathbf{1}_Y^\top$, we recall that $n = \mathbf{\Sigma}_Y \mu$ is the vectorized form of $n_x = \sum_y \mu_{xy}$.[1] If we introduce the value u_x of each state x, which is $u_x = G_x(\Phi\lambda)$, we recall that $\mathbf{\Sigma}_Y^\top u$ is the doubly indexed vector of size XY whose xy term is u_x, and $u_x = G_x(\Phi\lambda)$ can be equivalently reexpressed as $G_x(\Phi\lambda - \mathbf{\Sigma}_Y^\top u) = 0$. Hence the objective function in problem (8.1.1) is $\hat{\mu}^\top\left(\Phi\lambda - \mathbf{\Sigma}_X^\top u\right)$, and the constraints are $G_x(\Phi\lambda - \mathbf{\Sigma}_Y^\top u) = 0$. Problem (8.1.1) can therefore be reexpressed as

$$\max_{\lambda\in\mathbb{R}^K, u\in\mathbb{R}^X} \{\hat{\mu}^\top\left(\Phi\lambda - \mathbf{\Sigma}_Y^\top u\right)\} \tag{8.1.6}$$

$$s.t.\quad G_x(\Phi\lambda - \mathbf{\Sigma}_Y^\top u) = 0 \ \forall x \in [X]. \tag{8.1.7}$$

[1]The notation $\mathbf{\Sigma}_Y$ means to recall that the effect of that matrix is to sum over the y indices.

Letting $\tilde{n}_x$ be the Lagrange multiplier associated to constraint (8.1.7), the saddle point formulation of the problem becomes

$$\max_{\lambda \in \mathbb{R}^K, u \in \mathbb{R}^X} \min_{\tilde{n} \in \mathbb{R}^X} \left\{ \hat{\mu}^\top \left(\Phi\lambda - \boldsymbol{\Sigma}_Y^\top u \right) - G\left(\Phi\lambda - \boldsymbol{\Sigma}_Y^\top u; \tilde{n} \right) \right\}$$

$$\text{where } G(U; \tilde{n}) := \sum_{x \in [X]} \tilde{n}_x G_x(U)$$

We have by Karush–Kuhn–Tucker conditions (see appendix A.1.6) with respect to u that

$$\left(\boldsymbol{\Sigma}_Y \hat{\mu} \right)_x = \tilde{n}_x \sum_{y \in [Y]} \partial G_x / \partial U_y (\Phi\lambda - u_x) = \tilde{n}_x,$$

and therefore at the optimum, $\tilde{n} = n$ and the problem boils down to

$$\max_{\lambda \in \mathbb{R}^K, u \in \mathbb{R}^X} \{ \hat{\mu}^\top \left(\Phi\lambda - \boldsymbol{\Sigma}_Y^\top u \right) \}$$

$$s.t. \quad G(\Phi\lambda - \boldsymbol{\Sigma}_Y^\top u; n) = 0.$$

Introducing the Z function. We now investigate another reformulation of problem (8.1.1) that will be meaningful in the logistic case. Another way to express the constraint $u_x = G_x(\Phi\lambda)$ is $1 = \exp(-u_x + G_x(\Phi\lambda))$, and letting N_x be the corresponding Lagrange multiplier, problem (8.1.1) can be rewritten in a saddle point formulation as

$$\max_{\lambda \in \mathbb{R}^K, u \in \mathbb{R}^X} \min_{N \in \mathbb{R}^X} \{ \hat{\mu}^\top \left(\Phi\lambda - \boldsymbol{\Sigma}_Y^\top u \right) + \sum_{x \in [X]} Z_x(\lambda, u_x; N_x),$$

$$\text{where } Z_x(\lambda, u_x; N_x) := N_x - N_x \exp(-u_x + G_x(\Phi\lambda)) \}.$$

While the interest of that formulation is questionable in general, considering the particular case of the logit random utility reveals its usefulness in that case. Indeed, the logit structure implies $\exp(G_x(\Phi\lambda)) = \sum_y \exp((\Phi\lambda)_{xy})$, and therefore

$$Z_x(\lambda, u; N) = N_x - N_x \sum_{y \in [Y]} \exp((\Phi\lambda)_{xy} - u_x),$$

where we recover the function Z introduced in chapter 6 and described in particular in section 6.4.

8.2 Empirical Matching Models

We now explore models of one-to-one matching similar to the one described in section 5.4, but with a fully general heterogeneity structure (although we shall maintain the continuous distribution and full support assumptions). We take again the example of the heterosexual marriage market, and we assume that if a man of type $x \in [X]$ and a woman of type $y \in [Y]$ match, they get respective surpluses $U_{xy} + \varepsilon_y$ and $V_{xy} + \eta_x$, where the vector $\varepsilon \sim \mathcal{P}_x$ and the vector $\eta \sim \mathcal{Q}_y$, and $U_{xy} + V_{xy} = \phi_{xy}$. It is assumed that $\mathcal{P}_x$ and $\mathcal{Q}_y$ satisfy assumptions 1.1 and 1.2 for all x and y. The expected indirect utility of a man of type x is $u_x = \mathbb{E}[\max_y\{U_{xy} + \varepsilon_y, \varepsilon_0\}] = G_x(U)$, while the expected indirect utility of a woman of type y is $v_y = \mathbb{E}[\max_x\{V_{xy} + \eta_x, \eta_0\}] = H_y(V)$. We let $G(U) = \sum_x n_x G_x(U)$ and $H(V) = \sum_y m_y H_y(V)$ be the total welfare of men and women, respectively, when their systematic utilities associated with matches of type xy are respectively U_{xy} and V_{xy}.

Equilibrium characterization. It follows from the Daly–Zachary–Williams theorem (theorem 1.2.1) that the number of men of type x who prefer women of type y is given by $\mu_{xy} = \partial G(U)/\partial U_{xy}$, and the number of women of type y who prefer men of type x is given by $\mu_{xy} = \partial H(V)/\partial V_{xy}$. As $U + V = \phi$, equilibrium is therefore given by

$$\frac{\partial G}{\partial U_{xy}}(U) = \frac{\partial H}{\partial V_{xy}}(V), \tag{8.2.1}$$

and we have:

Theorem 8.2.1. *If $\mathcal{P}_x$ and $\mathcal{Q}_y$ satisfy assumptions 1.1 and 1.2, the equilibrium matching μ satisfies*

$$\max_{\mu \in \mathbb{R}_+^{X \times Y}} \left\{ \sum_{xy \in [X \times Y]} \mu_{xy}\phi_{xy} - \mathcal{E}(\mu) \right\}, \tag{8.2.2}$$

where $\mathcal{E}(\mu) := G^*(\mu) + H^*(\mu)$ defines the entropy of matching associated with the random utility structure, and the systematic utilities U_{xy} and V_{xy} arise as the solution to

$$\min_{U,V \in \mathbb{R}^{X \times Y}} \{G(U) + H(V)\} \tag{8.2.3}$$
$$s.t. \quad U_{xy} + V_{xy} = \phi_{xy} \; \forall xy \in [X \times Y].$$

Proof. The equilibrium conditions (8.2.1) are the optimality conditions associated with the convex optimization problem (8.2.3). The Lagrangian associated with this problem is $\mathcal{L}(U,V;\mu)=\mu^\top\phi-\mu^\top U+G(U)-\mu^\top V+H(V)$, and therefore the dual of problem (8.2.1) is $\max_\mu \min_{U,V}\mathcal{L}(U,V;\mu)$; but $\min_{U,V}\mathcal{L}(U,V;\mu)=\mu^\top\phi-G^*(\mu)-H^*(\mu)$, from which the formulation (8.2.2) follows. ■

We turn to the logit model for an example.

Example 1.2 (Logit model, continued). In the case when $\mathcal{P}_x$ and $\mathcal{Q}_y$ are the distributions of i.i.d. Gumbel random vectors, one has

$$\begin{cases} G^*(\mu)=\sum_{x\in[X]}\left(\sum_{y\in[Y]}\mu_{xy}\log\frac{\mu_{xy}}{n_x}+\mu_{x0}\log\frac{\mu_{x0}}{n_x}\right),\\ H^*(\mu)=\sum_{y\in[Y]}\left(\sum_{x\in[X]}\mu_{xy}\log\frac{\mu_{xy}}{m_y}+\mu_{0y}\log\frac{\mu_{0y}}{m_y}\right),\end{cases}$$

where μ_{x0} is defined from the other entries of (μ_{xy}) by $\mu_{x0}=n_x-\sum_{y\in[Y]}\mu_{xy}$, and, similarly, μ_{0y} is defined as $\mu_{0y}=m_y-\sum_{x\in[X]}\mu_{xy}$, which implies

$$\mathcal{E}(\mu)=\sum_{xy\in[X\times Y]}\mu_{xy}\ln\frac{\mu_{xy}^2}{n_x m_y}+\sum_{x\in[X]}\mu_{x0}\log\frac{\mu_{x0}}{n_x}+\sum_{y\in[Y]}\mu_{0y}\log\frac{\mu_{0y}}{m_y},$$

so that problem (8.2.2) becomes

$$\max_{\mu\in\mathbb{R}_+^{X\times Y}}\left\{\begin{array}{l}\sum_{xy\in[X\times Y]}\mu_{xy}\phi_{xy}-\sum_{xy\in[X\times Y]}\mu_{xy}\ln\frac{\mu_{xy}^2}{n_x m_y}\\ -\sum_{x\in[X]}(n_x-\sum_{y\in[Y]}\mu_{xy})\log\left(1-\sum_{y\in[Y]}\frac{\mu_{xy}}{n_x}\right)\\ -\sum_{y\in[Y]}(m_x-\sum_{x\in[X]}\mu_{xy})\log\left(1-\sum_{x\in[X]}\frac{\mu_{xy}}{m_y}\right)\end{array}\right\}. \tag{8.2.4}$$

This complex expression illustrates that, despite the apparent simplicity of the abstract formulation (8.2.2), its application to concrete cases can be intricate. However, as seen in section 6.4, the TU-logit matching model admits a much simpler representation, which the following discussion will show how to derive.

Note that in problem (8.2.2), the constraints $\sum_{y\in[Y]}\mu_{xy}\le n_x$ and $\sum_{x\in[X]}\mu_{xy}\le m_y$ are imposed in the presence of $\mathcal{E}(\mu)$ in the objective function, as $\mathcal{E}(\mu)=G^*(\mu)+H^*(\mu)$, and $G^*(\mu)<+\infty$ implies that $\mu_{xy}\ge 0$ and that $\sum_{y\in[Y]}\mu_{xy}\le n_x$ should hold for all $x\in[X]$, while $H^*(\mu)<+\infty$ also implies that $\mu_{xy}\ge 0$, but now implies that $\sum_{x\in[X]}\mu_{xy}\le m_y$ should hold for all $y\in[Y]$. The number of individuals of each type who choose not to match are given by $\mu_{x0}=n_x-\sum_{y\in[Y]}\mu_{xy}$ and $\mu_{0y}=m_y-\sum_{x\in[X]}\mu_{xy}$; they are therefore defined implicitly from the number of matched individuals μ_{xy}. Inspired by the logit example that we just saw, we would like to introduce a convex function E defined on $\mathbb{R}_+^{XY+X+Y}$ whose argument is the vector $(\mu_{xy}, \mu_{x0}, \mu_{0y})$ that is such that

$$\mathcal{E}(\mu) = E((\mu_{xy}), (n_x - \sum_{y \in [Y]} \mu_{xy}), (m_y - \sum_{x \in [X]} \mu_{xy})). \tag{8.2.5}$$

We shall now regard μ as the vector of size $XY + X + Y$ obtained by the concatenation of $vec_R(\mu_{xy})$, (μ_{x0}), and (μ_{0y}), and consider Φ as the vector of size $XY + X + Y$ obtained by the concatenation of ϕ_{xy} and the zero vector of size $X + Y$ (given that $\phi_{x0} = \phi_{0y} = 0$), and let

$$\begin{cases} \mathbf{\Sigma}_X := \begin{pmatrix} \mathbf{I}_X \otimes \mathbf{1}_Y^\top & \mathbf{I}_X & \mathbf{0}_{X \times Y} \end{pmatrix}, \\ \mathbf{\Sigma}_Y := \begin{pmatrix} \mathbf{1}_X^\top \otimes \mathbf{I}_Y & \mathbf{0}_{Y \times X} & \mathbf{I}_Y \end{pmatrix}, \end{cases}$$

so that the x-margin constraints $\sum_{y \in [Y]} \mu_{xy} + \mu_{x0} = n_x$ read $\mathbf{\Sigma}_X \mu = n$ and the y-margin constraints $\sum_{x \in [X]} \mu_{xy} + \mu_{0y} = m_y$ read $\mathbf{\Sigma}_Y \mu = m$. With this in mind, we rewrite problem (8.2.2) as

$$\begin{aligned} \max_{\mu \in \mathbb{R}_+^{XY+X+Y}} \quad & \{\mu^\top \Phi - E(\mu)\} \\ s.t. \quad & \mathbf{\Sigma}_X \mu = n, \\ & \mathbf{\Sigma}_Y \mu = m. \end{aligned} \tag{8.2.6}$$

Introducing u_x and v_y as the Lagrange multipliers of the constraints in problem (8.2.6), the Lagrangian formulation is

$$\max_{\mu \in \mathbb{R}_+^{XY+X+Y}} \min_{u \in \mathbb{R}^X, v \in \mathbb{R}^Y} \left\{ n^\top u + m^\top v + \mu^\top \left(\Phi - \mathbf{\Sigma}_X^\top u - \mathbf{\Sigma}_X^\top v \right) - E(\mu) \right\},$$

which, by strong duality, leads to the following result:

Theorem 8.2.2. *Under the assumptions of theorem 8.2.1, an equivalent formulation of problem* (8.2.3) *is*

$$\min_{(u,v) \in \mathbb{R}^{X+Y}} \{n^\top u + m^\top v + E^* \left(\Phi - \mathbf{\Sigma}_X^\top u - \mathbf{\Sigma}_X^\top v \right)\}, \tag{8.2.7}$$

where E^*, defined on $\mathbb{R}^{XY+X+Y}$, is the Legendre–Fenchel transform of E, expressed as $E^*(a) = \max_\mu \{\mu^\top a - E(\mu) : \mu \in \mathbb{R}_+^{XY+X+Y}\}$.

Example 1.2 (Logit model, continued). Continuing the logit example above, we can take $E(\mu) = \sum_{xy} \mu_{xy} \ln(\mu_{xy}^2/(n_x m_y)) + \sum_x \mu_{x0} \ln(\mu_{x0}/n_x)$

Algorithm 8.2.1 Generalized IPFP algorithm for solving problem (8.2.7)

Input : F, v_y^0
Parameter: η (tolerance)
while $\max_{y\in[Y]} \|v_y^t - v_y^{t-1}\| \geq \eta$ **do**
  $u_x^t = \arg\max_u F(u, v^{t-1})$
  $v_y^t = \arg\max_v F(u^t, v)$
end
Return : u^t, v^t

$+\sum_y \mu_{0y} \ln(\mu_{0y}/m_y) - 2\sum_{xy} \mu_{xy} - \sum_x \mu_{x0} - \sum_y \mu_{0y} + \sum_x n_x + \sum_y m_y$, and a straightforward calculation shows that

$$E^*(a) = 2 \sum_{xy\in[X\times Y]} \sqrt{n_x m_y} e^{\frac{a_{xy}}{2}} + \sum_{x\in[X]} n_x e^{a_{x0}} + \sum_{y\in[Y]} m_y e^{a_{0y}} - \sum_{x\in[X]} n_x - \sum_{y\in[Y]} m_y, \tag{8.2.8}$$

so that $E^*\left(\Phi - \mathbf{\Sigma}_X^\top u - \mathbf{\Sigma}_X^\top v\right)$ is precisely the function Z introduced in section 6.4. This leads to recovering the expression (5.4.9) obtained in chapter 5, that is,

$$\min_{(u,v)\in\mathbb{R}^{X+Y}} \left\{ \begin{array}{c} \sum_{x\in[X]} n_x u_x + \sum_{y\in[Y]} m_y v_y \\ +2\sum_{xy\in[X\times Y]} \sqrt{n_x m_y} \exp\left(\frac{\phi_{xy}-u_x-v_y}{2}\right) \\ +\sum_{x\in[X]} n_x \exp(-u_x) + \sum_{y\in[Y]} m_y \exp(-v_y) \end{array} \right\}.$$

Equilibrium computation. With formulation (8.2.7) of the problem in mind, we can provide a general formulation of the iterated proportional fitting procedure (IPFP), described as algorithm 5.2.1. Remember, that algorithm consists of coordinate descent on the dual problem. Here, we introduce

$$F(u, v) = n^\top u + m^\top v + E^*\left(\Phi - \mathbf{\Sigma}_X^\top u - \mathbf{\Sigma}_X^\top v\right), \tag{8.2.9}$$

and we can formally describe the generalized IPFP algorithm (still under assumptions 1.1 and 1.2) as algorithm 8.2.1.

It is important to note, however, that computational benefits of using the coordinate gradient descent as one does in algorithm 8.2.1 are uncertain in the general case. As we have seen in section 5.4, there are in the logit case very simple closed-form expressions for u^t and v^t; the existence of these closed-form expressions is intrinsically tied to the logistic framework and does extend to the general case. Therefore, one may prefer using gradient descent in order to compute the optimal (u, v) pair that minimizes F, which would lead to iterations of the form

$$\begin{cases} \mu^t = \nabla E\left(\Phi - \mathbf{\Sigma}_X^\top u - \mathbf{\Sigma}_Y^\top v\right), \\ u_x^{t+1} = u_x^t - \epsilon\left(n - \left(\mathbf{\Sigma}_X^\top \mu^t\right)\right)_x, \ x \in [X], \\ v_y^{t+1} = v_y^t - \epsilon\left(m - \left(\mathbf{\Sigma}_Y^\top \mu^t\right)\right)_y, \ y \in [Y], \end{cases} \tag{8.2.10}$$

and we see that the updating rule (8.2.10) has an interpretation in terms of a *tâtonnement algorithm*: u_x is interpreted as the price of x and v_y as the price of y; if the sum of μ_{xy}^t over y is below the number of x in the population, it means that x is underdemanded, and therefore too expensive, and one needs to decrease its price u_x; a similar argument applies to y. Convergence of the procedure consisting of iterating (8.2.10) follows from standard arguments since $F(u, v)$ is a convex and smooth function.

However, the computation of E itself may be complex in some cases, and the cases when E is available in closed form are relatively rare. For instance, when all the random utilities have multivariate extreme value distributions, it was seen in section 2.2 that the only objects that can be computed in closed form are G and H and their gradients. In that case, we can base our computational approach directly on the initial formulation (8.2.3) of this section, and we can use the gradient descent method

$$U_{xy}^{t+1} = U_{xy}^t - \epsilon\left(\frac{\partial G}{\partial U_{xy}}\left(U^t\right) - \frac{\partial H}{\partial V_{xy}}\left(\Phi - U^t\right)\right), \ xy \in [X \times Y], \tag{8.2.11}$$

which can again be interpreted as a tâtonnement procedure: U_{xy} is the systematic share of utility that goes to x in a match with y; at equilibrium, the demand for matches of type xy that originates from individuals of type x, which is $\partial G / \partial U_{xy}(U_{xy})$, should coincide with the demand for matches of type xy that originates from individuals of type y; if the former quantity exceeds the latter, then the terms of the match should be readjusted in favor of the y, and therefore U_{xy} should decrease, as implied by the updating formula (8.2.11). A Python implementation of this gradient descent procedure is provided in the computational appendix in section B.8.1.

Estimation. Given the previous discussion, the estimation strategy will come as no surprise. We assume the following parametrization of the surplus function $\Phi_{xy}^\lambda = \sum_k \phi_{xyk} \lambda_k$, which we write in a vectorized manner as $\Phi^\lambda = \mathbf{\Phi} \lambda$, where, as in section 5.4, $\mathbf{\Phi}$ is a matrix of size $XY \times K$ whose term is ϕ_{xyk}. Letting $\hat{\mu}_{xy}$ be the number of matches of type xy observed in the sample, we can base an estimator of λ on the following convex optimization problem:

$$\max_{\lambda,U,V} \left\{ \hat{\mu}^\top \mathbf{\Phi}\lambda - G(U) - H(V) \right\} \tag{8.2.12}$$

$$s.t. \quad U + V = \mathbf{\Phi}\lambda. \tag{8.2.13}$$

Denoting by μ the vector of Lagrange multipliers associated with the system of constraints (8.2.13), we have the following optimality conditions with respect to U and V:

$$\nabla G(U) = \mu = \nabla H(V),$$

which express that μ is an equilibrium matching associated with splitting the systematic surplus $\mathbf{\Phi}\lambda$ at the optimal value λ into U and V according to (8.2.13), while the optimality conditions with respect to λ yield

$$\mathbf{\Phi}^\top \mu = \mathbf{\Phi}^\top \hat{\mu},$$

or, more explicitly, $\sum_{xy} \mu_{xy}\phi_{xyk} = \sum_{xy} \hat{\mu}_{xy}\phi_{xyk}$, which express that the K moments predicted by μ coincide with the K moments observed in the sample.

Inference. We can reformulate the estimation of λ from problem (8.2.12) as an M-estimation procedure (see appendix A.2.5). Letting N be the total number of households, and $\hat{f}_{xy} = \hat{\mu}_{xy}/N$ be the empirical frequency of a match of type xy for $xy \in [X \times Y + X \times 0 + 0 \times Y]$, we can rescale problem (8.2.12) by a factor N to obtain that the estimator $\hat{\lambda}$ of λ is given by

$$\max_{\lambda \in \mathbb{R}^K} \{ \hat{f}^\top \mathbf{\Phi}\lambda - W(\lambda, \hat{f}) \},$$

where

$$W(\lambda, \mu) := \min_{U \in \mathbb{R}^{X \times Y}} \{ G(U; \mathbf{\Sigma}_X \mu) + H(\mathbf{\Phi}\lambda - U; \mathbf{\Sigma}_Y \mu) \}$$

and the notation $G(U; n)$ indicates the dependence in n, that is, $G(U; n) = \sum_x n_x G_x(U)$, and similarly for $H(V; m) = \sum_y m_y H_y(V)$.

Assuming that the true value of λ is $\lambda^\star$, and that, in the population, the frequency of a match of type xy is $f^\star_{xy}$, we have $\varphi(\hat{\lambda}, \hat{f}) = 0$ and $\varphi(\lambda^\star, f^\star)$, where $\varphi(\lambda, f) = \nabla_\lambda W(\lambda, f) - \mathbf{\Phi}^\top f$. Assuming that the observations are drawn from the population in an i.i.d. manner, we have the convergence in distribution of the empirical frequency toward the population frequency

$$\sqrt{N}\left(\hat{f} - f^\star\right) \xrightarrow{d} N(0, V),$$

where $V = diag(f) - ff^\top$. By linearization, we have therefore the first-order approximation $D_\lambda \varphi(\lambda^\star, f^\star)(\hat{\lambda} - \lambda^\star) + D_f \varphi(\lambda^\star, f^\star)(\hat{f} - f^\star) + o_P(1/\sqrt{N})$, from which the following statement can be deduced:

Theorem 8.2.3. *Under assumptions 1.1 and 1.2, the estimator $\hat{\lambda}$ given by (8.2.12) is consistent, and the following convergence holds in distribution*

$$\sqrt{N}\left(\hat{\lambda} - \lambda^\star\right) \xrightarrow{d} N\left(0, H^{-1} BV BH^{-1}\right),$$

where $V = diag(f) - ff^\top$, $B = D_f \varphi\left(\lambda^\star, f^\star\right) = D^2_{\lambda f} W(\lambda^\star, f^\star) - \mathbf{\Phi}$, and $H = D_\lambda \varphi(\lambda^\star, f^\star) = D^2_{\lambda\lambda} W(\lambda^\star, f^\star)$. These quantities can be consistently estimated themselves by plugging in $\hat{\lambda}$ and $\hat{f}$ instead of $\lambda^\star$ and $f^\star$.

Simulation. When G and H cannot be obtained in closed form, we can resort to simulation. Let I be the number of terms ε_{ixy} and η_{ixy} drawn from the distributions $\mathcal{P}_x$ and $\mathcal{Q}_y$ of the random utilities on both sides of the market for each cell. The idea is to replace $G(U) = \sum_x n_x G_x(U)$ by $G_I(U)$, defined as

$$G_I(U) = \sum_{x \in [X]} \frac{n_x}{I} \sum_{i \in [I]} \max_y \left\{U_{xy} + \varepsilon_{ixy}, \varepsilon_{ix0}\right\},$$

and replace $H(V)$ by $H_I(V)$, which has a similar expression. Problem (8.2.12) then becomes the linear programming problem

$$\max_{\lambda, u, v, U, V} \left\{\hat{\mu}^\top \mathbf{\Phi}\lambda - \sum_{ix \in [I \times X]} \frac{n_x}{I} u_{ix} - \sum_{iy \in [I \times Y]} \frac{m_y}{I} v_{iy}\right\} \tag{8.2.14}$$
$$\begin{aligned} s.t.\quad & U_{xy} + V_{xy} \geq (\mathbf{\Phi}\lambda)_{xy} \; \forall xy \in [X \times Y] \\ & u_{ix} \geq U_{xy} + \varepsilon_{ixy} \; \forall ixy \in [I \times X \times Y] \\ & u_{ix} \geq \varepsilon_{ix0} \; \forall ix \in [I \times X] \\ & v_{iy} \geq V_{xy} + \eta_{ixy} \; \forall ixy \in [I \times X \times Y] \\ & v_{iy} \geq \eta_{i0y} \; \forall iy \in [I \times Y], \end{aligned}$$

which vectorizes into

$$\begin{aligned} \max_\theta \quad & b^\top \theta \\ s.t.\; & C\theta \geq d, \end{aligned}$$

where the optimization variables are concatenated into $\theta = (\lambda^\top, u^\top, v^\top, U^\top, V^\top)^\top$, the objective function is determined by $b = (\hat{\mu}^\top \mathbf{\Phi}, -\mathbf{1}_I^\top$

$\otimes n^\top/I, -\mathbf{1}_I^\top \otimes m^\top/I, \mathbf{0}_{2XY}^\top)^\top$, the right-hand side of the constraint is given by $d = (\mathbf{0}_{XY}^\top, \varepsilon^\top, \varepsilon_0^\top, \eta^\top, \eta_0^\top)^\top$, and the constraint matrix is given by

$$C = \begin{pmatrix} -\mathbf{\Phi} & \cdot & \cdot & \mathbf{I}_{XY} & \mathbf{I}_{XY} \\ \cdot & \mathbf{I}_{IX} \otimes \mathbf{1}_Y & \cdot & -\mathbf{1}_I \otimes \mathbf{I}_{XY} & \cdot \\ \cdot & \mathbf{I}_{IX} & \cdot & \cdot & \cdot \\ \cdot & \cdot & \mathbf{I}_I \otimes \mathbf{1}_X \otimes \mathbf{I}_Y & \cdot & -\mathbf{1}_I \otimes \mathbf{I}_{XY} \\ \cdot & \cdot & \mathbf{I}_{IY} & \cdot & \cdot \end{pmatrix},$$

where the dots stand for zero block matrices. This approach is implemented in Python in section B.8.2 in the appendix, where a matching market with probit heterogeneity is estimated.

CCP approach. Just as in section 6.8, an alternative approach consists in running the CCP inversion in order to retrieve the joint surplus and then estimating the parameter vector via least squares. More specifically, given the observation of $\hat{\mu}_{xy}$ for $xy \in [X \times Y] \cup [X \times 0] \cup [0 \times Y]$, we may compute

$$\hat{\phi}_{xy} = \frac{\partial G_x^*}{\partial \pi_y}\left((\frac{\hat{\mu}_{xy}}{n_x})_{y \in [Y]} \right) + \frac{\partial H_y^*}{\partial \pi_x}\left((\frac{\hat{\mu}_{xy}}{m_y})_{x \in [X]} \right),$$

and, next, run generalized least squares of $\hat{\phi}$ so obtained on the basis functions that are given by the rows of $\mathbf{\Phi}$, looking for the parameter vector λ providing the minimal distance between the predicted surplus $\mathbf{\Phi}\lambda$ and the inferred surplus $\hat{\phi}$. We get

$$\min_\lambda \left| \hat{\phi} - \mathbf{\Phi}\lambda \right|_{\mathbf{W}},$$

where, as before, $|z|_{\mathbf{W}}^2 = z^\top \mathbf{W} z$.

8.3 Coalition Formation

We next turn to coalition formation models. As in section 5.5, $z \in [Z]$ are types of individuals, and there is a mass q_z of individuals of type z. A coalition indexed by $b \in [B]$ contains Σ_{zb} agents of type $z \in [Z]$, where Σ_{zb} is a nonnegative integer. We define B as the number of allowed coalitions and assume this number is finite. We recall that the size of the coalition b is $s_b = (\Sigma^\top \mathbf{1}_Z)_b$.

For $b \in [B]$, the quantity $\mu_b \in \mathbb{R}_+$ is the mass of coalitions of type b that are formed, and the market clearing condition is

$$q_z = \sum_{b \in [B]} \Sigma_{zb} \mu_b = (\Sigma \mu)_z, \tag{8.3.1}$$

which means that the total mass of individuals of type z is obtained by summing over the total number of coalitions the mass of a coalition times the number of individuals of type z contained by that coalition. We assume that if an individual i of type z is in coalition b, then this individual can obtain a utility equal to

$$U_{zb} + \varepsilon_{ib},$$

where the distribution of the B-dimensional random vector (ε_{ib}) follows a distribution $\mathcal{P}_z$, and U_{zb} is the share of the systematic surplus available to member z in coalition b. ϕ_b is the joint systematic surplus to coalition $b \in [B]$ to be shared between its members, so that

$$\sum_{z \in [Z]} \Sigma_{zb} U_{zb} = \phi_b.$$

Equilibrium definition. We let $G_z(U) = \mathbb{E}_{\mathcal{P}_z}[\max_b \{U_{zb} + \varepsilon_{ib}\}]$ and $G(U) = \sum_z q_z G_z(U)$ be the welfare of agents of type z and aggregate welfare, respectively. The Legendre–Fenchel transform of G_z, as usual denoted by G_z^*, is the entropy of choice associated with G, and G^*, the Legendre–Fenchel transform of G, is given by $G^*(\mu) = \sum_z q_z G_z^*(\mu / q_z)$. The mass of agents of type z who choose coalition b is given by $\partial G(U) / \partial U_{zb}$; this number should coincide with $\Sigma_{zb} \mu_b$, and, as a result, one may define a coalitional equilibrium in the following fashion:

Definition 8.3.1 (Coalitional equilibrium). *A coalitional equilibrium outcome is a vector of numbers $(\mu_b, U_{zb}) \in \mathbb{R}^B \times \mathbb{R}^{ZB}$ such that*

$$\begin{cases} \sum_{z \in [Z]} \Sigma_{zb} U_{zb} = \phi_b, \; \forall b \in [B], \\ \Sigma_{zb} \mu_b = \frac{\partial G(U)}{\partial U_{zb}}, \; \forall zb \in [Z \times B]. \end{cases}$$

Note that we are not explicitly requesting condition (8.3.1) to hold in our definition of a coalitional equilibrium; in fact, this condition is implicit, given that $\sum_b \Sigma_{zb} \mu_b = \sum_b \partial G(U) / \partial U_{zb} = q_z$.

Equilibrium computation. The following result, whose proof results directly from the Karush–Kuhn–Tucker conditions, provides a characterization of a coalitional equilibrium as a convex optimization problem:

Theorem 8.3.1. *A vector* $(\mu_b, U_{zb}) \in \mathbb{R}^B \times \mathbb{R}^{ZB}$ *is a coalitional equilibrium if and only if the vector of mass* (μ_b) *is a solution to the primal problem*

$$\max_{\mu_b \in \mathbb{R}_+^B} \left\{ \sum_{b \in [B]} \mu_b \phi_b - G^* \left((\Sigma_{zb} \mu_b)_{zb \in [Z \times B]} \right) \right\} \tag{8.3.2}$$

and the payoff vector U *is a solution to the corresponding dual*

$$\min_{U \in \mathbb{R}^{ZB}} G(U) \tag{8.3.3}$$

$$\text{s.t.} \sum_{z \in [Z]} \Sigma_{zb} U_{zb} = \phi_b. \tag{8.3.4}$$

Moreover, the value of the primal and the dual coincide, and the solution vector to the primal problem is the vector of Lagrange multipliers for the dual.

This theorem should by now have a familiar tone. The value of the primal $\sum_b \mu_b \phi_b - G^*(\Sigma_{zb}\mu_b)$ is made of two terms. The first term, $\sum_b \mu_b \phi_b$, is the weighted sum of the systematic joint surpluses associated with each coalition, weighted by the masses of the respective coalitions. The second term, $-G^*(\Sigma_{zb}\mu_b)$, is the total amount of utility resulting from the random utilities. Again, the value of the primal problem accounts for the origination of the surplus, which is generated by coalitions. The value of the dual, which coincides with that of the primal, accounts for the destination of the utilities, which is distributed among individuals: indeed, $G(U)$ is a sum of utilities across individuals, the condition $\sum_z \Sigma_{zb} U_{zb} = \phi_b$ ensuring that these utilities remain feasible. Note that if μ_b is the Lagrange multiplier in constraint (8.3.4), the optimality condition with respect to U_{zb} yields the equilibrium relation

$$\frac{\partial G(U)}{\partial U_{zb}} = \Sigma_{zb} \mu_b.$$

Estimation and inference. We now turn to estimation and inference issues, and in the spirit of the previous developments, we assume that our data consists of a vector $\hat{\mu}_b$, which counts the number of times a coalition of type b is formed. We parametrize the surplus function linearly by a K-dimensional vector λ, so that $\phi_b^\lambda = (\mathbf{\Phi}\lambda)_b$, and, given the previous developments, it should come as no surprise if the proposed estimator $\hat{\lambda}$ of the parameter arose as the solution of

$$\max_{\lambda \in \mathbb{R}^K, U \in \mathbb{R}^{\mathcal{B}}} \left\{ \hat{\mu}^\top \mathbf{\Phi} \lambda - G(U) \right\} \tag{8.3.5}$$

$$\text{s.t.} \sum_{z \in [Z]} \Sigma_{zb} U_{zb} = (\mathbf{\Phi}\lambda)_b.$$

Denoting μ_b the Lagrange multiplier of the constraint, the optimality conditions with respect to U yield

$$\frac{\partial G(U)}{\partial U_{zb}} = \Sigma_{zb} \mu_b,$$

which expresses that (μ, U) is a coalitional equilibrium outcome, and the optimality conditions with respect to λ give

$$\hat{\mu}^\top \mathbf{\Phi} = \mu^\top \mathbf{\Phi},$$

which we summarize into:

Theorem 8.3.2. *The solution of problem* (8.3.5) *provides a coalitional equilibrium matching that has the same K $\mathbf{\Phi}$-moments as those of the observed matching $\hat{\mu}$.*

Theorem 8.2.3 readily extends to the general case of coalitional equilibria to provide the consistency and the asymptotic normality of the estimator $\hat{\lambda}$.

8.4 Dynamic Discrete Choice

We now turn to dynamic discrete choice models, and we extend the framework to random utilities that are not necessarily logit. Define $\mathcal{P}_{tx}$ as the distribution of the Y-dimensional random vector of the utility shocks ε_{ity} that an agent i of type x associates to various options y at time t. In chapter 6, starting from section 6.2, we assumed that we were in the logistic framework, or, more precisely, that $\mathcal{P}_{tx}$ was the distribution of i.i.d. Gumbel random vectors for all t and x; however, in the present section, we shall relax that assumption and solely assume that $\mathcal{P}_{tx}$ satisfies assumptions 1.1 and 1.2 for all t and x. Throughout the section, G_{tx} shall stand for $G_{\mathcal{P}_{tx}}$. Just as in chapter 6, we will start with the finite horizon case, before moving to the infinite horizon case.

Finite horizon, computation. We consider the model described in section 6.2, with the only difference that the structure of the random utilities is not necessarily Gumbel. Assume that the policy variable μ is subject to a capacity constraint that can be expressed as $A\mu \leq c$. In this case, the Bellman equation is $u_{tx} = G_{tx}((\phi - A^\top \tau)_{txy} + \sum_{x'} \mathbf{P}_{x',xy} u_{(t+1)x'})_y)$. Recalling the definition of $\mathbf{J}_T$ as the $T \times T$ lower shift matrix with 1s on the subdiagonal and 0s elsewhere, i.e., $(\mathbf{J}_T)_{tt'} = \mathbf{1}\{t = t' + 1\}$, this constraint can be written in a vectorized way as $u_{tx} = G_{tx}((\phi - A^\top \tau + \mathbf{J}_T^\top \otimes \mathbf{P}^\top u)_{txy, y \in [Y]})$, and problem 6.2.5 generalizes into

$$\begin{aligned}
\min_{u \in \mathbb{R}^{TX}, \tau \in \mathbb{R}^L} \quad & \sum_{x \in [X]} N_x u_x^1 + \sum_{l \in [L]} c_l \tau_l \\
\text{s.t.} \quad & G_{tx}\left(\left(\phi - A^\top \tau + \left(\mathbf{J}_T^\top \otimes \mathbf{P}^\top - \mathbf{I}_T \otimes \mathbf{\Sigma}_Y^\top\right) u\right)_{txy, y \in [Y]}\right) \\
& = 0 \ \forall tx \in [T \times X],
\end{aligned}$$

where we recall the usual definition $\mathbf{\Sigma}_Y = \mathbf{I}_X^\top \otimes \mathbf{1}_Y$.

Finite horizon, estimation. We now turn to the estimation problem, relaxing the framework of section 6.3. For concision we shall assume away any capacity constraint, and we shall assume the usual linear parametrization of the surplus function $\phi = \Phi\lambda$, where λ is a K-dimensional parameter vector to be estimated. A natural generalization of problem (6.3.2) leads to

$$\begin{aligned}
\max_{u \in \mathbb{R}^{TX}, \lambda \in \mathbb{R}^K} \quad & \hat{\mu}^\top \left(\Phi\lambda + \left(\mathbf{J}_T^\top \otimes \mathbf{P}^\top - \mathbf{I}_T \otimes \mathbf{\Sigma}_Y^\top\right) u\right) \\
\text{s.t.} \quad & G_{tx}\left(\left(\Phi\lambda + \left(\mathbf{J}_T^\top \otimes \mathbf{P}^\top - \mathbf{I}_T \otimes \mathbf{\Sigma}_Y^\top\right) u\right)_{txy, y \in [Y]}\right) \\
& = 0 \ \forall tx \in [T \times X].
\end{aligned} \tag{8.4.1}$$

The optimality conditions have natural interpretations. Let n_{tx}^* be the Lagrange multiplier of the constraint, let $\pi_{txy}^* = \partial G_{tx} / \partial U_y$, and let $\mu_{txy}^* = n_{tx}^* \pi_{txy}^*$ be the predicted policy variable. We have by optimality with respect to λ that

$$\hat{\mu}^\top \Phi = \mu^{*\top} \Phi,$$

which has the usual interpretation that the predicted moments of the K basis functions of Φ match the observed ones, and the optimality conditions with respect to u yield

$$\left(\mathbf{J}_T \otimes \mathbf{P} - \mathbf{I}_T \otimes \mathbf{\Sigma}_Y\right) \hat{\mu} = \left(\mathbf{J}_T \otimes \mathbf{P} - \mathbf{I}_T \otimes \mathbf{\Sigma}_Y\right) \mu^*,$$

which has the same interpretation as the one given in remark 6.3.1 in the logit case: at the initial time ($t = 1$), $\sum_y \mu_{1xy}^*$, which is the number of units in state

x implied by the predicted policy variable μ^*, matches the observed one N_x; and in each later state ($t \geq 2$), the predicted change in the number of units matches with the observed one.

Finite horizon, simulation. We can simulate this model and estimate it using linear programming by

$$\max_{\lambda_k, u_{tx}, \tilde{u}_{itx}} \left\{ \hat{\mu}^\top \left(\Phi\lambda + \left(\mathbf{J}_T^\top \otimes \mathbf{P}^\top - \mathbf{I}_T \otimes \boldsymbol{\Sigma}_Y^\top \right) u \right) \right\} \tag{8.4.2}$$

$$s.t. \quad u_{tx} = \frac{1}{I} \sum_{i \in [I]} \tilde{u}_{itx} \; \forall x \in [T \times X],$$

$$\tilde{u}_{itx} \geq \left(\Phi\lambda + \mathbf{P}^\top u_{t+1} \right)_{xy} + \varepsilon_{itxy} \; \forall itxy \in [I \times T \times X \times Y],$$

which leads to the following vectorized formulation:

$$\max_{\lambda \in \mathbb{R}^K, u \in \mathbb{R}^{TX}, \tilde{u} \in \mathbb{R}^{ITX}} \hat{\mu}^\top \Phi\lambda + \hat{\mu}^\top \left(\mathbf{J}_T^\top \otimes \mathbf{P}^\top - \mathbf{I}_T \otimes \boldsymbol{\Sigma}_Y^\top \right) u$$

$$s.t.$$

$$\begin{array}{rcccccl} \mathbf{0}_{TX \times K}\lambda & + & \mathbf{I}_{TX} u & - & \frac{1}{I}\left(\mathbf{1}_I^\top \otimes \mathbf{I}_{TX}\right) \tilde{u} & = & \mathbf{0}_{TX}, \\ -\left(\mathbf{1}_I \otimes \Phi\right)\lambda & - & \left(\mathbf{1}_I \otimes \mathbf{J}_T^\top \otimes \mathbf{P}^\top\right) u & + & \left(\mathbf{I}_{ITX} \otimes \mathbf{1}_Y\right) \tilde{u} & \geq & \varepsilon. \end{array}$$

This approach is implemented in Python in section B.8.3 in the appendix, where a finite horizon dynamic discrete choice model with probit heterogeneity is estimated.

Infinite horizon, estimation. We now turn to the model with infinite horizon described in section 6.6, again with the only difference that we are no longer in the framework of logit random utilities. Based on data consisting of $\hat{\mu}_{xy}$, the number of times decision y was made for a unit in state x, we seek to estimate the surplus ϕ_{xy} based on a parametric family $\phi_{xy}^\lambda = \sum_k \phi_{xyk}\lambda_k$, or $\phi^\lambda = \Phi\lambda$ where Φ is the matrix of size $XY \times K$ of term ϕ_{xyk}. The stationary Bellman equation is

$$u_x = G_x((\phi + \beta\mathbf{P}^\top u)_{xy, y \in [Y]}), \tag{8.4.3}$$

which can be rewritten in a vectorized way as $G_x((\phi + (\beta\mathbf{P}^\top - \Sigma_Y^\top)u)_{xy, y \in [Y]}) = 0$, and the natural generalization of problem (6.6.5) is

$$\max_{u \in \mathbb{R}^X, \lambda \in \mathbb{R}^K} \hat{\mu}^\top \left(\Phi\lambda + \left(\beta\mathbf{P}^\top - \boldsymbol{\Sigma}_Y^\top \right) u \right) \tag{8.4.4}$$

$$s.t. \quad G_x\left(\left(\Phi\lambda + \left(\beta\mathbf{P}^\top - \boldsymbol{\Sigma}_Y^\top \right) u \right)_{xy, y \in [Y]} \right) = 0 \; \forall x \in [X]. \tag{8.4.5}$$

If we let (n_x^*) be the vector of Lagrange multipliers associated with the constraints, $\pi_{xy}^* = \partial G_x / \partial U_y$ (evaluated at the optimal value of u and λ) be the vector of conditional choice probabilities, and $\mu_{xy}^* = n_x^* \pi_{xy}^*$ be the predicted policy variable, the optimality conditions with respect to λ in problem (8.4.4) lead to

$$\hat{\mu}^\top \Phi = \mu^{*\top} \Phi,$$

which has the usual interpretation that the predicted moments of the K basis functions of Φ match the observed ones, and the optimality conditions with respect to u yield

$$(\beta \mathbf{P} - \mathbf{\Sigma}_Y)\, \hat{\mu} = (\beta \mathbf{P} - \mathbf{\Sigma}_Y)\, \mu^*,$$

which again means that the predicted policy variable μ^* assigns the same change of mass in units in each state as the one observed in the sample.

Infinite horizon, computation. Again, we can simulate the random utilities ε_y to approximate problem (8.4.4). Consider the linear programming problem

$$\begin{aligned} \max_{\lambda_k, u_x, \tilde{u}_{ix}} \quad & \left\{ \hat{\mu}^\top \left(\Phi\lambda + \left(\beta \mathbf{P}^\top - \mathbf{\Sigma}_Y^\top \right) u \right) \right\} \\ s.t. \quad & u_x = \frac{1}{I} \sum_{i \in [I]} \tilde{u}_{ix} \; \forall x \in [X], \\ & \tilde{u}_{ix} \geq \left(\Phi\lambda + \beta \mathbf{P}^\top u \right)_{xy} + \varepsilon_{ixy} \; \forall ixy \in [I \times X \times Y], \end{aligned} \tag{8.4.6}$$

which vectorizes into

$$\max_{\lambda \in \mathbb{R}^K, u \in \mathbb{R}^X, \tilde{u} \in \mathbb{R}^{IX}} \quad \hat{\mu}^\top \Phi \lambda + \left(\beta \hat{\mu}^\top \mathbf{P}^\top - \hat{\mu}^\top \mathbf{\Sigma}_Y^\top \right) u$$

$$s.t.$$

$$\begin{array}{rcrcrl} \mathbf{0}_{X \times K} \lambda & + & \mathbf{I}_X u & - & \frac{1}{I} \left(\mathbf{1}_I^\top \otimes \mathbf{I}_X \right) \tilde{u} & = \mathbf{0}_X, \\ -\left(\mathbf{1}_I \otimes \Phi \right) \lambda & - & \beta \left(\mathbf{1}_I \otimes \mathbf{P}^\top \right) u & + & \left(\mathbf{I}_{IX} \otimes \mathbf{1}_Y \right) \tilde{u} & \geq \varepsilon. \end{array}$$

This approach is implemented in Python in section B.8.4 in the appendix, where an infinite horizon dynamic discrete choice model with probit heterogeneity is estimated.

CCP inversion approach. Last, we show that the approach to dynamic discrete choice inversion based on the inversion of conditional choice probability systems, as described in section 6.8 for the logit case, also extends well beyond that framework. As in section 6.8, we are in the infinite-horizon setting, where we seek to estimate the per-period surplus ϕ. To this end, we take a linear parametrization $\phi^{\lambda} = \Phi\lambda$, where, as before, Φ is an $XY \times K$ matrix whose (xy, k) entry is ϕ_{xyk}. Equation (8.4.3) then inverts into

$$\left(\nabla G_x^* (\pi)\right)_{xy\in[X\times Y]} = \Phi\lambda + (\beta \mathbf{P}^\top - \mathbf{\Sigma}_Y^\top) u, \tag{8.4.7}$$

which is the generalization of equation (6.8.1), and given $\mathbf{W}$ a symmetric, positive definite square matrix of weights of size XY, we may mimic what we did in section 6.8 and estimate (λ, u) by generalized least squares using the following program:

$$\min_{\lambda,u} |(\nabla G_x^* (\hat{\pi}))_{xy\in[X\times Y]} - \Phi\lambda - (\beta \mathbf{P}^\top - \mathbf{\Sigma}_Y^\top) u|_{\mathbf{W}}, \tag{8.4.8}$$

where we recall $|z|^2_{\mathbf{W}} = z^\top \mathbf{W} z$.

8.5 Exercises

Exercise 8.1. *Consider the regret function $R(\lambda)$ defined as in* (8.1.5), *and recall that the proposed estimator $\hat{\lambda}$ minimizes that function. Show that $R(\hat{\lambda}) \geq 0$. What can be said when $R(\hat{\lambda}) = 0$?*

Exercise 8.2. *What becomes of the model of section* (8.2) *when $\sum_x n_x = \sum_y m_y$ and when agents are forced to match?*

Exercise 8.3. *Show that the estimation problems in the matching models of section 8.2, the coalition formation models of section 8.3, and the dynamic discrete choice models of section 8.4 can be reformulated as a minimax regret procedure in the spirit of the estimation of one-sided choice problems in section 8.1.*

Exercise 8.4. *Work out explicitly the updating steps of algorithm 8.2.1 when the random utilities on both sides of the market are distributed according to a nested logit distribution.*

Exercise 8.5. *Extend the coalition formation model in section 8.3 to incorporate a dynamic component in the spirit of section 6.9.*

Exercise 8.6. *Implement an augmented Lagrangian algorithm for problem* (8.4.1) *when the random utilities follow a nested logit distribution.*

Exercise 8.7. *Making use of lemma 2.3.1, show that the nested logit model (with one nest) can be recast as a (two-period) finite horizon dynamic discrete choice model.*

8.6 Problems

Problem 8.1. *Take the same data and the same model as in problem 3.1, but with the difference that* ε_{iy} *are now supposed to be drawn from i.i.d.* $\mathcal{N}(0,1)$ *random variables. You may use the GHK simulator of appendix B.4.3.*

(i) Compute the moment matching estimator of λ.

(ii) Compute the maximum likelihood estimator of λ.

(iii) Comment on the results.

8.7 References and Notes

The main reference for sections 8.1 and 8.2 is Galichon and Salanié (2022). Multivariate generalized linear models are discussed in Agresti (2012), section 6.1.2, p. 204, and chapter 3 of Fahrmeir, Tutz, Hennevogl, and Salem (1994). A good reference for minimax regret learning is Bubeck and Cesa-Bianchi (2012). The CCP inversion approach in matching models was proposed in Galichon and Salanié (2024). In section 8.3 the material on estimation and simulation in section 8.4 is original; the part that relies on matching estimation via CCP inversion is based on Chiong, Galichon, and Shum (2016). The paper by Mauerer and Tutz (2023) handles heterogeneity in general discrete choice models.

8.8 Summary

SUMMARY OF CHAPTER 8

Chapter 8 extends the models of choice, allocation, and equilibrium from logistic frameworks to more general structures of random utilities.

Section 8.1 introduces a generalized one-sided choice model. In this model, a decision-maker of type x associates a random utility $U_{xy} + \varepsilon_y$ to option y, where ε follows a distribution $\mathcal{P}_x$. The concepts of welfare function $G_x(U)$ and entropy of choice $G_x^*(\pi)$ are recalled. The generalization of logistic regression leads to a multivariate generalized linear model, providing a broader framework for choice modeling beyond logistic regression.

Section 8.2 explores one-to-one matching models with general heterogeneity. It discusses equilibrium conditions using welfare functions $G(U)$ and $H(V)$ for two-sided markets. The equilibrium matching is characterized through a convex optimization problem, whose primal formulation provides the equilibrium matching, while its dual formulation predicts the sharing rules for the joint surplus at equilibrium. Detailed computational methods such as gradient descent and a formal generalization of the iterative proportional fitting procedure (IPFP) for finding equilibria are investigated.

Section 8.3 addresses coalition formation models where agents of different types form coalitions that generalize matching models, going beyond the logistic framework. The utility for an individual in a coalition includes a random component with a general distribution. The chapter defines coalitional equilibrium and provides a characterization via convex optimization. Estimation methods ensure consistency and asymptotic normality of the estimator for the surplus function's parameters.

Section 8.4 extends dynamic discrete choice models to accommodate general random utility structures. For finite horizons, it formulates the Bellman equation and describes estimation through linear programming. For infinite horizons, it provides an estimation procedure and simulation methods. The conditional choice probability inversion approach generalizes the estimation method by using conditional choice probabilities, offering a robust framework for non-logit models of dynamic discrete choice.

APPENDIX A

Mathematical Toolbox

A.1 Optimization

A.1.1 CONVEX ANALYSIS

Basic notions. We define a function $f : \mathbb{R}^n \to \mathbb{R} \cup \{+\infty\}$ to be convex if (i) f is not identically equal to $+\infty$; (ii) f is *lower semicontinuous*, meaning that its sublevel sets $\{x \in \mathbb{R}^n : f(x) \leq t\}$ are closed for all $t \in \mathbb{R}$; and (iii) the inequality

$$\lambda f(x) + (1-\lambda) f(y) \geq f(\lambda x + (1-\lambda) y) \tag{A.1.1}$$

holds for all $\lambda \in [0,1]$ and for all $x, y \in \mathbb{R}^n$. A function f is *concave* if $-f$ is convex. We define f to be *strictly convex* if, in addition, inequality (A.1.1) holds strictly whenever $\lambda \in (0,1)$ and $x \neq y$. The domain of a convex function f, denoted by *dom* (f), is the set of $x \in \mathbb{R}^n$ such that $f(x) \in \mathbb{R}$. It is easy to see that the domain of a convex function is a convex set.

Compared with the usual definition of convex functions, we impose the additional requirements (i) of non-empty domain and (ii) of lower semicontinuity, which will allow us to write shorter statements. Our notion of convexity is sometimes called *proper convexity* by some authors. These requirements will be met in any practical situations we'll encounter, and are not very demanding, as seen in the following property.

Proposition A.1.1. *If $f : \mathbb{R}^n \to \mathbb{R} \cup \{+\infty\}$ satisfies (A.1.1), then it is continuous on the interior of its domain.*

We also have the following immediate property:

Proposition A.1.2. *If $f : \mathbb{R}^n \to \mathbb{R} \cup \{+\infty\}$ is a convex function and if the set of its minimizers is non-empty, then that set is a convex set.*

The following straightforward characterization of convexity is quite useful:

Proposition A.1.3. *If f is twice differentiable, then f is convex if and only if $dom\left(f\right)$ is convex and $D^2 f\,(x) \succeq_{\text{psd}} 0$, $\forall x \in dom\left(f\right)$.*

Strict convexity has a similar characterization by replacing positive semidefinite by positive definite.

Legendre–Fenchel transform. We now introduce the first important concept in this section, the *Legendre–Fenchel transform.*

Definition A.1.1 (Legendre–Fenchel transform). *Given a function $f: \mathbb{R}^n \to \mathbb{R} \cup \{+\infty\}$ not identically equal to $+\infty$, one may define its Legendre–Fenchel transform, also known as the* convex conjugate *of f, as*

$$f^*\left(y\right) = \sup_{x \in \mathbb{R}^n} \left\{x^\top y - f\,(x)\right\}. \tag{A.1.2}$$

Note that the Legendre–Fenchel transform applies to a function f that is not necessarily convex. It is easy to verify that the Legendre–Fenchel transform f^* is always a convex function unless it is identically equal to $+\infty$. It follows from the definition that:

Proposition A.1.4. *For any x and y in $\mathbb{R}^n$, one has $f(x) + f^*(y) \geq x^\top y$.*

We now provide two useful properties. The first one says that the Legendre–Fenchel transform reverses the order between functions. The second one investigates how the Legendre–Fenchel transform responds under a location-scale transform. We state:

Proposition A.1.5. *$f(x) \leq g(x)$ for all x implies $f^*(y) \geq g^*(y)$ for all y.*

Proposition A.1.6. *If $g\,(x) = \lambda f\,(x) + \gamma$ for $f: \mathbb{R}^n \to \mathbb{R}$, $\lambda > 0$ and $\gamma \in \mathbb{R}$, then $g^*\left(y\right) = \lambda f^*\left(y/\lambda\right) - \gamma$.*

We have $f^{**} \leq f$, with equality if and only if f is convex. The following claim relates the strict convexity of f and the smoothness of f^*.

Proposition A.1.7. *Given a convex function $f: \mathbb{R}^n \to \mathbb{R} \cup \{+\infty\}$:*

(i) if f is strictly convex, then f^* is differentiable on the interior of its domain;

(ii) *if f is differentiable on the interior of its domain, and if $\|\nabla f(x)\| \to +\infty$ as x tends to a point on the frontier of the domain of f, then f^* is strictly convex on its domain.*

Coercivity. Convexity alone is not enough to yield the existence of a minimizer, even when the function is bounded below. For example, $f(x) = \exp(x)$ is convex and bounded below by 0, yet has no minimizer on $\mathbb{R}$. In order to guarantee the existence of a minimizer, we shall need an additional assumption called "coercivity," which rules out the existence of a minimizing sequence whose norm tends to infinity. Precisely, we state:

Definition A.1.2. *Let $f : \mathbb{R}^n \to \mathbb{R}$ be a convex function. Then f is* coercive *if $f(x) \to +\infty$ as $\|x\| \to +\infty$.*

Coercivity ensures the existence of a minimizer for a convex function:

Proposition A.1.8. *Consider $f : \mathbb{R}^n \to \mathbb{R}$, a convex and coercive function. Then f has a minimizer on $\mathbb{R}^n$.*

Subdifferential. We next move on to the second main concept of this section, the *subdifferential* of a function.

Definition A.1.3 (Subdifferential). *Given a function $f : \mathbb{R}^n \to \mathbb{R} \cup \{+\infty\}$ not identically equal to $+\infty$, one defines the subdifferential of f as*

$$\partial f(x) = \left\{ y \in \mathbb{R}^n : f(z) \geq f(x) + y^\top (z - x)\ \forall z \in \mathbb{R}^n \right\}.$$

The subdifferential is therefore a *correspondence* of $\mathbb{R}^n$, which is a function from $\mathbb{R}^n$ to the power set of $\mathbb{R}^n$. Note that the subdifferential can be defined over more general functions than just the convex ones; but the notion will be particularly useful for convex functions. Indeed, the set $\partial f(x)$ may in general be empty, for instance, in the case when $f(x) = -|x|^2$. However, when f is convex, the set $\partial f(x)$ is never empty, and we have an alternative formula for the subdifferential:

Proposition A.1.9. *If f is convex, one has*

$$\partial f(x) = \arg\max_{y \in \mathbb{R}^n} \left\{ x^\top y - f^*(y) \right\} = \left\{ y \in \mathbb{R}^n : f(x) + f^*(y) = x^\top y \right\}. \quad \text{(A.1.3)}$$

This implies:

Proposition A.1.10. *Consider a convex function $f : \mathbb{R}^n \to \mathbb{R} \cup \{+\infty\}$. Then the following three statements are equivalent:*

(i) $y \in \partial f(x)$; (ii) $x \in \partial f^*(y)$; and (iii) $f(x) + f^*(y) = x^\top y$.

Gradient of a convex function. The following property relates the notion of subdifferential to classical derivation.

Proposition A.1.11. *A convex function f is differentiable at $x \in dom(f)$ if and only if $\partial f(x)$ has a single element. In that case, $\partial f(x) = \{\nabla f(x)\}$.*

Under mild requirements, a differentiable convex function is continuously differentiable:

Proposition A.1.12. *Let $f : \mathbb{R}^n \to \mathbb{R} \cup \{+\infty\}$ be convex and differentiable on an open set X. Then f is C^1 on X.*

If f is strictly convex and C^1, then f^* is C^1, and thus proposition A.1.10 implies:

Proposition A.1.13. *If f is strictly convex and C^1, then*

$$\nabla f^* = (\nabla f)^{-1}, \tag{A.1.4}$$

where the inverse is understood in the sense of the composition of functions.

Hessian of a convex function. If f and f^* are twice differentiable, then one may derive relation (A.1.4) to obtain:

Proposition A.1.14. *Let $f : \mathbb{R}^n \to \mathbb{R}$ be a convex function. If f and f^* are twice differentiable respectively at x and $y = \nabla f(x)$, then*

$$D^2 f^*(y) = (D^2 f)^{-1}(x), \tag{A.1.5}$$

where the inverse is understood in the sense of matrix algebra.

Differentiability of a convex function. Finally, note that convex functions are Lebesque differentiable almost everywhere.

Proposition A.1.15. *Let $f : \mathbb{R}^n \to \mathbb{R}$ be a convex function. Then:*

(i) the set $\{x \in \mathbb{R}^n : f \text{ is not differentiable at } x\}$ is of zero Lebesgue measure,
(ii) the set $\{x \in \mathbb{R}^n : f \text{ is not twice differentiable at } x\}$ is of zero Lebesgue measure.

The first claim is a consequence of an important mathematical result called Rademacher's theorem. The second one (which implies the first one) is a consequence of another landmark result called Alexandrov's theorem.

Both these results are significantly more general than the finite-dimensional statements given here.

References

Hiriart-Urruty and Lemaréchal (2001) is a gentle introduction to the topic, while Rockafellar and Wets (1998) and Ekeland and Temam (1976) are more comprehensive presentations.

A.1.2 OPTIMIZATION BY GRADIENT DESCENT

Consider the following minimization problem:

$$\min_{x\in\mathbb{R}^d} f(x), \tag{A.1.6}$$

where f is a convex function. In this appendix, we provide conditions on f that imply the convergence of the gradient descent algorithm, defined by an initial point $x^0 \in \mathbb{R}^d$ and

$$x^{t+1} = x^t - \alpha \nabla f(x^t), \tag{A.1.7}$$

where $\alpha > 0$ is a scalar that does not depend on time, and that is called the *learning rate*. Equivalently, we could consider the maximization of a concave function; the results of this section apply after a simple change of the sign of f, so the gradient descent for maximizing a concave function f is changed into $x^{t+1} = x^t + \alpha \nabla f(x^t)$.

It is easy to see that the convexity of f alone does not suffice to guarantee the convergence of (A.1.7) for any value of x_0 and α, as exemplified in dimension 1 by taking $f(x) = x^2/2$ and $\alpha = 2$. What we need in fact is to ensure that f is smooth enough, in the sense that it should not change direction too fast. In dimension 1, one would impose an upper bound β on the second derivative of f. More generally, this intuition is formalized by the notion of β-smoothness, which we define as:

Definition A.1.4 (β-smooth convex or concave function). *A convex function f is* β-smooth *for $\beta > 0$ if for all $x, x' \in dom(f)$, and $t \in [0,1]$, one has*

$$f\left(tx + (1-t)x'\right) \geq tf(x) + (1-t)f\left(x'\right) - \frac{\beta t(1-t)}{2}|x - x'|^2.$$

A concave function f is β-smooth if $-f$ is β-smooth in the sense of convex functions defined above.

Note that a (convex or concave) β-smooth function f is always continuously differentiable, and its gradient is β-Lipschitz. When the function is C^2, we have a very simple characterization of β-smoothness:

Proposition A.1.16. *When a (convex or concave) function f is C^2, it is β-smooth if and only if for all $x \in dom\left(f\right)$, the largest absolute value of the eigenvalues of $D^2 f\left(x\right)$ is less than or equal to β.*

We have the following convergence result for the gradient descent of β-smooth functions:

Theorem A.1.1. *Let $f : \mathbb{R}^n \to \mathbb{R}$ be a C^1, convex, β-smooth function. Assume that f has a minimizer x^* on $\mathbb{R}^n$. Consider the sequence x^t obtained by the gradient descent with constant learning rate $\alpha = \beta^{-1}$. Then, for all $t \geq 1$, one has*

$$\left|\nabla f\left(x^t\right)\right| \leq \frac{\beta^2 \left|x^0 - x^*\right|}{\sqrt{\left(1+t\right) t}} \text{ and } 0 \leq f\left(x^t\right) - f\left(x^*\right) \leq \frac{\beta \left|x^0 - x^*\right|^2}{2\left(t+1\right)}.$$

References

References for this section are Luenberger and Ye (2015) and Carlier (2022).

A.1.3 PROXIMAL GRADIENT DESCENT

Proximal gradient descent applies to settings where one seeks to minimize $f\left(x\right) = g\left(x\right) + \varphi\left(x\right)$, where g is smooth but φ is not necessarily smooth. Standard gradient descent would lead to

$$x^{t+1} = x^t - \alpha\left(\nabla g\left(x^t\right) + \nabla\varphi\left(x^t\right)\right),$$

but, instead, we shall apply a semi-implicit scheme, where we replace $\nabla\varphi\left(x^t\right)$ by $\nabla\varphi\left(x^{t+1}\right)$ and we seek to have

$$x^{t+1} = x^t - \alpha\left(\nabla g\left(x^t\right) + \nabla\varphi\left(x^{t+1}\right)\right).$$

We can rewrite this as looking for x^{t+1} such that $x^{t+1} + \alpha\nabla\varphi\left(x^{t+1}\right) = \tilde{x}^{t+1}$, where $\tilde{x}^{t+1} := x^t - \alpha\nabla g\left(x^t\right)$. The expression of x^{t+1} is no longer given explicitly as a function of x^t, but implicitly, under the form of an equation to solve. However, this equation arises as the first-order condition of the convex minimization problem:

$$x^{t+1} = \arg\min_{x \in \mathbb{R}^d} \left\{\frac{1}{2}|x|^2 + \alpha\varphi\left(x\right) - x^\top \tilde{x}^{t+1}\right\},$$

or, equivalently, $x^{t+1} = prox_{\alpha\varphi}\left(\tilde{x}^{t+1}\right)$, where the *proximal operator* associated with function φ is defined as

$$prox_{\varphi}\left(\tilde{x}\right) = \arg\min_{x \in \mathbb{R}^d} \left\{ \frac{1}{2} |x - \tilde{x}|^2 + \varphi\left(x\right) \right\}.$$

The advantage of the semi-implicit scheme just described is that it does not require that φ be smooth. Further, the proximal operator has closed-form expressions in a number of important cases:

Example A.1. The following are commonly used proximal operators:

(a) When $\varphi\left(x\right) = 0$, $prox_{\varphi}\left(\tilde{x}\right) = \tilde{x}$.
(b) When $\varphi\left(x\right) = x^{\top}\Sigma x/2$, $prox_{\varphi}\left(\tilde{x}\right) = \left(I + \Sigma\right)^{-1}\tilde{x}$.
(c) When $\varphi\left(x\right) = \lambda|x|_1$, we have $\left(prox_{\varphi}\left(\tilde{x}\right)\right)_i = S_{\lambda}\left(x_i\right)$, where $S_{\lambda}\left(u\right) := \left(u - \lambda\right)^{+} - \left(u + \lambda\right)^{-}$ is called the *soft thresholding function*.

The proximal gradient algorithm is formally described in algorithm A.1.1:

Algorithm A.1.1 Proximal gradient descent

Input : the smooth part g of the function to be minimized, rules for picking the descent direction $d(g, x)$, the step size $\alpha(g, x, d)$, and the proximal operator $prox_{\alpha\varphi}$
Parameter: the starting point x^0, the tolerance ξ
Set $t = 0$, $x^t = x^0$
repeat
  $d^t = d(g, x^t)$, $\alpha^t = \alpha(g, x^t, d^t)$
  $\tilde{x}^{t+1} := x^t - \alpha^t d^t$
  $x^{t+1} := prox_{\alpha\varphi}(\tilde{x}^{t+1})$
  $t = t + 1$
until $\|x^{t+1} - x^t\| \leq \xi$;
Return : x^t

References

A reference for proximal gradient descent algorithms is Bauschke and Combettes (2017).

A.1.4 MINIMAX THEORY

Duality. Let $\mathcal{X}$ and $\mathcal{Y}$ be two subsets of $\mathbb{R}^n$, and consider $L : \mathcal{X} \times \mathcal{Y} \to \mathbb{R}$. For any $x' \in \mathcal{X}$ and $y' \in \mathcal{Y}$, we have $\inf_{x\in\mathcal{X}} L(x, y') \leq L(x', y') \leq \sup_{y\in\mathcal{Y}} L(x', y)$, thus

$$\varphi(y') := \inf_{x\in\mathcal{X}} L(x, y') \leq \sup_{y\in\mathcal{Y}} L(x', y) =: \psi(x'),$$

thus $\sup \varphi \leq \inf \psi$, and therefore

$$\sup_{y\in\mathcal{Y}} \inf_{x\in\mathcal{X}} L(x, y) \leq \inf_{x\in\mathcal{X}} \sup_{y\in\mathcal{Y}} L(x, y),$$

which leads to:

Lemma A.1.1 (Weak duality). *We have the* weak duality inequality

$$V_D := \sup_{y\in\mathcal{Y}} \inf_{x\in\mathcal{X}} L(x, y) \leq \inf_{x\in\mathcal{X}} \sup_{y\in\mathcal{Y}} L(x, y) =: V_P.$$

The problems V_P and V_D are called the *primal problem* and the *dual problem*, respectively, and the quantity $V_P - V_D \geq 0$ is called the *duality gap*. When the duality gap is zero, one says that *strong duality* holds.

Saddle points. A *saddle point* for L is a pair $(x^*, y^*) \in \mathcal{X} \times \mathcal{Y}$ such that

$$L(x^*, y) \leq L(x^*, y^*) \leq L(x, y^*)$$

holds for all $(x, y) \in \mathcal{X} \times \mathcal{Y}$.

Lemma A.1.2. *The following are equivalent:*

(i) $(x^*, y^*) \in \mathcal{X} \times \mathcal{Y}$ *is a saddle point, and*
(ii) x^* *is optimal for the primal problem,* y^* *is optimal for the dual problem, and there is no duality gap, that is,*

$$x^* \in \arg\min_{x\in\mathcal{X}} \sup_{y\in\mathcal{Y}} L(x, y), \quad y^* \in \arg\max_{y\in\mathcal{Y}} \inf_{x\in\mathcal{X}} L(x, y), \text{ and}$$

$$\sup_{y\in\mathcal{Y}} \inf_{x\in\mathcal{X}} L(x, y) = \inf_{x\in\mathcal{X}} \sup_{y\in\mathcal{Y}} L(x, y).$$

Proof. Let us show (i) implies (ii). Assume (x^*, y^*) is a saddle point. Then we have $\sup_{y\in\mathcal{Y}} L(x^*, y) \leq \inf_{x\in\mathcal{X}} L(x, y^*)$, and therefore

$$\inf_{x\in\mathcal{X}} \sup_{y\in\mathcal{Y}} L(x, y) \leq \sup_{y\in\mathcal{Y}} L(x^*, y) \leq L(x^*, y^*) \leq \inf_{x\in\mathcal{X}} L(x, y^*) \leq \sup_{y\in\mathcal{Y}} \inf_{x\in\mathcal{X}} L(x, y),$$

i.e., $V_P \leq V_D$. However, weak duality implies the converse inequality $V_D \leq V_P$, so all these inequalities hold as equalities, and therefore

$$L\left(x^*, y^*\right) = \sup_{y\in\mathcal{Y}} L\left(x^*, y\right) = \inf_{x\in\mathcal{X}} \sup_{y\in\mathcal{Y}} L\left(x, y\right) = \inf_{x\in\mathcal{X}} L\left(x, y^*\right)$$
$$= \sup_{y\in\mathcal{Y}} \inf_{x\in\mathcal{X}} L\left(x, y\right), \text{ QED.}$$

Conversely, we show (ii) implies (i). Assume $x^* \in \mathcal{X}$ is optimal for the primal, $y^* \in \mathcal{Y}$ is optimal for the dual, and that there is no duality gap. We have

$$\inf_{x\in\mathcal{X}} L\left(x, y^*\right) \leq L\left(x^*, y^*\right) \leq \sup_{y\in\mathcal{Y}} L\left(x^*, y\right).$$

But using the optimality of x^* and y^* respectively for the primal and the dual, one has

$$\sup_{y\in\mathcal{Y}} \inf_{x\in\mathcal{X}} L\left(x, y\right) = \inf_{x\in\mathcal{X}} L\left(x, y^*\right) \leq L\left(x^*, y^*\right) \leq \sup_{y\in\mathcal{Y}} L\left(x^*, y\right) = \inf_{x\in\mathcal{X}} \sup_{y\in\mathcal{Y}} L\left(x, y\right),$$

but by the absence of a duality gap, the upper bound and the lower bound coincide, hence for any $(x, y) \in \mathcal{X} \times \mathcal{Y}$, we get that

$$L\left(x^*, y\right) \leq \sup_{y\in\mathcal{Y}} L\left(x^*, y\right) = L\left(x^*, y^*\right) = \inf_{x\in\mathcal{X}} L\left(x, y^*\right) \leq L\left(x, y^*\right)$$

and therefore (x^*, y^*) is a saddle point. ■

References

A standard reference for minimax problems in the convex-concave case is Rockafellar and Wets (1998).

A.1.5 LINEAR PROGRAMMING

A linear programming problem consists in optimizing (maximizing or minimizing) a linear objective function under constraints that are all linear. Any nontrivial linear programming problem can be reformulated as a problem *in standard form*, which is

$$V_P = \max_{x\in\mathbb{R}^n_+} c^\top x \tag{A.1.8}$$
$$\text{s.t. } Ax = d,$$

where A is an $I \times J$ matrix of full rank such that $I \leq J$. Thus I is the number of constraints and J is the number of variables. Linear programming solvers, such as Gurobi, solve programs such as (A.1.8) with tens of thousands of variables and constraints within seconds on a standard laptop.

Feasible and optimal solutions. Problem (A.1.8) is called the *primal program*. If $x \in \mathbb{R}^J$ satisfies the constraints $x \geq 0$ and $Ax = d$, it is called a *feasible solution*; if it is a solution to problem (A.1.8), it is called an *optimal solution*. Note that the set of feasible solutions may be empty; if $A = 1$ and $d = -1$, for instance, then there is no feasible solution. In this case, because a maximum over an empty set should have a smaller value than any finite number, we assign value $V_P = -\infty$ to problem (A.1.8). Even when the set of feasible solutions is non-empty, the set of optimal solutions may still be empty because the objective function may not be bounded above on the feasible set; this will be the case, for example, when $A = 0$ and $d = 1$ and $c = 1$, in which case the value of problem (A.1.8) is $V_P = +\infty$.

Duality. Duality is a powerful tool which provides insight into the analysis of problem (A.1.8). One can rewrite the primal problem as

$$V_P = \max_{(x_j) \geq 0} \min_{(y_i)} \left\{ c^\top x + (d - Ax)^\top y \right\}.$$

Indeed, the solution to the inner minimization problem will be $-\infty$ as soon as one of the primal constraints is not satisfied. By the minimax inequality,

$$\begin{aligned} V_P \leq V_D &:= \min_{(y_i)} \max_{(x_j) \geq 0} \left\{ c^\top x + (d - Ax)^\top y \right\} \\ &= \min_{(y_i)} \left\{ d^\top y + \max_{(x_j) \geq 0} \left\{ x^\top (c - A^\top y) \right\} \right\}, \end{aligned}$$

and we see that $\max_{(x_j) \geq 0} \left\{ x^\top w \right\}$ is equal to 0 if all the w_j are nonpositive, and to $+\infty$ if some w_j is positive. Thus V_D is the value of the *dual program*, defined by

$$\begin{aligned} V_D := \min_{(y_i)} \ & d^\top y \\ \text{s.t. } & A^\top y \geq c, \end{aligned} \tag{A.1.9}$$

and we call y a *feasible dual solution* if it satisfies the constraints of (A.1.9), and an *optimal dual solution* if it is a solution to (A.1.9).

Duality gap. We can characterize the cases when the *duality gap*, defined as the difference $V_D - V_P$ between the values of the dual and the primal problem, is not zero. As we shall see, this will depend upon the existence of primal or dual feasible solutions. The linear programming problem is called primal (respectively dual) feasible if its primal (respectively dual) problem has feasible solutions. We have:

Proposition A.1.17. *The following are the only possible cases regarding the values of the primal and the dual:*

- $V_P = V_D = +\infty$*: the primal is unbounded and the dual is unfeasible;*
- $V_P = V_D \in \mathbb{R}$*: both the primal and the dual are feasible and bounded;*
- $V_P = V_D = -\infty$*: the primal is unfeasible and the dual is unbounded;*
- $V_P = -\infty < V_D = +\infty$*: both the primal and the dual are unfeasible.*

Note that the case when both the primal and dual programs are unfeasible is the only case when strong duality does not hold. This can arise, for instance, if

$$A = \begin{pmatrix} -1 & 1 \\ 1 & -1 \end{pmatrix}, \quad c = \begin{pmatrix} 0 \\ 1 \end{pmatrix}, \quad d = \begin{pmatrix} -1 \\ 2 \end{pmatrix}.$$

Complementary slackness. If $(x_j^\star) \geq 0$ is an optimal primal solution, and $(y_i^\star)$ is an optimal dual solution, then complementary slackness holds, that is, $x_j^\star > 0$ implies $(A^\top y^\star)_j = c_j$. We summarize these results into the following statement.

Theorem A.1.2 (Finite-dimensional linear programming). *In the setting described above:*

(i) *The weak duality inequality holds:* $V_P \leq V_D$.

(ii) *As soon as the primal or the dual program has an optimal solution, both programs have an optimal solution, and strong duality holds:* $V_P = V_D$.

(iii) *If* $(x_j^\star) \geq 0$ *is an optimal primal solution, and* $(y_i^\star)$ *is an optimal dual solution, then the complementary slackness conditions hold:*

$$x_j^\star > 0 \text{ implies } (A^\top y^\star)_j = c_j. \tag{A.1.10}$$

(iv) *Conversely, if* $(x_j^\star) \geq 0$ *is a feasible primal solution, and* $(y_i^\star)$ *is a feasible dual solution such that the complementary slackness conditions (A.1.10) hold, then* $(x_j^\star) \geq 0$ *is an optimal primal solution, and* $(y_i^\star)$ *is an optimal dual solution.*

References

A good reference for linear programming, which goes well beyond the present summary, is Bertsimas and Tsitsiklis (1997). See a nice introduction aimed at economists in Vohra (2005), chapter 4. Gurobi (2021) optimization is a state-of-the-art large scale linear optimization engine. It is a commercial product but is provided freely to the academic community. A good introductory source for linear programming is Luenberger and Ye (2015).

A.1.6 NONLINEAR PROGRAMMING

Consider a problem of the form

$$V_P := \min_{x \in \mathbb{R}^{\mathcal{I}}} f(x) \quad \text{s.t. } g_j(x) \leq 0,\ \forall j \in \mathcal{J}, \tag{A.1.11}$$

where f and the g_j are convex functions, and hence the set over which one is optimizing is a convex set. Problem (A.1.11) is then called a "convex" program.

Theorem A.1.3 (Weak duality). *One has $V_P \geq V_D$, where*

$$V_D := \sup_{y \in \mathbb{R}_+^{\mathcal{J}}} \varphi(y), \tag{A.1.12}$$

where

$$\varphi(y) := \inf_{x \in \mathbb{R}^{\mathcal{I}}} \left\{ f(x) + y^\top g(x) \right\} \tag{A.1.13}$$

is a concave function of y.

Karush–Kuhn–Tucker conditions. The following result provides necessary and sufficient conditions to express that (x^*, y^*) is a saddle point for L, the inner function in (A.1.13) defined by $L(x, y) = f(x) + y^\top g(x)$:

Theorem A.1.4. *Assume the problem (A.1.11) is convex. Then, given $(x^*, y^*) \in \mathbb{R}^{\mathcal{I}} \times \mathbb{R}^{\mathcal{J}}$, the following two statements are equivalent:*

(i) *(x^*, y^*) is a saddle point for L on $\mathbb{R}^{\mathcal{I}} \times \mathbb{R}_+^{\mathcal{J}}$, and*
(ii) *(x^*, y^*) satisfies the Karush–Kuhn–Tucker (KKT) conditions:*
   (a) primal feasibility: $g_j(x^*) \leq 0,\ \forall j \in \mathcal{J}$,
   (b) dual feasibility: $y_j^* \geq 0,\ \forall j \in \mathcal{J}$,
   (c) complementary slackness: $y_j^* > 0 \implies g_j(x^*) = 0,\ \forall j \in \mathcal{J}$,
   (d) stationarity: $0_{\mathcal{I}} \in \partial_x L(x^*, y^*)$.

Slater's theorem. We now introduce Slater's condition, which is a sufficient condition for strong duality. It allows linear constraints to be treated separately from nonlinear constraints, as the requirement imposed on them will be weaker. Therefore, we let $\mathcal{J}_A \subseteq \mathcal{J}$ such that $g_j(x) = a_j^\top x - b_j$ is an affine function for $j \in \mathcal{J}_A$. Hence, we may assume that g_j is not affine for $j \notin \mathcal{J}_A$, and we are able to state Slater's condition:

Assumption A.1 (Slater's condition). *There exists $x^* \in \mathbb{R}^{\mathcal{J}}$ such that*

$$g_j\left(x^*\right) < 0, \ \forall j \notin \mathcal{J}_A.$$

Theorem A.1.5 (Slater's theorem). *If problem (A.1.11) is convex and Slater's condition (A.1) is met, then strong duality holds: the value of the primal problem (A.1.11) coincides with the value of the dual problem (A.1.12).*

References

References include Boyd and Vandenberghe (2004), Luenberger and Ye (2015), and Carlier (2022).

A.1.7 AUGMENTED LAGRANGIAN METHOD

Consider a variant of problem (A.1.11) with equality constraint:

$$V_P := \min_{x \in \mathbb{R}^{\mathcal{I}}} f(x) \qquad \text{s.t. } g(x) = 0, \tag{A.1.14}$$

where $f : \mathbb{R}^{\mathcal{I}} \to \mathbb{R}$ and $g : \mathbb{R}^{\mathcal{I}} \to \mathbb{R}^{\mathcal{J}}$ are C^2 functions. The (classical) Lagrangian for this constrained minimization problem is:

$$L\left(x, y\right) = f(x) + y^\top g(x). \tag{A.1.15}$$

The augmented Lagrangian method, also known as the method of multipliers, adds a penalty term that helps stabilize the convergence of the method when dealing with constraint violations. We define the augmented Lagrangian as

$$L_A\left(x, y, \gamma\right) = f(x) + y^\top g(x) + \frac{\gamma}{2} \|g(x)\|^2, \tag{A.1.16}$$

where $\gamma > 0$ increases the cost of violating the constraint.

The following result is the conceptual basis for the convergence of algorithm A.1.2:

Algorithm A.1.2 Augmented Lagrangian method

Input : Objective function $f(x)$, constraint functions $g(x)$
Parameter: Initial variables (x^0, y^0), increasing sequence $\gamma^t \to +\infty$, tolerance ξ
Output : Approximate primal solution x^*
Set $t = 0$
repeat
 $x^{t+1} = \arg\min_x L_A\left(x, y^t, \gamma^t\right)$
 $y^{t+1} = y^t + \gamma^t g\left(x^{t+1}\right)$
 $t = t + 1$
until $\|g\left(x^t\right)\| \leq \xi$;

Theorem A.1.6. *Assume f and g are continuous functions. Assume (γ^t) is an increasing real-valued sequence that tends to $+\infty$. Let (y^t) be a bounded sequence of elements of $\mathbb{R}^{\mathcal{J}}$, and consider x^t a global minimum of $L\left(x, y^t, \gamma^t\right)$ over $x \in \mathbb{R}^{\mathcal{I}}$. Then every limit point of (x^t) is a solution to* (A.1.14).

References

The augmented Lagrangian method, and, more generally, the Lagrange multiplier method, is the focus of the textbook Bertsekas (2014). The method was invented independently by Powell and Hestenes and appeared in 1969 simultaneously in Powell (1969) and in Hestenes (1969). Our theorem A.1.6 appears in Bertsekas (2014) as proposition 2.1; we refer to that text for the proof and more refined results.

A.1.8 OPTIMAL TRANSPORT

Assume that $\mathcal{X}$ and $\mathcal{Y}$ are closed subsets of $\mathbb{R}^d$ and $\mathbb{R}^{d'}$, respectively. Let n and m be two probability measures of $\mathcal{X}$ and $\mathcal{Y}$, respectively, and let $\mathcal{M}(n, m)$ be the set of joint distributions μ over $\mathcal{X} \times \mathcal{Y}$ such that if $(X, Y) \sim \mu$, then $X \sim n$ and $Y \sim m$. Let $\Phi : \mathcal{X} \times \mathcal{Y} \to \mathbb{R} \cup \{-\infty\}$ and consider the so-called *primal problem*

$$\sup_{\mu \in \mathcal{M}(n,m)} \mathbb{E}_\mu[\Phi(X, Y)], \tag{A.1.17}$$

which is an infinite-dimensional linear programming problem whose dual is expressed as

$$\inf_{u(x)+v(y) \geq \Phi(x,y)} \mathbb{E}_n[u(X)] + \mathbb{E}_m[v(Y)], \tag{A.1.18}$$

where it is understood that the infimum is taken over all n- and m-integrable functions u and v.

Monge–Kantorovich duality. The Monge–Kantorovich theorem provides conditions for the absence of a duality gap (equality between the values of the primal and the dual problems) and for the existence of solutions to the primal and dual problems.

Theorem A.1.7 (Monge–Kantorovich duality theorem). *Assume that Φ is upper semicontinuous and that there are $\overline{a} \in L^1(n)$ and $\overline{b} \in L^1(m)$ such that $\Phi(x, y) \leq \overline{a}(x) + \overline{b}(y)$ for all $x \in \mathcal{X}$ and $y \in \mathcal{Y}$. Then:*

(i) The values of the primal and the dual problems coincide, i.e.,

$$\sup_{\mu \in \mathcal{M}(n,m)} \mathbb{E}_\mu[\Phi(X, Y)] = \inf_{u(X)+v(Y) \geq \Phi(X,Y)} \mathbb{E}_n[u(X)] + \mathbb{E}_m[v(Y)]. \tag{A.1.19}$$

(ii) The primal problem on the left-hand side of (A.1.19) has an optimal solution μ.

(iii) Assume further that there are $\underline{a} \in L^1(n)$ and $\underline{b} \in L^1(m)$ such that $\Phi(x, y) \geq \underline{a}(x) + \underline{b}(y)$ for all $x \in \mathcal{X}$ and $y \in \mathcal{Y}$. Then the dual problem on the right-hand side of (A.1.19) has an optimal solution (u, v) as well.

The proof of this result is given in Villani (2009), Theorem 5.10. The solution μ of the primal problem is sometimes called an optimal transportation plan, optimal coupling, or optimal matching. The solutions u and v of the dual problem are sometimes called Kantorovich potentials.

Discrete setting. In the discrete setting, when $\mathcal{X}$ and $\mathcal{Y}$ are finite sets, n and m are two nonnegative vectors of $\mathbb{R}^{\mathcal{X}}$ and $\mathbb{R}^{\mathcal{Y}}$ respectively such that $\sum_x n_x = \sum_y m_y = 1$, and the set of transport plans is given as

$$\mathcal{M}(n, m) := \left\{ \mu \in \mathbb{R}_+^{\mathcal{X} \times \mathcal{Y}} : \begin{array}{c} \sum_{y \in \mathcal{Y}} \mu_{xy} = n_x, \ \forall x \in \mathcal{X} \\ \sum_{x \in \mathcal{X}} \mu_{xy} = m_y, \ \forall y \in \mathcal{Y} \end{array} \right\}. \tag{A.1.20}$$

The primal problem is then obtained as the maximization of the surplus $\sum_{xy} \mu_{xy} \Phi_{xy}$ over $\mu \in \mathcal{M}(n, m)$, that is,

$$\max_{\mu_{xy} \geq 0} \sum_{xy \in \mathcal{X} \times \mathcal{Y}} \mu_{xy} \Phi_{xy} \tag{A.1.21}$$

$$\text{s.t. } \mu \in \mathcal{M}(n, m),$$

and its dual is

$$\min_{u_x, v_y} \sum_{x \in \mathcal{X}} n_x u_x + \sum_{y \in \mathcal{Y}} m_y v_y \qquad \text{(A.1.22)}$$
$$\text{s.t. } u_x + v_y \geq \Phi_{xy}.$$

Both the primal and the dual problems being feasible, strong duality holds and the value of the program is finite. Complementary slackness holds and ensures that:

Proposition A.1.18 (Complementary slackness). *Let $\mu \geq 0$ be primal feasible and (u, v) be dual feasible. Then the following two assertions are equivalent:*

(i) μ is optimal for the primal problem (A.1.21) and (u, v) is optimal for the dual problem (A.1.22).
(ii) The complementarity relation holds:

$$\forall xy \in \mathcal{X} \times \mathcal{Y},\ \mu_{xy} > 0 \Longrightarrow u_x + v_y = \Phi_{xy}. \qquad \text{(A.1.23)}$$

Entropic regularization. We now discuss a popular variant of the optimal transport problem, namely, optimal transport with entropic regularization. Let us describe it in the discrete case. It simply consists of adding an entropic penalty term to the objective function in the primal problem (A.1.21), which becomes

$$\max_{\mu \in \mathcal{M}(n,m)} \sum_{xy \in \mathcal{X} \times \mathcal{Y}} \mu_{xy} \Phi_{xy} - T \sum_{xy \in \mathcal{X} \times \mathcal{Y}} \mu_{xy} \log \mu_{xy}, \qquad \text{(A.1.24)}$$

where $T > 0$ is a temperature parameter that controls how much we regularize. The case $T = 0$ boils down to the unregularized problem (A.1.21), while setting T large amounts to giving a predominant role to the entropy in the objective function. Calculation shows that the dual to problem (A.1.24) is

$$\min_{u,v} \sum_{x \in \mathcal{X}} n_x u_x + \sum_{y \in \mathcal{Y}} m_y v_y + T \sum_{xy \in \mathcal{X} \times \mathcal{Y}} \exp\left(\frac{\Phi_{xy} - u_x - v_y}{T} \right), \qquad \text{(A.1.25)}$$

whose first-order conditions determine u and v by what is called a *Bernstein–Schrödinger system* of equations:

$$\begin{cases} n_x = \sum_{y \in \mathcal{Y}} \exp\left((\Phi_{xy} - u_x - v_y)/T \right), \\ m_y = \sum_{x \in \mathcal{X}} \exp\left((\Phi_{xy} - u_x - v_y)/T \right). \end{cases} \qquad \text{(A.1.26)}$$

We claim:

Theorem A.1.8. *Consider the Bernstein–Schrödinger system* (A.1.26). *Then:*

(i) *The system has a unique solution* (u, v) *up to a constant—that is, if* (u, v) *and* (u', v') *are solutions, then there is a constant* $c \in \mathbb{R}$ *such that* $u'_x = u_x + c$ *and* $v'_y = v_y - c$.

(ii) *The set of solutions to the system coincides with the set of minimizers of the dual problem* (A.1.25).

(iii) *If* (u, v) *is a solution to the system, then* μ_{xy} *defined by*

$$\mu_{xy} := \exp\left(\frac{\Phi_{xy} - u_x - v_y}{T}\right) \tag{A.1.27}$$

is a solution to the primal regularized problem (A.1.24). *Conversely, every primal solution arises in that way.*

The proof is constructive, and involves the description of the *iterated proportional fitting procedure* (IPFP), also called *Sinkhorn's algorithm.* Let $a_x = \exp(u_x/T)$ and $b_y = \exp(v_y/T)$. Given b^0_y, we construct iterative estimates (a^t, b^t) of the solutions (a, b) as follows:

$$a_x^{t+1} = \frac{1}{n_x} \sum_{y \in \mathcal{Y}} \exp\left(\frac{\Phi_{xy}}{T}\right) \frac{1}{b_y^t}, \tag{A.1.28}$$

and next, in the same loop, we look for v_y^{t+1} to satisfy the second set of equations of system (A.1.26), so that

$$b_y^{t+1} = \frac{1}{m_y} \sum_{x \in \mathcal{X}} \exp\left(\frac{\Phi_{xy}}{T}\right) \frac{1}{a_x^{t+1}}. \tag{A.1.29}$$

We have the following important convergence result, which establishes the convergence of (u^t, v^t), as defined by (A.1.28)–(A.1.29) above, toward a pair $(u^\star, v^\star)$ solution to (A.1.26) using the Hilbert metric defined in (A.3.18) below. Let A be the $|\mathcal{X}| \times |\mathcal{Y}|$ matrix of term $A_{xy} = \exp(\Phi_{xy}/T)/n_x$ and let B be the $|\mathcal{Y}| \times |\mathcal{X}|$ matrix of term $B_{yx} = \exp(\Phi_{xy}/T)/m_y$. Let $K_A = \kappa(A)$ and $K_B = \kappa(B)$, where $\kappa(\cdot)$ is defined in (A.3.20), and let $K = K_A K_B$, which is smaller than 1. We have:

Theorem A.1.9 (Convergence of the IPFP). *The solution* $(u^\star, v^\star)$ *to system* (A.1.26) *exists and is unique up to a constant. Then, letting* $a^\star_x = \exp(u^\star_x/T)$ *and* $b^\star_y = \exp(v^\star_y/T)$, *we have for any* $t \geq 0$, $d_H(a^t, a^\star) \leq K^t d_H(a^0, a^\star)$, *and, similarly,* $d_H(b^t, b^\star) \leq K^t d_H(b^0, b^\star)$.

Proof. Let F and G be the maps such that $F_x(a) = 1/a_x$ and $G_y(b) = 1/b_y$. These preserve the Hilbert metric. Equation (A.1.28) can be rewritten as $a^{t+1} = AG(b^t)$, and equation (A.1.29) can be rewritten as $b^{t+1} = BF(a^{t+1})$. Combined together, these facts imply the result. ∎

References

References for this section include Villani (2009), Santambrogio (2015) for a mathematical treatment, my book Galichon (2016) for economic applications, and Peyré and Cuturi (2019) for applications to data science. The proof of theorem A.1.9 is taken from the paper Franklin and Lorenz (1989).

A.2 Probability and Statistics

A.2.1 GEOMETRY OF THE SIMPLEX

A probability vector $\pi \in \mathbb{R}^{\mathrm{J}}$ can be thought of as an element of the $(\mathrm{J}-1)$-dimensional *standard simplex* $\{\pi \in \mathbb{R}_{+}^{\mathrm{J}} : \sum_j \pi_j = 1\}$, whose J extreme points are $\mathbf{e}_{\mathrm{J}}^{j}$, the elements of the canonical basis of $\mathbb{R}^{\mathrm{J}}$. More generally, a $(\mathrm{J}-1)$-dimensional simplex is the convex hull of J points $(a_1, \ldots, a_{\mathrm{J}})$ in $\mathbb{R}^{\mathrm{J}}$. The simplex is said to be "generated" by the vectors a_j. A one-dimensional simplex $(\mathrm{J} = 2)$ is a segment and a two-dimensional simplex $(\mathrm{J} = 3)$ is a triangle.

Simplical subdivision. Any point a of the simplex $\Delta = conv(a_1, \ldots, a_{\mathrm{J}})$ can be written as a barycenter of the points that generate this simplex. In particular, a point a_0 in the interior of Δ has expression $a_0 = \sum_j t_j a_j$ with $t_j > 0$ and $\sum_j t_j = 1$. For $j \in [\mathrm{J}]$, we define the *simplical subdivision* Δ_j with respect to a_0 by $\Delta_j = conv(\{a_k : k \neq j\} \cup \{a_0\})$. This is illustrated in figure A.1a for $\mathrm{J} = 3$. The following result, whose proof is straightforward, is a simple criterion for deciding in which simplical subdivision a point described in barycentric coordinates belongs. We claim:

Proposition A.2.1. *Let $b \in \Delta$ so that b can be written $b = \sum_j \theta_j a_j$ with $\theta_j \geq 0$ and $\sum_j \theta_j = 1$. Then $b \in \Delta_j$ if and only if $\theta_j t_k \leq t_j \theta_k$ for all $k \in [\mathrm{J}]$.*

The $(\mathrm{J}-1)$-dimensional volume of $\Delta = conv(a_1, \ldots, a_{\mathrm{J}})$ is given by

$$vol(\Delta) = \frac{1}{(\mathrm{J}-1)!} \sqrt{G(\{a_j - a_{\mathrm{J}}, j \in [\mathrm{J}-1]\})}, \tag{A.2.1}$$

where $G(\{v_j, j \in [\mathrm{J}-1]\})$ is the Gram determinant of the vectors (v_j) for $j \in [\mathrm{J}-1]$, which is the determinant of the *Gram matrix* of these vectors, itself

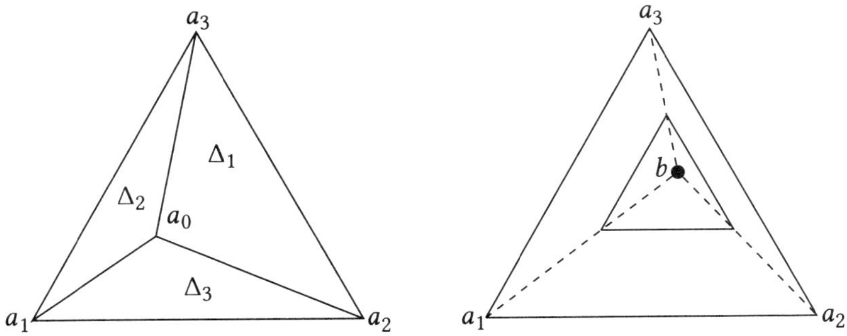

(a) Simplical subdivisions of a simplex. (b) Parallel subsimplex of a simplex.

Figure A.1: Left: Simplical subdivisions. Right: Parallel subsimplex.

defined as the $(\mathrm{J}-1)$ by $(\mathrm{J}-1)$ matrix whose (j,j')th term for $1 \leq j, j' \leq \mathrm{J}-1$ is the scalar product $(v_j)(v_{j'})$. If we substitute $a_0 = \sum_j t_j a_j$ for a_J in (A.2.1), and, more generally, for any a_j, we get an expression for the volume of the simplical subdivision Δ_j, namely:

Proposition A.2.2 (Volume of the simplical subdivision). *We have*

$$vol(\Delta_j) = t_j \; vol(\Delta). \tag{A.2.2}$$

Parallel subsimplex. Consider $\Delta = conv(a_1, \ldots, a_\mathrm{J})$, a $(\mathrm{J}-1)$-dimensional simplex, and for $\tau \in \mathbb{R}^\mathrm{J}$, let the *parallel subsimplex* associated with τ be the subset $\Delta(\tau)$ of Δ made of the vectors a that can be written $a = \sum_j t_j a_j$ for $t_j \geq 0$ such that $t_j \geq \tau_j$ and $\sum_j t_j = 1$.

One has $\Delta(0) = \Delta$. Consider $\tau \in \mathbb{R}^\mathrm{J}_+$. We have:

- If $\sum_j \tau_j > 1$, then $\Delta(\tau)$ is empty.
- If $\sum_j \tau_j = 1$, then $\Delta(\tau) = \{b\}$, where $b = \sum_j \tau_j a_j$.
- If $\sum_j \tau_j < 1$, then we let $k = \sum_j \tau_j$ and define $\theta_j = \tau_j / k$, and $b = \sum_j \theta_j a_j$. We can show that $\Delta(\tau)$ is obtained from Δ by a homothety of ratio $1-k$ and with center b.

We have therefore established:

Proposition A.2.3 (Volume of the parallel subsimplex). *We have*

$$vol\big(\Delta(\tau)\big) = \Big(\Big(1 - \sum_{j \in [\mathrm{J}]} \tau_j\Big)^+\Big)^{\mathrm{J}-1} vol(\Delta). \tag{A.2.3}$$

Dirichlet distribution. We call *flat Dirichlet distribution* of order J the uniform distribution on the $\mathrm{J}-1$ subsimplex $\{a \in \mathbb{R}^{\mathrm{J}} : a_j \geq 0, \sum_j a_j = 1\}$. This distribution is denoted by $\mathcal{D}_{\mathrm{J}}$. The flat Dirichlet distribution on the one-dimensional standard simplex is thus the distribution of (D_1, D_2) given by $D_1 = U$, $D_2 = 1 - U$ for $U \sim \mathcal{U}$, a uniform random variable over $[0,1]$. More generally, it follows from proposition A.2.3 that the Dirichlet distribution of order I is characterized by the fact that if $D \sim \mathcal{D}_{\mathrm{J}}$, then

$$\Pr\left(D_j \geq x_j,\ \forall j \in [\mathrm{J}]\right) = \left(\left(1 - \sum_{j \in [\mathrm{J}]} x_j\right)^+\right)^{\mathrm{J}-1}. \tag{A.2.4}$$

It follows from equation (A.2.4) that if $D \sim \mathcal{D}_{\mathrm{J}}$, then for any $j \in [\mathrm{J}]$, D_j has a c.d.f. at $z \in [0,1]$ equal to $1 - (1-z)^{\mathrm{J}-1}$. As a result, we can set $x_j = b_j/z$, and we have:

Proposition A.2.4. *If $D \sim \mathcal{D}_{\mathrm{J}}$ has a flat Dirichlet distribution of order* J, *then*

$$\max_{j \in [\mathrm{J}]} \frac{b_j}{D_j} \stackrel{d}{=} \frac{\sum_{j \in [\mathrm{J}]} b_j}{D_0}, \tag{A.2.5}$$

where D_0 has the same distribution as each of the components D_j of the random vector D.

Finally, we have:

Proposition A.2.5. *If $X_1, \ldots, X_I$ are i.i.d. exponential random variables, then $D_i = X_i / (\sum_i X_i)$ is flat Dirichlet of order I.*

References

An early reference is the classic paper Dempster (1968). For something more recent, see Jacob, Gong, Edlefsen, and Dempster (2021).

A.2.2 POSITIVE STABLE DISTRIBUTIONS

A probability distribution with support contained in $\mathbb{R}_+$ is *positive (add-) stable* if for $Z_1, \ldots, Z_I$ i.i.d. variables with that distribution, there exists a constant $c_I \in \mathbb{R}_+$ such that one has

$$Z_1 + \cdots + Z_I \stackrel{d}{=} c_I Z, \text{ where } Z \stackrel{d}{=} Z_i\ \forall i \in [I]. \tag{A.2.6}$$

It is clear that (A.2.6) holds if and only if the Laplace transform of the distribution, denoted by $L(t) := \mathbb{E}[\exp(-tZ)]$, satisfies $L^I(t) = L(c_I t)$ for all $t \geq 0$ and $I \in \mathbb{N}$. We can thus apply the following result to $\varphi(t) = -\log L(t)$:

Lemma A.2.1. *Let $\varphi : \mathbb{R}_+ \to \mathbb{R}_+$ be a continuous function such that for each I there exists some $c_I \in \mathbb{R}_+$ such that $I\varphi(t) = \varphi(c_I t)$ holds for all $t \in \mathbb{R}_+$. Then $\varphi(t) = t^\alpha$ for some $\alpha \in \mathbb{R}$.*

This result implies that a distribution is positive add-stable if and only if its Laplace transform satisfies $L(t) = \exp(-t^\alpha)$. But one can show that the map $t \mapsto \exp(-t^\alpha)$ is the Laplace transform of a probability distribution if and only if $\alpha \in (0, 1]$, which establishes:

Theorem A.2.1. *A probability distribution is positive add-stable if and only if there is a scalar parameter $\alpha \in (0, 1]$, called the* stability parameter, *such that the Laplace transform of the distribution is equal to*

$$\mathbb{E}\left[e^{-tZ}\right] = \exp\left(-t^\alpha\right), \tag{A.2.7}$$

in which case the distribution is called the positive stable distribution of parameter α, *denoted by $\mathcal{PS}_\alpha$, and $c_I = I^{1/\alpha}$ in* (A.2.6), *which yields*

$$Z_1 + \cdots + Z_I \stackrel{d}{=} I^{1/\alpha} Z. \tag{A.2.8}$$

Of course, if $\alpha = 1$, then $\mathcal{PS}_1$ is the degenerate distribution associated with the constant $X = 1$. When $\alpha < 1$, it is easy to see that a positive stable distribution does not have a first moment. Indeed, if that were the case, the expectation of Z would be positive; yet, taking expectation in (A.2.6) yields $I\mathbb{E}[Z] = I^{1/\alpha}\mathbb{E}[Z]$.

Note that for $\alpha < 1$, the domain of $\mathcal{PS}_\alpha$ is $\mathbb{R}_+$, and its mean is $+\infty$. There is no simple formula for the density of a positive add-stable distribution, but there is a compact probabilistic representation of these distributions using two independent factors. Indeed, if U and V are independent $\mathcal{U}([0, 1])$ random variables, then

$$\left(\frac{\sin\left((1-\alpha)\pi U\right)}{-\log V}\right)^{\frac{1-\alpha}{\alpha}} \frac{\sin(\alpha\pi U)}{\left(\sin(\pi U)\right)^{1/\alpha}} \sim \mathcal{PS}_\alpha. \tag{A.2.9}$$

The following result holds:

Proposition A.2.6. *If* $Z_1 \sim \mathcal{PS}_{\alpha_1}$ *and* $Z_2 \sim \mathcal{PS}_{\alpha_2}$ *are independent random variables, then* $Z_1 Z_2^{1/\alpha_1} \sim \mathcal{PS}_{\alpha_1 \alpha_2}$.

There is an interesting relation between the positive add-stable distributions just introduced and the positive "max-stable" distributions, which are stable by the maximum rather than by the sum. One can see that if $X_1, X_2, \ldots, X_n$ are independent variables that all have the same distribution as $\alpha \log Z$ with $Z \sim \mathcal{PS}_\alpha$, then

$$\text{softmax}^{\alpha}_{i\in[I]} \{X_i + x_i\} \overset{d}{=} \log \Big(\sum_{i\in[I]} \exp x_i \Big) + X, \tag{A.2.10}$$

where X has the same distribution as each term X_i, and the *soft-maximum operator* is defined by $\text{softmax}^{\alpha}_i \{z_i\} = \alpha \log \sum_i \exp (z_i/\alpha)$. Since $\text{softmax}^{\alpha}_i \{z_i\} \to \max_i\{z_i\}$ as $\alpha \to 0$, we may expect for $\alpha \log Z$ to converge in some sense to the standard Gumbel distribution, which is max-stable. This is made precise by the following statement:

Theorem A.2.2. *Let* $\mathcal{S}_\alpha$ *be the distribution of* $\alpha \log Z$*, where* $Z \sim \mathcal{PS}_\alpha$*. Then*

$$W_2 (\mathcal{S}_\alpha, \mathcal{G}) \leq \alpha \sqrt{\gamma^2 + \frac{\pi^2}{6}},$$

where $W_2 (\mathcal{S}_\alpha, \mathcal{G})$ *is the 2-Wasserstein distance, and* $\mathcal{G}$ *is the standard Gumbel distribution.*

References

A reference for theorem A.2.1 is Feller (1991), chapter XIII, section 6, and chapter VI, section 3 for the other claims in the section, in particular, proposition A.2.6. Formula (A.2.9) was derived in Kanter (1975).

A.2.3 ROSENBLATT QUANTILES AND IMPORTANCE SAMPLING

Knothe and Rosenblatt independently discovered an important multivariate extension of the notion of univariate quantile. Let $\mathcal{Q}$ be a continuous distribution over $\mathbb{R}^J$, and let Z be a J-dimensional random vector with distribution $\mathcal{Q}$. We shall denote by F_{Z_1} the c.d.f. of Z_1, by $F_{Z_2|Z_1}$ the conditional c.d.f. of Z_2 given Z_1, and so on.

Definition A.2.1. *The Rosenblatt quantile map associated with the distribution $\mathcal{Q}$ is the map $T : \mathbb{R}^{\mathrm{J}} \to \mathbb{R}^{\mathrm{J}}$ defined by*

$$\begin{cases} T_1(x) = F_{Z_1}^{-1}(x_1), \\ T_2(x) = F_{Z_2|Z_1}^{-1}(x_2 \mid Z_1 = T_1(x_1)), \\ \vdots \\ T_{\mathrm{J}}(x) = F_{Z_{\mathrm{J}}|Z_1,\ldots,Z_{\mathrm{J}-1}}^{-1}\left(x_{\mathrm{J}} \mid Z_j = T_j(x)\ \forall j \in [\mathrm{J}-1]\right). \end{cases}$$

It is clear from the definition that T_1 depends only on x_1, T_2 depends on x_1 and x_2, and so on. As a result, the Jacobian of T is lower triangular with positive diagonal terms. The term "quantile" is motivated by the following fact, whose proof is left to the reader, and that extends a characterization of the quantile map when $\mathrm{J} = 1$ to the general case $\mathrm{J} \geq 1$:

Lemma A.2.2. *If T is the Rosenblatt quantile associated with distribution $\mathcal{Q}$, then $X \sim \mathcal{U}([0,1]^{\mathrm{J}})$ implies $T(X) \sim \mathcal{Q}$.*

In fact, one can characterize T as the only such mapping that is in addition increasing in the lexicographic order of $\mathbb{R}^{\mathrm{J}}$.

Example A.2. When $\mathcal{Q} = \mathcal{N}(0, \Sigma)$, the Rosenblatt quantile of $\mathcal{Q}$ is given by $T(x) = L\Phi^{-1}(x)$, where L is the Choleski factorization of Σ, i.e., the unique lower triangular matrix with a positive diagonal such that $LL^{\top} = \Sigma$, and Φ is the c.d.f. of the standard normal distribution, with Φ^{-1} its inverse, the quantile map. Here Φ^{-1} applies component-wise on the entries of $x \in \mathbb{R}^{\mathrm{J}}$, so that $\left(\Phi^{-1}(x)\right)_j = \Phi^{-1}(x_j)$.

Importance sampling. The Rosenblatt quantile is useful for a simulation technique called *importance sampling*, for which we provide intuition in the bivariate case ($\mathrm{J} = 2$). Assume we want to compute

$$\pi = \Pr(Z_1 \leq z_1, Z_2 \leq z_2), \tag{A.2.11}$$

and suppose we know the Rosenblatt quantile for the distribution $\mathcal{Q}$ of Z, which is $T(x) = (T_1(x_1), T_2(x_1, x_2))$. We denote by $T_1^{-1}(z_1)$ the inverse map of T_1, and by $T_2^{-1}(x_1, z_2)$ the inverse map of $x_2 \mapsto T_2(x_1, x_2)$, holding x_1 fixed. We assume that both T_1^{-1} and T_2^{-1} are either known in closed form, or can easily be obtained by numerical computation.

For $ij \in [I \times \mathrm{J}]$ ($\mathrm{J} = 2$), we draw x_{ij}, an i.i.d. sample of uniform random variables over $[0,1]$. One may compute π using the *naive accept-reject simulator*

$$\hat{\pi}_I^{NAR} = \frac{1}{I} \sum_{i \in [I]} \mathbf{1}\{T_1(x_{i1}) \le z_1\} \mathbf{1}\{T_2(x_{i1}, x_{i2}) \le z_2\} \tag{A.2.12}$$

whose variance is $\pi(1-\pi)/I$. However, note that one does not have to sample two dimensions of (X_1, X_2); indeed, $\pi = \mathbb{E}[\mathbf{1}\{X_1 \le T_1^{-1}(z_1)\} T_2^{-1}(X_1, z_2)]$, and therefore one may draw I samples x_i from the $\mathcal{U}([0,1])$ distribution and define the *integrated accept-reject simulator*

$$\hat{\pi}_I^{IAR} = \frac{1}{I} \sum_{i \in [I]} \mathbf{1}\{x_i \le T_1^{-1}(z_1)\} T_2^{-1}(x_i, z_2). \tag{A.2.13}$$

Alternatively, one may use an importance sampling idea, and argue that $\pi = \pi_1 \pi_{2|1}$, where $\pi_1 = \Pr(Z_1 \le z_1)$ and $\pi_{2|1} = \Pr(Z_2 \le z_2 \mid Z_1 \le z_1)$. Letting $X \sim \mathcal{U}([0,1]^2)$, and denoting $Z_1 = T_1(X_1)$ and $Z_2 = T_2(X_1, X_2)$, we have

$$\pi_1 = T_1^{-1}(z_1), \tag{A.2.14}$$

as well as $\Pr(Z_2 \le z_2 \mid Z_1 \le z_1) = \Pr\left(T_2(X_1, X_2) \le z_2 \mid X_1 \le T_1^{-1}(z_1)\right)$. However, the distribution of $X_1 \sim \mathcal{U}([0,1])$ conditional on $X_1 \le T_1^{-1}(z_1)$ is simply the uniform distribution over $[0, T_1^{-1}(z_1)]$, which is the unconditional distribution of $\tilde{X}_1 T_1^{-1}(z_1)$ for $\tilde{X}_1 \sim \mathcal{U}([0,1])$. Therefore $\Pr(Z_2 \le z_2 \mid Z_1 \le z_1) = \Pr(T_2(\tilde{X}_1 T_1^{-1}(z_1), \tilde{X}_2) \le z_2)$, where $(\tilde{X}_1, \tilde{X}_2) \sim \mathcal{U}([0,1]^2)$, which is equal to $\Pr(\tilde{X}_2 \le T_2^{-1}(\tilde{X}_1 T_1^{-1}(z_1), z_2))$, and therefore

$$\pi_{2|1} = \mathbb{E}\left[T_2^{-1}(\tilde{X}_1 T_1^{-1}(z_1), z_2))\right], \tag{A.2.15}$$

and, as a result, combining (A.2.14) and (A.2.15) and still using the i.i.d. sample of I draws $\tilde{x}_i$, $i \in [I]$ from the uniform distribution over $[0,1]$, we get the *importance sampling simulator*

$$\hat{\pi}_I^{IS} = \frac{\pi_1}{I} \sum_{i \in [I]} T_2^{-1}(\tilde{x}_i T_1^{-1}(z_1), z_2)). \tag{A.2.16}$$

More generally, for $J \ge 2$, we denote by $T_j^{-1}\left(x_1, \ldots, x_{j-1}, z_j\right)$ the inverse of $x_j \mapsto T_j\left(x_1, \ldots, x_{j-1}, x_j\right)$ at z_j. In the Gaussian case, when $\mathcal{Q} = \mathcal{N}(0, \Sigma)$ and L is the Cholesky lower triangular matrix such that $LL^\top = \Sigma$, we have

$$\begin{cases} T_j(x) = L_{j1}\Phi^{-1}(x_1) + \cdots + L_{j,j-1}\Phi^{-1}(x_{j-1}) + L_{jj}\Phi^{-1}(x_j), \\ T_j^{-1}\left(x_1, \ldots, x_{j-1}, z_j\right) = \Phi(-\frac{L_{j1}}{L_{jj}}\Phi^{-1}(x_1) - \cdots - \frac{L_{j,j-1}}{L_{jj}}\Phi^{-1}(x_{j-1}) + \frac{z_j}{L_{jj}}), \end{cases}$$

and the following lemma generalizes the above discussion:

Lemma A.2.3. *We have*

$$\Pr(Z_1 \le z_1, \ldots, Z_J \le z_J) = \mathbb{E}[\prod_{j \in [J]} T_j^{-1}\left(\hat{X}_1, \hat{X}_2, \ldots, \hat{X}_{j-1}, z_j\right)], \tag{A.2.17}$$

where, given $\tilde{X} \sim \mathcal{U}([0,1]^J)$, *the* J*-dimensional random vector* $\hat{X}$ *is defined recursively by*

$$\begin{cases} \hat{X}_1 = \tilde{X}_1 \; T_1^{-1}(z_1), \\ \hat{X}_2 = \tilde{X}_2 \; T_2^{-1}\left(\hat{X}_1, z_2\right), \\ \vdots \\ \hat{X}_J = \tilde{X}_J \; T_J^{-1}\left(\hat{X}_1, \ldots, \hat{X}_{J-1}, z_J\right). \end{cases} \tag{A.2.18}$$

This lemma is the basis of the GHK (Geweke–Hajivassiliou–Keane) algorithm, described in example 4.1. Numerical studies show that the variance of the importance sampling simulator (that is, in the context of discrete choice, the GHK simulator) is typically much lower than the variance of the accept-reject simulator.

References

References for the Rosenblatt quantile are the paper Rosenblatt (1952) and Knothe (1957). For the importance sampling result, see Geweke (1989), Hajivassiliou and McFadden (1998), and Keane (1994).

A.2.4 NUMERICAL INTEGRATION BY GAUSS–HERMITE QUADRATURE

The (physicist's) Hermite polynomials are the polynomials H_N such that

$$H_N(x) = (-1)^N e^{x^2} \frac{d^N}{dx^N} e^{-x^2},$$

from which we can deduce the recurrence relation

$$H_{N+1}(x) = 2xH_N(x) - 2NH_{N-1}(x),$$

and one can show that $H_N(x)$ is a polynomial of degree N with 2^N as its leading coefficient (that is, the coefficient of x^N). We can easily compute the first Hermite polynomials $H_0(x)=1$, $H_1(x)=2x$, $H_2(x)=4x^2-2$, $H_3(x)=8x^3-12x$, and so on. We have

$$\int H_N(x)\, H_M(x)\, \frac{e^{-x^2}}{\sqrt{\pi}}\, dx = 2^N N! \text{ if } N=M \text{ and } =0 \text{ otherwise.}$$

Thus the Hermite polynomials $H_0, H_1, ..., H_N$ form an orthogonal basis of the space of polynomials of degree N.

Gauss–Hermite quadrature. If x_n, $n \in [N]$, are the roots of H_N, then a good approximation of the integral of $f(x)\exp(-x^2)$ over the real line is provided by

$$\int_{-\infty}^{+\infty} f(x)\, e^{-x^2} dx \approx \sum_{n\in[N]} w_n f(x_n), \text{ where } w_n = \frac{2^{N-1}N!}{N^2 \left(H_{N-1}(x_n)\right)^2},$$

and the approximation is exact if f is a polynomial of degree at most $2N-1$. As a result, if $X \sim \mathcal{N}(0,1)$, the expectation $\mathbb{E}[f(X)]$ is approximated by

$$\int_{-\infty}^{+\infty} f(x)\, \frac{e^{-x^2/2}}{\sqrt{2\pi}} dx = \int_{-\infty}^{+\infty} f\left(\sqrt{2}x\right) \frac{e^{-x^2}}{\sqrt{\pi}} dx \approx \frac{1}{\sqrt{\pi}} \sum_{n\in[N]} w_n f\left(\sqrt{2}x_n\right). \tag{A.2.19}$$

References

A standard reference is the classic text Abramowitz and Stegun (1965). An in-depth mathematical treatment of orthogonal polynomials is the book by Szegő (1939). For economic applications, see the handbook Judd (1998).

A.2.5 M-ESTIMATION

M-estimation is a general class of estimation methods that includes, as special cases, maximum likelihood estimation, least squares, and many others. An M-estimator of a parameter vector $\theta \in \mathbb{R}^K$ is defined as the solution to the following optimization problem:

$$\hat{\theta} = \arg\min_{\theta\in\mathbb{R}^K} \sum_{i\in[I]} m(Z_i, \theta), \tag{A.2.20}$$

where $\{Z_i\}_{i\in[I]}$, which are vectors of $\mathbb{R}^P$, are observed data, and $m(z,\theta)$ is a real-valued function often referred to as the objective function or criterion function. We introduce several assumptions:

(A1) The data $\{Z_i\}_{i\in[I]}$ are independent and identically distributed from a probability distribution $\mathcal{P}$.
(A2) The function $\theta \mapsto m(z,\theta)$ is twice continuously differentiable with respect to θ for every $z \in \mathbb{R}^P$.
(A3) The function $z \mapsto m(z,\theta)$ is Borel measurable for every $\theta \in \mathbb{R}^K$.
(A4) One has $\mathbb{E}_{\mathcal{P}}[\sup_{\theta\in\Theta} m(Z,\theta)] < +\infty$ for every compact subset Θ of $\mathbb{R}^K$.
(A5) The true parameter value $\theta_0 \in \mathbb{R}^K$ is the only minimizer of the population analog of problem (A.2.20), namely, $\theta \mapsto \mathbb{E}_{\mathcal{P}}[m(Z,\theta)]$.
(A6) One has $\mathbb{E}_{\mathcal{P}}\big[\sup_{\theta\in\Theta} |\partial^2 m(Z,\theta)/\partial\theta_k\partial\theta_l|\big] < +\infty$ for every compact subset Θ of $\mathbb{R}^K$ and every k and l in $[K]$.
(A7) Each random variable $\partial m(Z,\theta_0)/\partial\theta_k$ has finite second moments for $k \in [K]$.

Given the assumptions above, we can state the following theorem regarding the asymptotic distribution of the M-estimator $\hat{\theta}$. Intuitively, we have $\mathbb{E}_{\hat{\mathcal{P}}}[\nabla_\theta m(Z,\hat{\theta})] = 0$, where $\hat{\mathcal{P}}$ is the sample distribution, and $\mathbb{E}_{\mathcal{P}}[\nabla_\theta m(Z,\theta_0)] = 0$, where $\mathcal{P}$ is the population distribution. By linearization, we get $0 = (\mathbb{E}_{\hat{\mathcal{P}}}[\nabla_\theta m(Z,\theta_0)] - \mathbb{E}_{\mathcal{P}}[\nabla_\theta m(Z,\hat{\theta})]) + \mathbb{E}_{\mathcal{P}}[D^2_{\theta\theta} m(Z,\theta_0)]$ $(\hat{\theta} - \theta_0) + o_p(\hat{\theta} - \theta_0)$, and thus

$$\hat{\theta} - \theta_0 = -\mathbb{E}_{\mathcal{P}}[D^2_{\theta\theta} m(Z,\theta_0)]^{-1}\left(\mathbb{E}_{\hat{\mathcal{P}}}[\nabla_\theta m(Z,\hat{\theta})] - \mathbb{E}_{\mathcal{P}}[\nabla_\theta m(Z,\theta_0)]\right) + o_p\left(I^{-1/2}\right).$$

As we have $\mathbb{E}_{\hat{\mathcal{P}}}[\nabla_\theta m(Z,\hat{\theta})] - \mathbb{E}_{\mathcal{P}}[\nabla_\theta m(Z,\theta_0)] \sim \mathcal{N}(0, \mathbb{E}_{\mathcal{P}}[\nabla_\theta m(Z,\theta_0)\nabla_\theta m(Z,\theta_0)^\top])$, we are able to conclude that:

Theorem A.2.3. *Let $\hat{\theta}$ be an M-estimator of θ based on a sample of size I under the assumptions stated above. Then:*

(i) Under assumptions (A1) to (A4), the uniform law of large numbers holds: the sample criterion function converges uniformly in probability to its expectation, i.e.,

$$\sup_{\theta\in\Theta}\left|\frac{1}{I}\sum_{i\in[I]} m(Z_i,\theta) - \mathbb{E}_{\mathcal{P}}[m(Z,\theta)]\right| \xrightarrow{p} 0 \tag{A.2.21}$$

holds as $I \to \infty$ for every compact subset Θ of $\mathbb{R}^K$.

(ii) Under assumptions (A1) to (A5), consistency holds: $\hat{\theta} \xrightarrow{p} \theta_0$ as $I \to \infty$.

(iii) Under assumptions (A1) to (A7), we have the asymptotic normality of $\hat{\theta}$, that is,

$$\sqrt{I}(\hat{\theta} - \theta_0) \xrightarrow{d} \mathcal{N}(0, H^{-1}CH^{-1}), \tag{A.2.22}$$

where the matrices H and C are given by

$$H_{kl} = \mathbb{E}_{\mathcal{P}}\left[\frac{\partial^2 m(Z,\theta_0)}{\partial\theta_k\partial\theta_l}\right] \quad \textit{and} \quad C_{kl} = \mathbb{E}_{\mathcal{P}}\left[\frac{\partial m(Z,\theta_0)}{\partial\theta_k}\frac{\partial m(Z,\theta_0)}{\partial\theta_l}\right]. \tag{A.2.23}$$

References

Our presentation follows that of Wooldridge (2010), chapter 12. See also van der Vaart (1998), chapter 5.

A.2.6 GENERALIZED METHOD OF MOMENTS

Generalized method of moments (GMM) consists of estimating $\theta^\star \in \mathbb{R}^K$ based on the L moment restrictions

$$\mathbb{E}\left[m_{l\tilde{\omega}}(\theta^\star)\right] = 0, \forall l \in [L] \tag{A.2.24}$$

for a data-generating process $\tilde{\omega} \sim \mathcal{P}$. Our data is assumed to be an i.i.d. sample $\{\omega_i, i \in [I]\}$ drawn from $\mathcal{P}$. Throughout, we shall denote by $\tilde{m}(\theta^\star)$ the L-dimensional random vector $(m_{l\tilde{\omega}})_l$, where $\tilde{\omega} \sim \mathcal{P}$.

The GMM estimator. Let's present the intuition in the case of discrete data, which is the case when the support of $\mathcal{P}$ is a finite set—the generalization will be straightforward. Assume that the support of $\mathcal{P}$ has cardinality $\Omega \in \mathbb{N}$ and can be described by $[\Omega] = \{1, \ldots, \Omega\}$. Then for $\omega \in [\Omega]$, we define $\hat{\pi}_\omega = I^{-1}\sum_i 1\{\omega_i = \omega\}$ to be the empirical frequency of outcome ω. Letting $\pi_\omega = \Pr_{\mathcal{P}}(\tilde{\omega} = \omega)$ be the population analog of that quantity, which forms a column vector π of size $\Omega \times 1$, and denoting by $M(\theta)$ the $L \times \Omega$ matrix of term $m_{l\omega}(\theta)$, we can rewrite (A.2.24) into

$$M\left(\theta^\star\right)\pi = 0. \tag{A.2.25}$$

Given a symmetric positive definite matrix W of size $L \times L$, we reformulate (A.2.25) as

$$\theta^\star = \arg\min_{\theta \in \mathbb{R}^K} F(\theta, \pi), \tag{A.2.26}$$

where, for $\theta \in \mathbb{R}^K$ and π an Ω-dimensional probability column vector, F is defined by

$$F(\theta,\pi) = \frac{1}{2}\pi^\top M(\theta)^\top WM(\theta)\pi, \tag{A.2.27}$$

allowing us to define the GMM estimator of $\hat{\theta}$ as the minimizer of $F(\theta, \hat{\pi})$ over $\theta \in \mathbb{R}^d$. The optimality conditions are $\partial F(\hat{\theta}, \hat{\pi})/\partial \theta_k = 0$ for $k \in [K]$, where

$$\frac{\partial F(\theta, \pi)}{\partial \theta_k} = \pi^\top M(\theta)^\top W \frac{\partial M}{\partial \theta_k}(\theta)\pi. \tag{A.2.28}$$

Large sample properties. Assume from now on that $M(\theta)$ is linear in θ, and let $\mathbf{G}$ be the $L \times K$ matrix of term $G_{lk} = \sum_\omega \frac{\partial M_{l\omega}}{\partial \theta_k} \pi_\omega$, which therefore does not depend on θ, and which is such that $\mathbb{E}[\tilde{m}(\theta)] = \mathbf{G}\theta$. Letting $\mathbf{M}$ be the $L \times \Omega$ matrix of term $M_{l\omega} = M_{l\omega}(\theta^\star)$, it follows from equation (A.2.25) that

$$D^2_{\theta\pi} F\left(\theta^\star, \pi\right) = \mathbf{G}^\top W \mathbf{M} \text{ and } D^2_{\theta\theta} F\left(\theta^\star, \pi\right) = \mathbf{G}^\top W \mathbf{G} \tag{A.2.29}$$

and a first-order expansion of $\partial_\theta F(\theta^\star, \pi) = 0$ around the true θ and the population π yields $\hat{\theta} - \theta^\star = -(\partial^2_{\theta\theta} F(\theta^\star, \pi))^{-1} \partial^2_{\theta\pi} F(\theta^\star, \pi)(\hat{\pi} - \pi) + o_P(I^{-1/2})$, that is, $\hat{\theta} - \theta^\star = -(\mathbf{G}^\top W \mathbf{G})^{-1}(\mathbf{G}^\top W \mathbf{M})(\hat{\pi} - \pi) + o_P(I^{-1/2})$. Now, with i.i.d. data, we have that the variance-covariance matrix of $\hat{\pi}$ is V_π / I, where $V_\pi := (\Delta_\pi - \pi\pi^\top)$. Therefore the variance-covariance matrix of $\mathbf{M}\hat{\pi} = I^{-1} \sum_i m_{l\omega_i}(\theta^\star)$ is Σ / I, where $\Sigma = \mathbf{M} V_\pi \mathbf{M}^\top$ is the $L \times L$ matrix of variance-covariance matrix of $\tilde{m}(\theta^\star)$, the L-dimensional random vector $(m_{l\tilde{\omega}})_l$. Thus

$$\sqrt{I}(\hat{\theta} - \theta^\star) \xrightarrow{d} \mathcal{N}\left(0, \left(\mathbf{G}^\top W \mathbf{G}\right)^{-1} \mathbf{G}^\top W \Sigma W^\top \mathbf{G} \left(\mathbf{G}^\top W \mathbf{G}\right)^{-1}\right). \tag{A.2.30}$$

Efficient weighting. Letting $\mathbf{F} = \Sigma^{-1/2}\mathbf{G}$, which is an $L \times K$ matrix of rank $K \leq L$, we have $\mathbf{I}_L \geq \mathbf{F}\left(\mathbf{F}^\top \mathbf{F}\right)^{-1} \mathbf{F}^\top$. Hence, for any symmetric definite positive matrix S, one has $\mathbf{F}^\top S^2 \mathbf{F} \geq \left(\mathbf{F}^\top S \mathbf{F}\right)\left(\mathbf{F}^\top \mathbf{F}\right)^{-1}\left(\mathbf{F}^\top S \mathbf{F}\right)$, and therefore $\left(\mathbf{F}^\top S \mathbf{F}\right)^{-1} \mathbf{F}^\top S^2 \mathbf{F} \left(\mathbf{F}^\top S \mathbf{F}\right)^{-1} \geq \left(\mathbf{F}^\top \mathbf{F}\right)^{-1}$. Taking $S = \Sigma^{1/2} W \Sigma^{1/2}$ and recalling the expression of $\mathbf{F}$ yields

$$\left(\mathbf{G}^\top W \mathbf{G}\right)^{-1} \mathbf{G}^\top W \Sigma W \mathbf{G} \left(\mathbf{G}^\top W \mathbf{G}\right)^{-1} \geq \left(\mathbf{G}^\top \Sigma^{-1} \mathbf{G}\right)^{-1},$$

which implies that the estimator $\hat{\theta}$ achieves minimum variance for $W = \Sigma^{-1}$, in which case it is called the *efficient GMM estimator*, and we have:

Theorem A.2.4. *The efficient GMM estimator is obtained for $W = \Sigma^{-1}$, in which case*

$$\sqrt{I}(\hat{\theta} - \theta^\star) \xrightarrow{d} \mathcal{N}\left(0, \left(\mathbf{G}^\top \Sigma^{-1} \mathbf{G}\right)^{-1}\right). \tag{A.2.31}$$

Note that Σ is the variance-covariance matrix of $\tilde{m}(\theta^\star)$, which requires knowledge of the true value of the parameter, and thus efficient GMM estimation is not directly feasible. In practice, a first consistent estimator of θ is obtained using a reasonable generic choice of the matrix W, and that parameter value is then plugged in to compute a consistent estimator of Σ. This approach is precisely illustrated in appendix A.2.7 below when dealing with instrumental variable estimation.

References

The generalized method of moments appeared in the seminal paper Hansen (1982).

A.2.7 INSTRUMENTAL VARIABLES

Our dataset consists, for each observation $i \in [I]$, of a dependent variable $y_i \in \mathbb{R}$; a vector of (potentially endogenous) regressors $x_i \in \mathbb{R}^K$; and a vector of instruments $z_i \in \mathbb{R}^L$. We assume that the data-generating process is $\tilde{y} = \tilde{x}^\top \beta^\star + \tilde{\epsilon}$, where the error $\tilde{\epsilon}$ is orthogonal to the vector of instruments, that is, $\mathbb{E}[\tilde{\epsilon}\tilde{z}] = 0$, or, in other words, $\beta^\star$ satisfies the moment restriction

$$\mathbb{E}[(\tilde{y} - \tilde{x}^\top \beta)\tilde{z}] = 0. \quad \text{(A.2.32)}$$

Denoting by Y the column vector of size I of entries y_i, by X the $I \times K$ matrix whose rows are the vectors x_i, and by Z the $I \times L$ matrix whose rows are the vectors z_i, we obtain the sample analog of (A.2.32), namely,

$$Z^\top (Y - X\beta) = 0. \quad \text{(A.2.33)}$$

Just identified case. Standard instrumental variable regression assumes that there are as many instruments as parameters: $L = K$, which is the just identified case. We assume that there is enough dependence between the instruments and the regressors for the $K \times K$ matrix $\mathbb{E}[\tilde{z}\tilde{x}^\top]$ to be invertible. In that case, restriction (A.2.33) yields the standard IV estimator, which is given by

$$\hat{\beta}^{IV} = (Z^\top X)^{-1} Z^\top Y.$$

Overidentified case. Often, however, there are more restrictions than parameters, so $L > K$. In that case, β provided by (A.2.33) is overidentified, and one needs to essentially pick K from among L instrumental restrictions. This is done by left-multiplying (A.2.33) by some $K \times L$ selection matrix P so that the $K \times K$ matrix $PZ^\top$ is invertible, and $PZ^\top (Y - X\beta) = 0$ yields $\beta = (PZ^\top X)^{-1} PZ^\top Y$.

Two-stage least squares estimator. The well-known two-stage least squares estimator provides an initial rationale for obtaining the selection matrix P. It consists of forming the best linear predictor of X from Z—call it $\hat{X}$—and regressing Y on $\hat{X}$. The best linear predictor is obtained by regressing X on Z: $X = Z\delta + \eta$, which gives $\delta = \left(Z^\top Z\right)^{-1} Z^\top X$, which yields $\hat{X} = Z\delta = Z\left(Z^\top Z\right)^{-1} Z^\top X$. Note that the $K \times K$ matrix $Z\left(Z^\top Z\right)^{-1} Z^\top$ is the orthogonal projection matrix on the linear space spanned by the rows of Z, and our selection matrix P introduced above has expression $P = X^\top Z\left(Z^\top Z\right)^{-1}$, which leads us to the two-stage least squares estimator

$$\hat{\beta}^{2SLS} = (X^\top Z\left(Z^\top Z\right)^{-1} Z^\top X)^{-1} X^\top Z\left(Z^\top Z\right)^{-1} Z^\top Y.$$

GMM estimation. A more efficient approach consists of using the generalized method of moments to estimate the moment restriction (A.2.32). In that case, the parameter denoted by θ in appendix A.2.6 is simply the regression parameter β, so that $\theta = \beta$, and we look for the value of β that minimizes

$$\mathbb{E}[\epsilon(\beta)z]^\top W \mathbb{E}[\epsilon(\beta)z], \tag{A.2.34}$$

where the weighting matrix W is the symmetric positive definite matrix of size $L \times L$ introduced in appendix A.2.6, and where $\epsilon(\beta) = y - x^\top \beta$. The sample analog of (A.2.34) is

$$\min_{\beta \in \mathbb{R}^K} (Y - X\beta)^\top Z W Z^\top (Y - X\beta), \tag{A.2.35}$$

and the objective function F introduced in (A.2.27) is $F(\beta) = (Y - X\beta)^\top Z W Z^\top (Y - X\beta)/I^2$. Optimality conditions yield

$$\beta^\star = \left(X^\top \Pi X\right)^{-1} \left(X^\top \Pi Y\right) \text{ and } F(\beta^\star) = Y^\top \Upsilon Y / I^2, \tag{A.2.36}$$

where $\Pi = Z W Z^\top$, and $\Upsilon = \Pi - \Pi^\top X\left(X^\top \Pi X\right)^{-1} X^\top \Pi$.

The $L \times \Omega$ matrix-valued function M introduced in appendix A.2.6 becomes

$$M_{l\omega}(\beta) = z_{l\omega}\left(y_\omega - \sum_{k \in [K]} x_{k\omega}\beta_k\right),$$

and the sample analog of $M(\beta)\mu$ is $M(\beta)\hat{\mu} = Z^\top (Y - X\beta)/I$. It is straightforward to verify that $G = \mathbb{E}[\tilde{z}\tilde{x}^\top]$, whose sample analog is $\hat{G} = Z^\top X/I$, and that $\Sigma = \mathbb{V}(\tilde{z}\tilde{\epsilon})$ is the variance-covariance matrix of the random vector $\tilde{z}\tilde{\epsilon}$, whose sample analog is

$$\Sigma = \hat{M}(\beta^\star)(\frac{\mathbf{I}_I}{I} - \frac{\mathbf{1}_I \mathbf{1}_I^\top}{I^2})\hat{M}(\beta^\star)^\top, \tag{A.2.37}$$

where $\hat{M}(\beta)$ is the $L \times I$ matrix whose (l, i) term is $\hat{M}_{li}(\beta) = z_{li} y_i - z_{li} \sum_k x_{k\omega} \beta_k$. Expression (A.2.37) is not feasible, as it requires the knowledge of $\beta^\star$. However, one can use instead a consistent estimator provided, for example, by the two-stage least squares estimator, and thus estimate Σ using

$$\hat{\Sigma} = \hat{M}(\hat{\beta}^{2SLS})(\frac{\mathbf{I}_I}{I} - \frac{\mathbf{1}_I \mathbf{1}_I^\top}{I^2})\hat{M}(\hat{\beta}^{2SLS})^\top. \tag{A.2.38}$$

The first-order conditions associated with the generalized method of moments estimator are $(Y - X\beta)^\top ZWG = 0$, which, given the expression of G, yields

$$\hat{\beta}^{GMM}(W) = (X^\top ZWZ^\top X)^{-1} X^\top ZWZ^\top Y.$$

In other words, the generalized method of moments consists of taking the selection matrix $P = X^\top ZW$. Note that the two-stage least squares estimator, which has $P = X^\top Z \left(Z^\top Z\right)^{-1}$, is obtained as a GMM estimator with weight matrix $W = \left(Z^\top Z\right)^{-1} / I$, which is feasible, as it does not require knowledge of the true value of the parameter, but not efficient. As seen in appendix A.2.6, efficient GMM is obtained instead for $W = \Sigma^{-1}$, where, again, Σ is consistently estimated using a feasible estimator, such as a two-stage least squares estimator, according to expression (A.2.38) above. The resulting estimator of β is called the IV-GMM estimator, and we therefore get the following procedure to compute the IV-GMM estimator:

$$\begin{cases} \hat{\beta}^{2SLS} = \hat{\beta}^{GMM}(W_1), \text{ where } W_1 = (\frac{Z^\top Z}{I})^{-1}, \\ \hat{\Sigma} = \hat{M}(\hat{\beta}^{2SLS})(\frac{\mathbf{I}_I}{I} - \frac{\mathbf{1}_I \mathbf{1}_I^\top}{I^2})\hat{M}(\hat{\beta}^{2SLS})^\top \\ \hat{\beta}^{IV-GMM} = \hat{\beta}^{GMM}(W_2), \text{ where } W_2 = \hat{\Sigma}^{-1}. \end{cases}$$

It follows from theorem A.2.4 that

$$\sqrt{I}(\hat{\beta}^{IV-GMM} - \beta^\star) \xrightarrow{d} \mathcal{N}\left(0, \left(\left(\frac{X^\top Z}{I}\right) \hat{\Sigma}^{-1} \left(\frac{Z^\top X}{I}\right)\right)^{-1}\right).$$

References

The method of instrumental variables can be traced back to the monograph Wright (1928), and later on appeared in the paper by Theil (1950), but was widely developed a few years after that in Sargan (1958). References for the IV-GMM approach include Hansen (1982), Hansen and Singleton (1982), and Arellano and Bond (1991).

A.2.8 GENERALIZED LINEAR MODELS

GLM specification. Generalized linear models (GLMs) extend linear regressions in the sense that if $\tilde{\mu}$ is the dependent variable and $\tilde{\rho}$ the P-dimensional column vector of explanatory variables (regressors), the linear specification $\mathbb{E}[\tilde{\mu} \mid \tilde{\rho}] = \tilde{\rho}^\top \theta$ is replaced by $f(\mathbb{E}[\tilde{\mu} \mid \tilde{\rho}]) = \tilde{\rho}^\top \theta$, where the index $\tilde{\rho}^\top \theta$ is often called the *score* associated with covariates $\tilde{\rho}$. The function f, which is called the *link function*, is increasing and continuous from an open (not necessarily bounded) interval $(\mu_l, \mu_u) \subseteq \mathbb{R}$ onto $\mathbb{R}$; it is therefore invertible, which leads to:

Definition A.2.2 (Generalized model specification). *The generalized linear model with link function f, with dependent variable $\tilde{\mu}$, and with vector of explanatory variables $\tilde{\rho}$ is given by the relationship*

$$\mathbb{E}[\tilde{\mu} \mid \tilde{\rho}] = f^{-1}\left(\tilde{\rho}^\top \theta\right). \tag{A.2.39}$$

Much of the appeal of generalized linear models comes from the possibility of computing them as a convex optimization problem. Indeed, letting F be the primitive of f, it follows from convex analysis (see appendix A.1.1 above) that the primitive of f^{-1} is F^*, the Legendre–Fenchel conjugate of F. As a result, the relation (A.2.39) recasts as $\mathbb{E}[\tilde{\mu} \mid \tilde{\rho}] = (F^*)'\left(\tilde{\rho}^\top \theta\right)$, from which it follows that the true parameter θ is the solution to

$$\max_{\theta \in \mathbb{R}^P} \mathbb{E}\left[\tilde{\mu}^\top \tilde{\rho}^\top \theta - F^*\left(\tilde{\rho}^\top \theta\right)\right]. \tag{A.2.40}$$

GLM estimator. Problem (A.2.40) has a natural sample analog, which leads to an estimator of θ. Let's assume we observe a sample $\left\{\left(\rho_\omega, \hat{\mu}_\omega\right), \omega \in [\Omega]\right\}$ of Ω realizations of $(\tilde{\rho}, \tilde{\mu})$, and let R be the *design matrix*, which is the matrix of size $\Omega \times P$ whose row ω is the vector ρ_ω, and which is assumed to be a full rank matrix. The sample analog of the objective function in (A.2.40) is $\theta \mapsto \hat{\mu}^\top R\theta - \sum_{\omega \in [\Omega]} F^*\left((R\theta)_\omega\right)$. Thus, we have:

Theorem A.2.5. *The GLM estimator $\hat{\theta}$ is obtained as the solution to*

$$\max_{\theta \in \mathbb{R}^P} \left\{\hat{\mu}^\top R\theta - \sum_{\omega \in [\Omega]} F^*\left((R\theta)_\omega\right)\right\}, \tag{A.2.41}$$

whose dual expression is

$$\min_{\mu \geq 0} \sum_{\omega \in [\Omega]} F(\mu_\omega) \tag{A.2.42}$$
$$\text{s.t. } \mu^\top R = \hat{\mu}^\top R.$$

The estimator $\hat{\theta}$ *solution to (A.2.41) satisfies*

$$\sum_{\omega \in [\Omega]} \left(f^{-1}((R\hat{\theta})_\omega) - \hat{\mu}_\omega \right) R_{\omega p} = 0 \quad \textit{for all } p \in [P]$$

and the solution $\bar{\mu}_\omega$ *to (A.2.42) satisfies* $\bar{\mu}_\omega = f^{-1}((R\hat{\theta})_\omega)$.

Weighted GLMs. We shall sometimes need to apply different weights w_ω to different observations ω. Letting $\mathbf{\Delta}_w = diag\,(w)$, the *weighted GLM estimator* now becomes

$$\max_{\theta \in \mathbb{R}^P} \left\{ \hat{\mu}^\top \mathbf{\Delta}_w R\theta - w^\top F^*\,(R\theta) \right\}, \tag{A.2.43}$$

whose first-order conditions are $\sum_\omega w_\omega \left(\hat{\mu}_\omega - f^{-1}\,((R\theta)_\omega) \right) R_{\omega p} = 0$, and whose dual formulation consists of minimizing $\sum_\omega w_\omega F(\mu_\omega)$ subject to $\mu^\top \mathbf{\Delta}_w R = \hat{\mu}^\top \mathbf{\Delta}_w R$. We obtain

$$\min_{\mu \geq 0} \sum_{\omega \in [\Omega]} w_\omega F(\mu_\omega) \tag{A.2.44}$$
$$\text{s.t. } \mu^\top \mathbf{\Delta}_w R = \hat{\mu}^\top \mathbf{\Delta}_w R.$$

Coercivity and existence of a GLM estimator. We have so far implicitly assumed that the optimization problem (A.2.41) has a solution. This, however, need not be the case. Take, for instance, $F^*(z) = \exp(z)$, which is the case in the Poisson regression discussed next in appendix A.2.9, with a single observation and a single parameter dimension, and assume $R = 1$. Then:

- if $\mu = -1$, the maximum in (A.2.41) is infinite, as the limit of the objective function when $\theta \to -\infty$ is $+\infty$;
- likewise, if $\mu = 0$, the value of problem (A.2.41) is zero;
- finally, if $\mu = 1$, the problem has a maximizer that is $\theta = 0$ (and hence the problem has a finite value).

The following theorem provides a necessary and sufficient condition for the existence of an estimator.

Theorem A.2.6 (Existence of a GLM estimator). *Assume that* $f^{-1}(w) > 0$ *for all* $w \in \mathbb{R}$, *and that* $F^*\,(w) \to 0$ *as* $w \to -\infty$ *and* $F^*\,(w)\,/w \to +\infty$ *as* $w \to +\infty$, *and consider problem* (A.2.41). *Then:*

(i) *the value of the problem is finite if and only if there is a vector* $\bar{\mu}$ *such that* $\bar{\mu}_\omega \geq 0$ *for all* $\omega \in [\Omega]$ *and* $\bar{\mu}^\top R = \hat{\mu}^\top R$.

(ii) *the problem has a maximizer if and only if there is a vector* $\bar{\mu}$ *such that* $\bar{\mu}_\omega > 0$ *for all* $\omega \in [\Omega]$ *and* $\bar{\mu}^\top R = \hat{\mu}^\top R$.

The assumption of theorem A.2.6 is satisfied in particular by the Poisson regression, which has $F^*(w) = \exp w$. While the proof of the theorem is left to the reader as an exercise, at least the direct implication in (ii) is easy to see: first-order conditions yield $f^{-1}((R\theta)_\omega)^\top R = \hat{\mu}^\top R$, so setting $\bar{\mu}_\omega = f^{-1}((R\theta)_\omega)$ leads to the result.

The following criterion, based on linear programming, allows for a simple test of conditions (i) and (ii) in theorem A.2.6. Consider the linear programming problem, called the *coercivity detector*,

$$V = \max_{t \geq 0, \bar{\mu}_\omega \geq 0} t \qquad \text{subject to } t\mathbf{1}_\Omega - \bar{\mu} \leq 0 \text{ and } \bar{\mu}^\top R = \hat{\mu}^\top R, \tag{A.2.45}$$

which is always feasible, as $t = 0$, $\bar{\mu} = \mu$ satisfies the constraints. By strong duality, the value of the coercivity detector coincides with the value of its dual expression

$$V = \min_{u_\omega \geq 0, \lambda_k} \lambda^\top Z\mu \qquad \text{subject to } \mathbf{1}_\Omega^\top u \geq 1 \text{ and } u \leq R\lambda, \tag{A.2.46}$$

and inspection of the dual problem reveals that it is not always feasible: for instance, when $R = 0$, there is no vector $u_\omega \geq 0$ such that $u \leq R\lambda$ and $\sum_\omega u_\omega \geq 1$. Hence V can take values in $\mathbb{R}_+ \cup \{+\infty\}$. The following theorem holds:

Theorem A.2.7 (Coercivity detector). *Consider* V*, the value of both the primal problem (A.2.45) and its dual (A.2.46). Then:*

(i) $V = +\infty$ *if and only if problem* (A.2.41) *is infinite.*
(ii) $0 < V < +\infty$ *if and only if problem* (A.2.41) *has a finite value and a maximizer.*
(iii) $V = 0$ *if and only if problem* (A.2.41) *has a finite value but no maximizer.*

References

References include Agresti (2012) and McCullagh and Nelder (1989).

A.2.9 POISSON REGRESSION

The random variable $\tilde{\mu}$ valued in $\mathbb{N}$ follows a *Poisson distribution* of parameter m if

$$\Pr(\tilde{\mu} = \mu) = e^{-m} \frac{m^\mu}{\mu!}.$$

The Poisson distribution has mean m and variance m. If $\rho \in \mathbb{R}^P$ is a vector of covariates, we assume that the conditional distribution of $\tilde{\mu}$ given $\tilde{\rho} = \rho$ is a Poisson distribution of parameter $m(\rho) = \exp\left(\rho^\top \theta\right)$, that is:

Definition A.2.3 (Poisson dependence). *Assume that the conditional distribution of μ given ρ is given by*

$$\Pr\left(\tilde{\mu} = \mu \mid \tilde{\rho} = \rho\right) = \exp(-\exp(\rho^\top \theta)) \frac{\exp(\mu \rho^\top \theta)}{\mu!}. \tag{A.2.47}$$

It follows from the Poisson dependence relation (A.2.47) that both the conditional expectation and the variance of $\tilde{\mu}$ given $\tilde{\rho} = \rho$ are equal to the parameter $m(\rho) = \exp\left(\rho^\top \theta\right)$, therefore

$$\mathbb{E}[\tilde{\mu} = \mu \mid \tilde{\rho} = \rho] = \exp\left(\rho^\top \theta\right) \quad \text{and} \quad \mathbb{V}\left[\tilde{\mu} = \mu \mid \tilde{\rho} = \rho\right] = \exp\left(\rho^\top \theta\right). \tag{A.2.48}$$

If we observe a sample of observations $(\hat{\mu}_\omega, \rho_\omega)_{\omega \in [\Omega]}$ indexed by $\omega \in [\Omega]$, the log-likelihood of observation ω is given by $\mu_\omega \rho_\omega^\top \theta - \exp\left(\rho_\omega^\top \theta\right)$, and therefore the log-likelihood of the sample is

$$\ell(\theta) = \hat{\mu}^\top R\theta - \sum_{\omega \in [\Omega]} \exp\left(\rho_\omega^\top \theta\right), \tag{A.2.49}$$

where the design matrix R is the $\Omega \times P$ matrix whose rows are the vectors ρ_ω. This yields the Poisson regression estimator:

Definition A.2.4. *The Poisson regression consists of the maximization of the log-likelihood function (A.2.49). The Poisson regression estimator $\hat{\theta}$ is thus the parameter vector θ solution to*

$$\max_{\theta \in \mathbb{R}^P} \left\{ \hat{\mu}^\top R\theta - \sum_{\omega \in [\Omega]} \exp\left((R\theta)_\omega\right) \right\}. \tag{A.2.50}$$

A Poisson regression can be recast as a *generalized linear model* (GLM) with the logarithm as link function. Then $f^{-1} = \exp$, and thus $F^* = \exp$ and expression (A.2.41) boils down to the Poisson regression. A primitive F is $F(m) = m \log m$, and by (A.2.42) the dual of the Poisson regression thus formulates as

$$\begin{aligned} &\min_{\mu \geq 0} \sum_{\omega \in [\Omega]} \mu_\omega \log \mu_\omega \\ &\text{s.t. } \mu^\top \Phi = \hat{\mu}^\top \Phi. \end{aligned} \tag{A.2.51}$$

Weighted Poisson regression. For the weighted Poisson regression with weight w_ω placed on observation ω, the primal formulation (A.2.50) becomes

$$\max_{\theta\in\mathbb{R}^P} \left\{ \hat{\mu}^\top \mathbf{\Delta}_w R\theta - w^\top \exp(R\theta) \right\}, \tag{A.2.52}$$

where, as in section A.2.8, $\mathbf{\Delta}_w = diag(w)$, and the dual formulation becomes

$$\min_{\mu\geq 0} \sum_{\omega\in[\Omega]} \mu_\omega \log \frac{\mu_\omega}{w_\omega} \tag{A.2.53}$$
$$\text{s.t. } \mu^\top \Phi = \hat{\mu}^\top \Phi.$$

Asymptotics. It will be useful to investigate the asymptotic distribution of $\hat{\theta}$ under a particular setting, namely, when Ω remains fixed, but when $N = \sum_\omega \hat{\mu}_\omega$ gets large. In that case, we need to introduce the concept of *homogeneous* weighted Poisson regression, as follows. We shall assume that

$$\hat{\pi} := \frac{\hat{\mu}}{N} \tag{A.2.54}$$

converges almost surely to π, and that the following convergence in distribution holds:

$$\sqrt{N}\left(\hat{\pi} - \pi\right) \xrightarrow{d} \mathcal{N}(0, \mathbf{V}_\pi). \tag{A.2.55}$$

Definition A.2.5. *A weighted Poisson regression with design matrix R and vector of weights $w \in \mathbb{R}_+^\Omega$ is called homogeneous if $\mathbf{1}_\Omega \in \operatorname{Im} R$, which implies that there is a unique vector $v \in \mathbb{R}^P$ such that $Rv = \mathbf{1}_\Omega$.*

In that case, we consider the Poisson regression of $\hat{\mu}$ on R with weights w, which formulates as problem (A.2.52) above. Letting $\mu^\theta = \exp(R\theta)$, the optimality conditions that characterize $\hat{\theta}$ are

$$R^\top \mathbf{\Delta}_w \mu^{\hat{\theta}} = R^\top \mathbf{\Delta}_w \hat{\mu}.$$

Now assume that we rescale the dependent variable, and we define $\hat{\mu}_K = \hat{\mu}/K$. Letting $\hat{\theta}_K$ be the resulting estimator, one has $\mu^{\hat{\theta}-(\ln K)v} = \mu^{\hat{\theta}}/K$, and thus $\hat{\theta}_K = \hat{\theta} - (\ln K)v$.

Under the assumption that we are in the homogeneous case, we shall replace $\hat{\mu}$ by $\hat{\pi} = \hat{\mu}/N$ in the Poisson regression. Indeed, the new estimator differs from the previous one by a $(\ln N)v$ term. We will thus consider the regression of $\hat{\pi}$ on R, and we shall from now on call $\hat{\theta}$ the regressor in that regression, so that $R^\top \mathbf{\Delta}_w \pi^{\hat{\theta}} = R^\top \mathbf{\Delta}_w \hat{\pi}$, where $\pi^\theta = \exp(R\theta)$. Letting θ_0 be the value of θ that satisfies the previous equation in the limit, that is, when $\hat{\pi}$ is replaced by π, we recall the asymptotic behavior of $\hat{\pi}$, which is given by (A.2.54) and (A.2.55). We have

$$R^\top \mathbf{\Delta}_w \exp(R\hat{\theta}) = R^\top \mathbf{\Delta}_w \hat{\pi}, \tag{A.2.56}$$

which, letting $\delta\hat{\theta} = \hat{\theta} - \theta_0$ and $\delta\hat{\pi} = \hat{\pi} - \pi$, gets linearized into $R^\top \mathbf{\Delta}_w \mathbf{\Delta}_{\pi^{\theta_0}} R\delta\hat{\theta} = R^\top \mathbf{\Delta}_w \delta\hat{\pi} + o_P(\sqrt{N})$, and therefore:

Theorem A.2.8. *Under the assumptions made above, and letting* $B = R^\top \mathbf{\Delta}_w \mathbf{\Delta}_{\pi^{\theta_0}} R$ *and* $C = R^\top \mathbf{\Delta}_w \mathbf{V}_\pi \mathbf{\Delta}_w R$, *the convergence* $\hat{\theta} \to \theta_0$ *holds almost surely as* $N \to +\infty$, *and*

$$\sqrt{N}(\hat{\theta} - \theta_0) \xrightarrow{d} \mathcal{N}\left(0, B^{-1}CB^{-1}\right). \tag{A.2.57}$$

References

In addition to the references on generalized linear models provided in the previous section, good sources for Poisson regressions and their applications are Winkelmann (2008) and Cameron and Trivedi (2013).

A.2.10 EXPONENTIAL FAMILIES

Exponential families are families of distributions for which maximum likelihood estimation is a convex optimization problem. More specifically, consider a parametric family $f(z;\theta)$ with support $\mathcal{Z} \subseteq \mathbb{R}^P$ and parameter $\theta \in \mathbb{R}^P$. Let $\Lambda : \mathbb{R}^P \to \mathbb{R}$ be a convex function, and let $\Lambda^* : \mathbb{R}^P \to \mathbb{R}$ be its Legendre transform, defined by $\Lambda^*(\theta) = \max_\zeta \{\zeta^\top \theta - \Lambda(\zeta)\}$. If we observe i.i.d. data $\{z_\omega : \omega \in [\Omega]\}$ sampled from $f(z;\theta^\star)$, we may define $\bar{z} = \Omega^{-1} \sum_\omega z_\omega$, and the problem

$$\arg\max_{\theta \in \mathbb{R}^P} \left\{ \frac{1}{\Omega} \sum_{\omega \in \Omega} z_\omega^\top \theta - \Lambda^*(\theta) \right\} \tag{A.2.58}$$

yields $\nabla\Lambda^*(\hat{\theta}) = \bar{z}$ for $\hat{\theta}$, the optimal value of the parameter θ in (A.2.58). In that case, the likelihood associated with observation z, $\log f(z;\theta)$, coincides with $z^\top \theta - \Lambda^*(\theta)$ up to an irrelevant function of z, denoted by $a(z)$, and thus

$$f(z;\theta) = \exp(z^\top \theta - \Lambda^*(\theta) - a(z)). \tag{A.2.59}$$

Further, as Λ is the Legendre–Fenchel transform of Λ^*, the value of the likelihood at the optimal value of the parameter is equal to $\Lambda(\bar{z}) - a(\bar{z})$. Obviously, imposing that $f(.;\theta)$ be a density requires that it sum to 1 when integrating over $z \in \mathcal{Z}$, and thus requires that

$$\Lambda^*(\theta) = \log \int_{\mathcal{Z}} \exp(z^\top \theta - a(z))dz. \tag{A.2.60}$$

It is time to make two remarks. First, note that while problem (A.2.58) bears close resemblance with a generalized linear model, it is not one in a strict sense—indeed, in our exposition of generalized linear models in section A.2.8, we have requested that the link function be univariate, while here, the analog of the link function, which is $\nabla\Lambda^*$, is a function from $\mathbb{R}^P$ to $\mathbb{R}^P$. Another distinction is that generalized linear models do not require Λ^* to be of the form (A.2.60), which means that generalized linear models do not necessarily fall within the maximum likelihood framework.

The last point brings us to a second remark: any function $a(z)$ on $\mathcal{Z}$ certainly defines a function Λ^* by (A.2.60) that is convex on its domain, but, conversely, not all convex functions Λ^* can be represented as (A.2.60). For instance, in dimension 1 and when $\mathcal{Z} = \mathbb{R}_-$, the Bernstein theorem imposes that the signs of the higher-order derivatives of $\theta \mapsto \exp(\Lambda^*(\theta))$ alternate in order for $a(z)$ to exist.

Examples. Many of the usual families of distributions are exponential families or can be recast as such. We mention only three of them, though many other examples exist.

Example A.3 (Example of exponential families). The following classical parametric families of distributions can be reformulated (often, after a nonlinear transformation of the variables) as exponential families:

(i) Gaussian distributions: In the univariate case, the p.d.f. of a Gaussian distribution with mean μ and variance σ^2 is proportional to $\exp(-(x-\mu)^2/(2\sigma^2))$, so this is (up to a transformation) an exponential family with $P=2$, $z_1 = x$, $z_2 = x^2$, $\theta_1 = \mu/\sigma^2$, $\theta_2 = -1/(2\sigma^2)$, $a(z) = 0$, and $\mathcal{Z} = \{z \in \mathbb{R}^2 : z_2 = z_1^2\}$. The extension to the multivariate case is straightforward.

(ii) Poisson distribution: A Poisson distribution with parameter λ assigns a probability mass proportional to $\exp(-\lambda)\lambda^x = \exp(x \ln \lambda - \lambda)$ to each $x \in \mathbb{N}$, so this is an exponential family with $P=1$, $\mathcal{Z} = \mathbb{N}$, $z = x$, $\theta = \ln \lambda$, and $\Lambda^*(\theta) = \exp \theta$.

(iii) Exponential distribution: An exponential distribution with parameter $\lambda \geq 0$ has p.d.f. $\lambda \exp(-\lambda x)$ for $x \geq 0$, which is an exponential family with $P=1$, $\mathcal{Z} = \mathbb{R}_+$, $\theta = -\lambda$, and $\Lambda^*(\theta) = -\ln(-\lambda)$.

We denote by $\mathbb{E}_\theta$ and $\mathbb{V}_\theta$ the expectation and variance-covariance matrix with respect to the distribution $f(.|\theta)$, and we have:

Proposition A.2.7. *If $\tilde{z} \sim f(.;\theta)$, then*

$$\nabla \Lambda^*(\theta) = \mathbb{E}_\theta[\tilde{z}] \text{ and } D^2\Lambda^*(\theta) = \mathbb{V}_\theta[\tilde{z}],$$

and if $\{z_\omega : \omega \in [\Omega]\}$ is an i.i.d. sample of random variables drawn from distribution $f(z;\theta^\star)$, the maximum likelihood estimator $\hat{\theta}$ of $\theta^\star$ is such that $\nabla\Lambda^*(\hat{\theta}) = \bar{z}$, where $\bar{z} = \Omega^{-1}\sum_\omega z_\omega$ has been defined as the sample average of the observations.

Latent data. Exponential families allow for a relatively simple way to handle missing data. Indeed, if $\mathcal{Z}' \subseteq \mathcal{Z}$ then it is easy to see that the distribution of $\tilde{z} \sim f(.;\theta)$ given $\tilde{z} \in \mathcal{Z}'$ is $f\left(z|\mathcal{Z}';\theta\right) = \exp(z^\top\theta - \Lambda^*_{\mathcal{Z}'}(\theta))$, where

$$\Lambda^*_{\mathcal{Z}'}(\theta) = \log \int_{\mathcal{Z}'} \exp(z^\top\theta)dz, \tag{A.2.61}$$

and one has

$$\nabla\Lambda^*_{\mathcal{Z}'}(\theta) = \mathbb{E}_\theta\left[\tilde{z}|\tilde{z} \in \mathcal{Z}'\right] \text{ and } \Pr\left(z \in \mathcal{Z}'\right) = \exp(\Lambda^*_{\mathcal{Z}'}(\theta) - \Lambda^*_{\mathcal{Z}}(\theta)). \tag{A.2.62}$$

As a result, if the data $z_1, \ldots, z_\Omega$ drawn from $f(.;\theta^\star)$ is latent, but if instead one observes the images $w_\omega = \varphi(z_\omega)$ of the sample points z_ω by a measurable map $\varphi : \mathcal{Z} \to \mathbb{R}^P$, we can use the previous results with $\mathcal{Z}' = \varphi^{-1}(w)$ in order to show that the maximum likelihood estimator of θ based on the sample $\{w_1, \ldots, w_\Omega\}$ is characterized by:

Proposition A.2.8. *The maximum likelihood estimator $\hat{\theta}$ based on $\{w_1, \ldots, w_\Omega\}$, where $w = \varphi(z)$ is characterized by*

$$\frac{1}{\Omega}\sum_{\omega\in[\Omega]} \mathbb{E}_{\hat{\theta}}\left[\tilde{z}|\varphi(\tilde{z}) = w_\omega\right] = \mathbb{E}_{\hat{\theta}}[\tilde{z}]. \tag{A.2.63}$$

A Bayesian point of view. If we put a prior distribution $\pi_0(\theta)$ on the parameter, it is easy to see that the posterior distribution of θ conditional on observing one observation $\hat{z}$ is

$$\pi\left(\theta|\hat{z}\right) = \frac{\pi_0(\theta)\exp(\hat{z}^\top\theta - \Lambda^*(\theta))}{\int \pi_0(\theta')\exp(\hat{z}^\top\theta' - \Lambda^*(\theta'))d\theta'}.$$

If one assumes a parametric form for the distribution π of θ—which is to say, if one parametrizes the distribution of the parameter, and if one lets $(\zeta, \nu) \in \mathbb{R}^P \times \mathbb{R}$ be that hyper-parameter—we may consider

$$\pi(\theta; \zeta, \nu) \propto \exp(\zeta^\top \theta - \nu \Lambda^*(\theta)),$$

where the $\propto$ sign indicates that one should divide by a normalization constant. Assuming that $\pi_0(.) = \pi(.; \zeta_0, \nu_0)$, we can show that $\pi_1(\theta) = \pi(.; \zeta_1, \nu_1)$, where $\zeta_1 = \zeta_0 + \hat{z}$ and $\nu_1 = \nu_0 + 1$.

References

A good and accessible reference for exponential families is the book Efron (2022). Formulas (A.2.60)–(A.2.63) are called Sundberg's formulas, and appeared in a paper by that author in Sundberg (1974).

A.2.11 THE EM ALGORITHM

The Expectation-Maximization (EM) algorithm is a powerful iterative method used for finding maximum likelihood estimates in statistical models where the data is incomplete or has missing values. The core idea of the EM algorithm is to iteratively improve parameter estimates for models with latent variables or incomplete data by alternating between two steps: the expectation (E) step and the maximization (M) step. In the E-step, the algorithm estimates the missing data based on the current parameters of the model, while in the M-step, it updates the model parameters to maximize the likelihood given the estimated data from the E-step. By iterating between these two steps, the EM algorithm gradually refines both the estimates of the missing data and the model parameters until it reaches a point where the changes are minimal, and the algorithm converges to a local maximum of the likelihood function.

To describe the algorithm, consider a latent variable Z whose distribution belongs to a parametric family $f_Z(z; \theta)$. However, Z is observed not fully but only partially, in the sense that only $Y = \varphi(Z)$ is, where the function φ is not invertible. For instance, there may be right-censoring, with $\varphi(Z) = \min\{Z, z_0\}$, i.e., we don't observe the values above the threshold but we observe when they exceed the threshold.[1] One observes an i.i.d. sample $z_\omega = \varphi(z_\omega)$ where the latent variables z_ω are drawn from the distribution $f_Z(\cdot; \theta^\star)$, for some true value $\theta^\star$ of the parameter, and one would like to find

[1] *Censoring* should be distinguished from *truncation*. In the presence of right-censoring, one observes the distribution of $\min\{Z, z_0\}$. In the presence of right-truncation, we observe the distribution of Z conditional on $Z \leq z_0$.

an estimator of $\theta^\star$. The most straightforward approach consists of denoting by $f_Y(y;\theta)$ the distribution of $Y=\varphi(Z)$ if $Z\sim f_Z(\cdot;\theta)$, and carrying out the maximum likelihood estimation, where the log-likelihood of the sample is $\sum_\omega \log f_Y(y_\omega;\theta)$. For instance, in the case of censored data, the distribution of Y has a continuous part on $(-\infty, z_0]$ with density $f_Z(y;\theta)$, and a mass point z_0 with probability $\Pr(Y=z_0)=\int_{z_0}^{+\infty} f_Z(z;\theta)\,dz$, and the log-likelihood of the sample is maximized accordingly, yielding

$$\max_\theta \left\{ \sum_{\omega\in[\Omega]} 1\{y_\omega < z_0\} \log f_Z(y_\omega;\theta) + \sum_{\omega\in[\Omega]} 1\{y_\omega = z_0\} \log \int_{z_0}^{+\infty} f_Z(z;\theta)\,dz \right\}. \tag{A.2.64}$$

Maximizing a quantity such as (A.2.64) may, however, not be easy in practice due to difficulties in evaluating $\int_{z_0}^{+\infty} f_Z(z;\theta)\,dz$, which may involve high-dimensional integration. A first way out of this problem consists of replacing the objective function in (A.2.64) by

$$\sum_{\omega\in[\Omega]} 1\{y_\omega < z_0\} \log f_Z(y_\omega;\theta) + \sum_{\omega\in[\Omega]} 1\{y_\omega = z_0\} \log(B^{-1} \sum_{b\in[B]} 1\{z_b \geq z_0\}), \tag{A.2.65}$$

where the terms z_b are B additional samples of simulated data drawn from $f_Z(\cdot;\theta)$, and maximizing over θ. However, this may not be a satisfactory solution. First, a maximization problem typically requires the gradient of the objective function; yet that information cannot be deduced from (A.2.65). A second, deeper problem is that it may be difficult to sample draws z_b from $f_Z(\cdot;\theta)$. Actually, the original data already contains a part of this information, through the points y_ω. In some cases, it is easy to sample Z conditional on $Y=y$ according to $f_{Z|Y}\left(\cdot|y;\theta\right)$, but difficult to sample the full random elements Z according to $f_Z(\cdot;\theta)$. If that is the case, one would want to reuse our sample $\{y_\omega\}$.

Following that insight, let us assume that for each observation ω, we may draw a sample of simulated points $z_{b\omega}$ from $f_{Z|Y}\left(\cdot|y_\omega;\theta\right)$, so that, by definition, we have $\varphi(z_{b\omega})=y_\omega$. The distribution of each $z_{b\omega}$ is $f_{Z|Y}\left(\cdot|y_\omega;\theta\right) f_Y(y_\omega;\theta^\star)$, which differs from $f_Z(\cdot;\theta^\star)$ unless $\theta=\theta^\star$. If we knew $\theta^\star$, we could use it in order to draw points $z_{b\omega}$ from $f_{Z|Y}\left(\cdot|y_\omega;\theta^\star\right)$, and thus the distribution of $z_{b\omega}$ would be $f_Z(\cdot;\theta^\star)$. Conversely, if the distribution of $z_{b\omega}$ were $f_Z(\cdot;\theta^\star)$, then maximum likelihood estimation would recover the parameter $\theta^\star$. The EM algorithm, described in display A.2.3, consists of alternating between these ideas. An initial guess of the parameter is taken. At period t, given the current parameter estimate θ^t, one draws in the E-step (expectation step), for each $\omega\in[\Omega]$, B simulated points $z_{b\omega}$ ($b\in[B]$) sampled in an i.i.d. fashion (conditionally on y_ω) from the distribution $f_{Z|Y}\left(\cdot|y_\omega;\theta^t\right)$. Each $z_{b\omega}$

Algorithm A.2.3 EM algorithm

Input : The observed sample $\{y_\omega\}$; the likelihood of the latent variable $f_Z(\cdot;\theta)$; a generator of $f_{Z|Y}(\cdot|y;\theta)$

Parameter: the starting point θ^0, the tolerance ξ

Set $t=0$, $\theta^t=\theta^0$

repeat

E-step: for each $\omega\in[\Omega]$ draw B i.i.d. points from the conditional distribution of Z given $Y=y_\omega$ at the current estimate of θ,

$$z_{b\omega}\sim f_{Z|Y}\left(\cdot|y_\omega;\theta^t\right),\; b\in[B]$$

M-step: update the estimate of θ by maximum likelihood estimation on the sample thus generated,

$$\theta^{t+1}=\arg\max_{\theta\in\mathbb{R}^p}\frac{1}{B}\sum_{b\omega\in[B\times\Omega]}\log f_Z(z_{b\omega};\theta)$$

$t=t+1$

until $\|\theta^{t+1}-\theta^t\|\leq\xi$;

Return : θ^t

has therefore distribution $f_{Z|Y}\left(\cdot|y_\omega;\theta^t\right)f_Y(y_\omega;\theta^\star)$. In the M-step (maximization step), we maximize the likelihood of the sample thus generated, and we update the parameter estimate to θ^{t+1}.

Large simulated pool. The sample uncertainty associated with the simulation in the E-step complicates the analysis, but it vanishes when B becomes large. For simplicity, we shall from now on place ourselves in the limit $B\to+\infty$ to make a claim about the algorithm. In that case, combining the E-step and the M-step, the algorithm becomes

$$\theta^{t+1}=\arg\max_{\theta\in\mathbb{R}^p}\sum_{\omega\in[\Omega]}\mathbb{E}_{f_{Z|Y}(\cdot|y_\omega,\theta^t)}\left[\log f_Z(Z;\theta)\right],$$

where the random term Z in the expectation has distribution $f_{Z|Y}(\cdot|y_\omega,\theta^t)$, which depends on y_ω. In the case when the parametric family of the latent variables belongs to the exponential family class, these update formulas take a particularly nice form.

Example A.4 (EM algorithm for exponential families). Assume that $f_Z(\cdot;\theta)$ is an exponential family as described in section A.2.10. Because of proposition

A.2.8, the EM algorithm consists of updating θ^{k+1} such that

$$\mathbb{E}_{\theta^{k+1}}[Z] = \mathbb{E}_{\theta^k}[Z|y_\omega]. \tag{A.2.66}$$

Interpretation as coordinate descent. There is a nice interpretation due to Neal and Hinton of the EM algorithm as a coordinate ascent algorithm, in which one maximizes alternatively over the space of conditional distributions of z given y and over the parameter space. Letting

$$\mathcal{L}(g,\theta) = \sum_{\omega\in[\Omega]} \mathbb{E}_{g(\cdot|y_\omega)}\left[\log \frac{f_Z(Z,\theta)}{g(Z|y_\omega)}\right],$$

we note that the expression $\mathcal{L}(g,\theta)$ can be rewritten as

$$\mathcal{L}(g,\theta) = \sum_{\omega\in[\Omega]} \mathbb{E}_{g(\cdot|y_\omega)}\left[\log \frac{f_{Z|Y}(Z|y_\omega;\theta)}{g(Z|y_\omega)}\right] + \sum_{\omega\in[\Omega]} \log f_Y(y_\omega;\theta) \tag{A.2.67}$$

$$\text{and } \mathcal{L}(g,\theta) = \sum_{\omega\in[\Omega]} \mathbb{E}_{g(\cdot|y_\omega)}\left[\log f_Z(Z,\theta)\right] - \sum_{\omega\in[\Omega]} \mathbb{E}_{g(\cdot|y_\omega)}\left[\log g(Z|y_\omega)\right], \tag{A.2.68}$$

from which we can deduce:

Proposition A.2.9. *One has*

$$g^t = \arg\max_g \mathcal{L}(g,\theta^t) \quad \text{and} \quad \theta^{t+1} = \arg\max_{\theta\in\mathbb{R}^P} \mathcal{L}(g^t,\theta). \tag{A.2.69}$$

As a result, we have $\mathcal{L}(g^{t+1},\theta^{t+1}) \geq \mathcal{L}(g^{t+1},\theta^t) \geq \mathcal{L}(g^t,\theta^t)$, but $\mathcal{L}(g^t,\theta^t) = \max_g \mathcal{L}(g,\theta^t)$, and because of expression (A.2.67), $\max_g \mathcal{L}(g,\theta^t)$ is obtained for $g = g^t = f_{Z|Y}(\cdot|y_\omega;\theta)$ and has therefore $\mathcal{L}(g^t,\theta^t) = \sum_{\omega\in[\Omega]} \log f_Y(y_\omega;\theta^t)$. We have shown that each step of the EM algorithm increases the value of the likelihood, which we state formally as:

Theorem A.2.9. *The value of the likelihood $\sum_\omega \log f_Y(y_\omega;\theta^t)$ is nondecreasing with respect to t.*

References

The first general analysis of the Expectation-Maximization (EM) algorithm was provided in the seminal paper Dempster, Laird, and Rubin (1977), where the authors formulated the general procedure for maximum likelihood estimation from incomplete data. Prior to this, Sundberg (1974) and Sundberg (1976) had made important early contributions based on ideas by Per and Anders Martin-Löf, laying the groundwork for the formalization of the EM algorithm and providing the formulation in the context of exponential families described in equation (A.2.66) above. A nice historical account is provided in the paper Laird (2014), which provided a historical account of Sundberg's contributions to the EM algorithm and its recognition in the statistical community. In a famous paper, Neal and Hinton (1998) offered an important reinterpretation of the EM algorithm as a coordinate ascent algorithm, providing insights into its convergence properties and leading to various extensions, such as incremental and sparse versions.

A.3 Numerical Analysis

A.3.1 VECTORS AND TENSORS

This section will deal with *tensors*, also called *multidimensional arrays*. Throughout this text, we shall work with tensors of various dimensions and sizes. We shall sometimes treat the same data as tensors of several different dimensions, and it is important to understand the mathematical properties of the operations consisting of reshaping these tensors. For instance, if we have data $T_{x,y,z}$ for $x \in [X]$, $y \in [Y]$, and $z \in [Z]$, we may naturally view T as a three-dimensional tensor of shape (X, Y, Z); but we may also lump the x and the y indices into $xy \in [X \times Y]$ and view T as a two-dimensional tensor (matrix) of shape (XY, Z); alternatively, we may also view the data as a matrix of shape (X, YZ); or we may stack all the data in a single vector and view T as a one-dimensional tensor of shape (XYZ).

Vectorization. Regardless of its shape, any tensor is represented as a sequence of numbers—the *vectorized* form—in the memory of the computer. In the case of a matrix, for instance, two obvious ways of storing its entries are stacking the rows and stacking the columns. In fact, programming languages are split regarding how they vectorize matrices (and, more generally, tensors).

- Stacking the rows of matrices (which is called the "row-major order") is used by `C`, and is the default convention for `NumPy`, `PyTorch`, and `TensorFlow`. According to this order, a matrix A is represented by varying the last index first, i.e., a 2×2 matrix is represented as $vec_R(A) = (A_{11}, A_{12}, A_{21}, A_{22})$. More generally, indices of a tensor are listed in the lexicographic (increasing) order, as they would be in a dictionary.

- Stacking the columns of matrices (or working with the "column-major order") is used by `Fortran`, `Matlab`, `R`, and most underlying core linear algebra libraries such as `BLAS`. The 2×2 matrix described above is represented under the column-major order as $vec_C(A) = (A_{11}, A_{21}, A_{12}, A_{22})$. In any dimension, a tensor is vectorized by listing the indices in reverse lexicographic order, by varying the first index first, then the second index, and so on.

Recall the standard notation $[N] = \{1, \ldots, N\}$, which is useful for single indices. In this text, I introduce a convenient notation for multiple indices. Given P integers $N_1, N_2, \ldots, N_P$, $[N_1 \times N_2 \times \cdots \times N_P]$ denotes the list of P-tuples $n_1 n_2 \ldots n_P$, where $n_p \in [N_p]$ for $p \in [P]$, and the P-tuples are ordered in the lexicographic increasing order.

Hadamard product. Given A and B, two matrices (or more generally tensors) of the same size, the Hadamard product $A \odot B$ is the tensor whose entries are the element-wise product of the entries of A and B. For instance, if A and B are two matrices of size $I \times J$, $(AB)_{ij} = A_{ij} B_{ij}$. In `NumPy`, the Hadamard product is implemented by the standard multiplication operator `*`.

Kronecker product. Given A a matrix of size $M \times N$ and B a matrix of size $P \times Q$, the *Kronecker product* $A \otimes B$ of A and B is the $MP \times NQ$ matrix defined by blocks as

$$A \otimes B = \begin{pmatrix} a_{11}B & \cdots & a_{1N}B \\ \vdots & \ddots & \vdots \\ a_{M1}B & \cdots & a_{MN}B \end{pmatrix}. \tag{A.3.1}$$

The Kronecker product is intimately connected with vectorization. Indeed, $A \otimes B$ has MP rows that are naturally doubly indexed by $m \in [M]$ and $p \in [P]$. The first column of $A \otimes B$, for instance, takes values from the first columns of both A and B; because it contains the repetition of the first column of B in a block-wise fashion, it lists the mp in the lexicographic increasing order, which is associated with the row-major order. Thus, the rows of $A \otimes B$ are indexed by $mp \in [M \times P]$. Similarly, each row of $A \otimes B$ contains repetitions of rows of B, and thus, the columns of $A \otimes B$ are indexed by $nq \in [N \times Q]$. As a result, we have

$$\text{for all } mp \in [M \times P] \text{ and } nq \in [N \times Q],\ (A \otimes B)_{mp,nq} = a_{m,n} b_{p,q}.$$

If $u \in \mathbb{R}^I$ and $v \in \mathbb{R}^J$ are both column vectors, then $u \otimes v^\top$ is a rank 1 matrix of size $I \times J$, and $vec_R(u \otimes v^\top) = u \otimes v$. We have the important identity

$$(A \otimes B)(C \otimes D) = (AC) \otimes (BD), \tag{A.3.2}$$

assuming that the matrix shapes are such that the products AC and BD exist.

Vectorization of matrix products. Next, we investigate the vectorization of matrix products. More precisely, if A is a matrix of size $M \times N$, if X is a matrix of size $N \times P$, and if B is a matrix of size $P \times Q$, we expect the vector of entries of the matrix product AXB to be a linear transform of the entries of the vector of entries of X. That is, there must be a relation of the form $vec_R(AXB) = K\, vec_R(X)$, where K is a matrix to be determined. We have

$$(AXB)_{m,q} = \sum_{n\in[N]} \sum_{p\in[P]} A_{m,n} X_{n,p} B_{p,q},$$

which we can rewrite in a vectorized way as $\sum_{np\in[N\times P]} (A_{m,n} B_{p,q}) X_{np}$. We note that $(A_{m,n} B_{p,q}) = (A_{m,n} B^{\top}_{q,p}) = (A \otimes B^{\top})_{mq,np}$, which implies

$$(AXB)_{m,q} = \sum_{np\in[N\times P]} (A \otimes B^{\top})_{mq,np} X_{np}.$$

As a result, the desired relation between the entries of AXB and those of X is

$$vec_R(AXB) = \left(A \otimes B^{\top}\right) vec_R(X). \tag{A.3.3}$$

Reduction and expansion. Kronecker products allow us to perform important operations on the entries of tensors, namely, reduction and expansion. Consider a tensor $\left(U_{x,y,z}\right)$ of shape (X, Y, Z). Its (sum) *reduction* with respect to y is the tensor S of shape (X, Z) whose (X, Y) entry is $S_{xz} = \sum_{y\in[Y]} U_{xyz}$. Recall that $\mathbf{1}_Y$ denotes the Y-dimensional column vector with all its entries equal to 1. The vector of entries of the reduced tensor S is given by

$$vec_R(S) = \left(\mathbf{I}_X \otimes \mathbf{1}_Y^{\top} \otimes \mathbf{I}_Z\right) vec_R(U). \tag{A.3.4}$$

Conversely, given a tensor $(V_{x,z})$ of shape (X, Z), its *expansion* of size Y between the first and second dimensions is the tensor T of shape (X, Y, Z) whose (x, y, z) entry is $T_{x,y,z} = V_{x,z}$. The vector of entries of the expanded tensor T is given by

$$vec_R(T) = (\mathbf{I}_X \otimes \mathbf{1}_Y \otimes \mathbf{I}_Z)\, vec_R(V). \tag{A.3.5}$$

Reductions and expansions are ubiquitous operations on tensors, and libraries that operate on tensors such as `NumPy`, `TensorFlow`, and `PyTorch`, have built-in versions of them. In `NumPy`, for instance, the tensor reduction operation (A.3.5) would be implemented using `S = U.sum(axis=1)`, and the tensor expansion operation would be implemented using `T = V[:,None,:]`.

Row-wise Kronecker product. We define the row-wise Kronecker product (also called the Khatri–Rao product) $\circledast$ as an operator acting on two matrices with the same number of rows. Let A be an $I \times J$ matrix and B be an

$I \times K$ matrix. The row-wise Kronecker product is defined as the $I \times JK$ matrix whose ith row is the Kronecker product of the ith row of A and the ith row of B. That is, for $i \in [I]$ and $jk \in [J \times K]$,

$$(A \circledast B)_{i,jk} = A_{ij} B_{ik}. \tag{A.3.6}$$

In particular, if a and b are column vectors (i.e., matrices with a single column), then $a \circledast b = a \otimes b$ and $a^\top \circledast b^\top = a^\top \odot b^\top$. Hence, the row-wise Kronecker product acts on column vectors as the standard Kronecker product, and on row vectors as the Hadamard product. The row-wise Kronecker product is implemented in Python in appendix B.9.2.

Diagonal selection matrix. For a positive integer I, we introduce the *diagonal selection matrix* $\mathfrak{D}_I$, a matrix of size $I \times I^2$ defined by

$$\mathfrak{D}_I := \sum_{i \in [I]} \mathbf{e}_I^i \otimes (\mathbf{e}_I^i)^\top \otimes (\mathbf{e}_I^i)^\top, \tag{A.3.7}$$

where we recall that e_I^i is the ith vector of the canonical basis of $\mathbb{R}^I$. This matrix has the important property that if A is a matrix of size $I^2 \times J$ whose rows are doubly indexed by $ii' \in [I^2]$ and whose columns are indexed by $j \in [J]$, so that the term of A is $(A_{ii',j})$, then $\mathfrak{D}_I A$ is a matrix of size $I \times J$ such that

$$(\mathfrak{D}_I A)_{ij} = A_{ii,j}. \tag{A.3.8}$$

In particular, if P is a $K \times K$ square matrix, then $vec_R(P)$ is a column vector of size K^2, and $\mathfrak{D}_K \, vec_R(P)$ is a column vector of size K such that $(\mathfrak{D}_K \, vec_R(P))_k = P_{k,k}$. Hence, calling $diagonal(P)$ the diagonal vector of the square matrix P, one has

$$\mathfrak{D}_K \, vec_R(P) = diagonal(P). \tag{A.3.9}$$

Conversely, if v is a column vector of size K, then, calling $\mathbf{\Delta}_v$ the $K \times K$ diagonal matrix whose diagonal is the vector v, we have $\mathfrak{D}_K^\top v = vec_R(\mathbf{\Delta}_v)$. We have the identity

$$A \circledast B = \mathfrak{D}_I (A \otimes B). \tag{A.3.10}$$

Appendix B.9.3 provides the Python code to construct the diagonal selection matrix.

Differentiation. If $f : \mathbb{R}^I \to \mathbb{R}^K$, one defines the first differential df at $x \in \mathbb{R}^I$ as the matrix of size $K \times I$ given by

$$df(x) = \sum_{ik \in [I \times K]} \frac{\partial f_k(x)}{\partial x_i} \mathbf{e}_K^k \otimes (\mathbf{e}_I^i)^\top,$$

so that df is the usual Jacobian of f. If $K=1$, $df(x)$ is a row vector of size I so that the usual gradient of f at x, which is a column vector, is the transpose of the differential $\nabla f(x) = \nabla df(x)^\top$. More generally, if $f(x)$ is a matrix of size $K \times L$, then the differential $df(x)$ is the matrix of size $K \times IL$ given by

$$df(x) = \sum_{ikl \in [I \times K \times L]} \frac{\partial f_{kl}(x)}{\partial x_i} \mathbf{e}_K^k \otimes (\mathbf{e}_I^i)^\top \otimes (\mathbf{e}_L^l)^\top.$$

It follows from this definition that, when $K=1$, $d^2 f = d(df)$ is a row vector of size I^2 whose expression is

$$d^2 f\,(x) = \sum_{ijk \in [I \times I \times K]} \frac{\partial^2 f_k\,(x)}{\partial x_i \partial x_j} \mathbf{e}_K^k \otimes (\mathbf{e}_I^i)^\top \otimes (\mathbf{e}_I^j)^\top,$$

and the usual Hessian of f at x, which is an $I \times I$ matrix, is given by $D^2 f(x) = d\nabla f(x) = d(df^\top)(x)$. The second differential $d^2 f(x)$ thus coincides with $vec(D^2(f))^\top$, the row vector of entries of the Hessian of f, indifferently taken in the row- or column-major order (because the Hessian is a symmetric matrix).

Let $f : \mathbb{R}^I \to \mathbb{R}^J$ be given by $f(x) = g[h(x)]$, where $h : \mathbb{R}^I \to \mathbb{R}^J$ and $g : \mathbb{R} \to \mathbb{R}$, and with the convention that if $y \in \mathbb{R}^J$, $g[y]$ is the vector of $\mathbb{R}^J$ whose jth entry is $g(y_j)$. If g and h are twice differentiable, we have

$$df\,(x) = \mathbf{\Delta}_{g'[h(x)]} dh\,(x), \tag{A.3.11}$$

where for any $y \in \mathbb{R}^J$, $\mathbf{\Delta}_y$ denotes the $J \times J$ diagonal matrix with y on the diagonal. Further, one has

$$d^2 f\,(x) = \mathbf{\Delta}_{g'[h(x)]} d^2 h\,(x) + \mathbf{\Delta}_{g''[h(x)]} dh\,(x)^{\circledast 2}, \tag{A.3.12}$$

where $dh(x)^{\circledast 2}$ denotes $dh(x) \circledast dh(x)$. As $dh(x)$ is a $J \times I$ matrix, $dh(x)^{\circledast 2}$ is a $J \times I^2$ matrix, the row-wise Kronecker product of $dh(x)$ times itself.

If $f : \mathbb{R}^P \to \mathbb{R}^I$, and if $\mathbf{\Delta}_f : \mathbb{R}^P \to \mathbb{R}^{I \times I}$ is the function that sends x to the diagonal matrix with $f(x)$ on its diagonal, then

$$d\mathbf{\Delta}_f(x) = df(x) \circledast \mathbf{I}_I. \tag{A.3.13}$$

If $f(x)$ and $g(x)$ are two matrix-valued functions of a parameter $x \in \mathbb{R}^P$ with respective shapes $I \times J$ and $J \times K$, the differential of their matrix product fg is given by

$$d\left(fg\right)(x) = df(x)\left(\mathbf{I}_P \otimes g(x)\right) + f(x)dg(x). \tag{A.3.14}$$

If $f(x)$ and $g(x)$ are two matrix-valued functions of a parameter $x \in \mathbb{R}^P$ with the same shape $I \times J$, the differential of their Hadamard product is given by

$$d\left(f \odot g\right)(x) = df(x) \odot \left(\mathbf{1}_P^\top \otimes g\right) + (\mathbf{1}_P^\top \otimes f) \odot dg(x). \qquad \text{(A.3.15)}$$

References

A good source for vectorization and Kronecker product is Magnus and Neudecker (2019); appropriate caution should, however, be applied, as that book takes the column-major order for vectorization, which makes it rather inconvenient when using a language based on the row-major order.

A.3.2 SCHUR COMPLEMENT

The Schur complement is a important concept in linear algebra that arises in the context of block matrices. Consider a block matrix M written in the form:

$$M = \begin{pmatrix} A & B \\ C & D \end{pmatrix},$$

where A, B, C, and D are submatrices of appropriate dimensions. The Schur complement provides a way to reduce or simplify the inversion of M under certain conditions. The Schur complement of the block A in the matrix M is defined as

$$S = D - CA^{-1}B,$$

provided that A is invertible. Similarly, the Schur complement of the block D is

$$T = A - BD^{-1}C,$$

provided that D is invertible.

Properties of the Schur complement. Key properties of the Schur complement include:

(i) If M is symmetric and positive definite, then both Schur complements S and T are also symmetric and positive definite.
(ii) The determinant of M can be expressed using the respective Schur complements as

$$\det(M) = \det(A)\det(S) = \det(T)\det(D)$$

when A (and respectively D) is invertible.

(iii) The inverse of M can be written as

$$M^{-1} = \begin{pmatrix} A^{-1} + A^{-1}BS^{-1}CA^{-1} & -A^{-1}BS^{-1} \\ -S^{-1}CA^{-1} & S^{-1} \end{pmatrix}$$

when both A and S are invertible, and

$$M^{-1} = \begin{pmatrix} T^{-1} & -T^{-1}BD^{-1} \\ -D^{-1}CT^{-1} & D^{-1} + D^{-1}CT^{-1}BD^{-1} \end{pmatrix}$$

if T and D are also invertible.

References

The Schur complement is discussed in standard textbooks on matrix analysis; see, e.g., Horn and Johnson (2012).

A.3.3 AUTOMATIC DIFFERENTIATION

Automatic differentiation is a set of techniques to numerically evaluate the derivative of a function specified by a computer program. It leverages the fact that every computer program, no matter how complex, executes a sequence of arithmetic operations and elementary functions (such as additions, multiplications, and trigonometric functions). By applying the chain rule repeatedly to these operations, automatic differentiation computes derivatives quickly and accurately, which is especially useful in optimization problems where gradient information is crucial.

Automatic differentiation is distinct from symbolic differentiation and numerical differentiation. Unlike symbolic differentiation, it does not manipulate mathematical expressions to obtain derivatives; instead, it works through the computational graph of the function. Compared to numerical differentiation, automatic differentiation provides exact derivatives (up to machine precision), without suffering from the issues of finite difference approximations. It operates by decomposing complex functions into elementary operations and applying the chain rule to these operations. It can be implemented in two successive steps:

- The **forward step** computes the values of each intermediate variable based on the input variables as it progresses through the computational graph from inputs to outputs.
- The **backward step** computes the derivative of the output variable with respect to each intermediate variable, working backward from

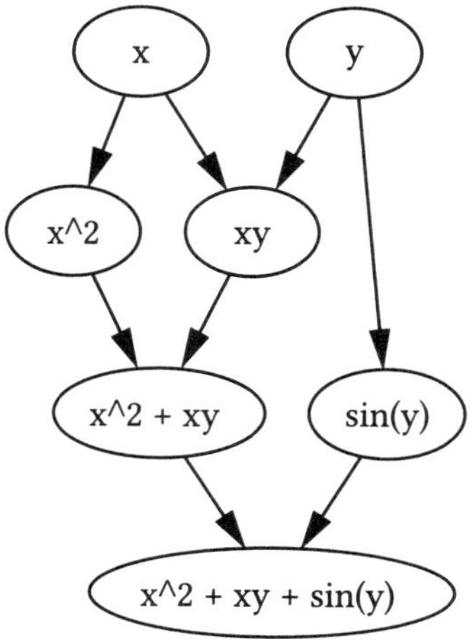

Figure A.2: The graph shows the forward pass, where the computation of the function $f(x, y)$ takes place. In the backward pass, the gradients are computed using the chain rule.

the output to the inputs. This mode is particularly efficient for functions with a large number of inputs and a small number of outputs, making it popular in machine learning.

As an example, consider a function $f(x, y) = x^2 + xy + \sin(y)$. To differentiate this function with respect to x and y, one constructs a computational graph representing the function, and then applies the automatic differentiation algorithm to compute the gradients. The derivatives of f with respect to x and y can be computed as follows:

$$\frac{\partial f}{\partial x}(x, y) = 2x + y, \qquad \frac{\partial f}{\partial y}(x, y) = x + \cos(y), \tag{A.3.16}$$

where these derivatives can be obtained by applying the chain rule to the computational graph of f, as seen in figure A.2.

The `PyTorch` library. The `torch.autograd` library in `PyTorch` automatically computes the gradients of tensors. It is a key component that allows developers to execute forward computations through neural networks while the library handles the backward pass automatically. This feature simplifies the development of deep learning models significantly. Consider the function $f(x, y) = x^2 + xy + \sin(y)$ from the above example, where x and y are input (scalar) tensors. The code displayed in appendix B.9.1 evaluates the gradient of

f at (x, y). In this example, the `backward()` method computes the gradients of f with respect to each tensor with `requires_grad=True`. After the computation, `x.grad` and `y.grad` hold the gradients of f with respect to x and y, respectively. Automatic differentiation in `PyTorch` is implemented using a computational graph. This graph is a directed acyclic graph (DAG) where nodes represent operations or variables, and edges represent the dependencies between these operations. Figure A.2 illustrates the computational graph for the function $f(x, y) = x^2 + xy + \sin(y)$ that helps explain how `PyTorch` computes gradients.

References

References on automatic differentiation include Baydin, Pearlmutter, Radul, and Siskind (2018), as well as the book Griewank and Walther (2008). See also Paszke et al. (2017) for a tutorial focused on Torch.

A.3.4 PERRON–FROBENIUS THEORY

In a broad sense, Perron–Frobenius theory studies positive matrices (i.e., matrices with positive entries), and, more generally, nonlinear operators that are monotone (i.e., order-preserving). We provide several results that are useful throughout the book.

Abstract Shapley operators. In the context of repeated zero-sum games, given a discounting factor $\beta \in [0, 1]$, the *Shapley operator* is a function $S^\beta : \mathbb{R}^X \to \mathbb{R}^X$ defined by

$$S_x^\beta(u) = \max_{y \in [Y]} \min_{z \in [Z]} \left\{ \Phi_{xyz} + \beta \sum_{x' \in [X]} P_{x'|xyz} u_{x'} \right\}, \tag{A.3.17}$$

where $x \in [X]$ is the state of the game, $y \in [Y]$ is the action of player 1, $z \in [Z]$ is the action of the opponent, and Φ_{xyz} is the short-term payoff that player 1 receives from player 2 if the game is in state x, if player 1 plays action y and player 2 plays action z. $P_{x'|xyz}$ is the probability of a transition from x to x' if the players' actions are y and z.

We start by introducing some important notations and properties. For $u \in \mathbb{R}^X$, let $|u|_\infty = \max_x |u_x|$ denote the *infinity norm*. Recall that the partial order on $\mathbb{R}^X$ is such that $u \le v$ if and only if $u_x \le v_x$ for all $x \in [X]$.

Definition A.3.1. *Let $F : \mathbb{R}^X \to \mathbb{R}^Y$. We say that*

(i) *F is unit-additive when $F(u + c\mathbf{1}_X) = F(u) + c\mathbf{1}_Y$;*

(ii) F *is unit sub-additive when* $c \in \mathbb{R}_+$ *implies* $F(u + c\mathbf{1}_X) \leq F(u) + c\mathbf{1}_Y$;
(iii) F *is order-preserving when* $u \leq u'$ *implies* $F(u) \leq F(u')$;
(iv) F *is nonexpansive for the infinity norm when* $|F(u) - F(v)|_\infty \leq |u - v|_\infty$.

The notion of *abstract Shapley operators*, which is a generalization of Shapley operators, plays a crucial role in dynamic programming. Its formal definition is as follows:

Definition A.3.2 (Abstract Shapley operator). *An abstract Shapley operator is a function* $F : \mathbb{R}^X \to \mathbb{R}^X$ *that is order-preserving and unit-additive.*

One can easily see that the operator S^1 defined by (A.3.17) is an abstract Shapley operator, while for $\beta < 1$, S^β is the composition of S^1 with the homothetic contraction $u \mapsto \beta u$.

Nonexpansiveness of order-preserving operators. The interest of abstract Shapley operators resides in the surprising fact that these operators are nonexpansive, which means that they are 1-Lipschitz with respect to a well-chosen norm. We have:

Theorem A.3.1 (Blackwell). *Assume* $F : \mathbb{R}^X \to \mathbb{R}^X$ *is unit-additive. Then the following two conditions are equivalent:*

(i) F *is order-preserving.*
(ii) F *is nonexpansive for the infinity norm.*

Proof. Assume that F is order-preserving and unit-additive. We have $|a|_\infty = \min\{\lambda \geq 0 : -\lambda \mathbf{1}_X \leq a \leq \lambda \mathbf{1}_X\}$. As a result, for u and u' in $\mathbb{R}^X$, we have $|u' - u|_\infty = \min\{\lambda \geq 0 : -\lambda \mathbf{1}_X \leq u' - u \leq \lambda \mathbf{1}_X\}$, but using the fact that F is order-preserving and unit-additive, $u' \leq u + \lambda \mathbf{1}_X$ implies $F(u') \leq F(u + \lambda \mathbf{1}_X) = F(u) + \lambda \mathbf{1}_X$ and thus $F(u') - F(u) \leq \lambda \mathbf{1}_X$. A similar argument shows that $F(u) - F(u') \leq \lambda \mathbf{1}_X$, which implies $|F(u') - F(u)|_\infty \leq \lambda$. Taking $\lambda = |u' - u|_\infty$ implies that F is nonexpansive for the infinity norm, which proves the direct implication. In order to prove the converse, we assume that F is at the same time nonexpansive for the infinity norm and unit-additive, and we seek to show that it is order-preserving. Consider a pair of vectors u and u' such that $u' \leq u$. Because F is nonexpansive, $|F(u' + \lambda) - F(u)|_\infty \leq |u' + \lambda - u|_\infty$ holds for any $\lambda \in \mathbb{R}$, which, as F is unit-additive, implies $|F(u') - F(u) + \lambda|_\infty \leq |u' + \lambda - u|_\infty$, hence $F(u') \leq F(u) - \lambda + |u' + \lambda - u|_\infty$. However, for $\lambda = |u - u'|_\infty$, we have $|u' + \lambda - u|_\infty \leq \lambda$, and thus $F(u') \leq F(u)$, which shows that F is order-preserving. ∎

Blackwell's theorem allows us to prove the following important consequence:

Theorem A.3.2. *Assume $F:\mathbb{R}^X \to \mathbb{R}^X$ is monotone. Then the following two conditions are equivalent:*

(i) F is unit sub-additive.
(ii) F is nonexpansive for the infinity norm.

Proof. Consider F monotone, which is unit sub-additive. Then define G: $\mathbb{R}^{X+1} \to \mathbb{R}^{X+1}$ by $G_x(u) = F_x(u_1 - u_{X+1}, \ldots, u_X - u_{X+1}) + u_{X+1}$ for $x \in [X]$, and $G_{X+1}(u) = u_{X+1}$. Clearly, G is unit-additive. If $u \le v$, then $G(u_1, \ldots, u_{X+1}) = F(u_1 - u_{X+1}, \ldots, u_X - u_{X+1}) + u_{X+1}$. By monotonicity of F, one has $F(u_1 - u_{X+1}, \ldots, u_X - u_{X+1}) \le F(v_1 - u_{X+1}, \ldots, v_X - u_{X+1})$. By unit sub-additivity, one has $F(v_1 - u_{X+1}, \ldots, v_X - u_{X+1}) \le F(v_1 - v_{X+1}, \ldots, v_X - v_{X+1}) + v_{X+1} - u_{X+1}$, which implies $G(u) \le G(v)$. It follows that G is monotone. By theorem A.3.1, this implies that G is nonexpansive for the infinity norm, which in turns implies that F is too, which shows the direct implication of the result. Conversely, assume F is monotone and nonexpansive for the infinity norm. Then for $c \in \mathbb{R}_+$, we have $|F(u + c\mathbf{1}_x) - F(u)|_\infty \le |u + c\mathbf{1}_x - u| = c$, and therefore $F(u + c\mathbf{1}_x) \le F(u) + c\mathbf{1}_x$, which implies that F is unit sub-additive. ■

Hilbert's projective metric. Given an $X \times X$ matrix A with positive coefficients, we cannot expect $u \mapsto Au$ to be a contraction mapping in general. For instance, if $A\mathbf{1}_X = \mathbf{1}_X$, then every multiple of $\mathbf{1}_X$ is a fixed point of A. However, we can show that A is a contraction mapping "up to a constant" for a well-chosen metric. We will make this statement precise using Hilbert's projective metric and Birkhoff's theorem.

Let a and b be two vectors of $\mathbb{R}^X_{++}$, the set of vectors with strictly positive entries. We define Hilbert's projective metric d_H by

$$d_H(a, b) = \log \max_{xx' \in [X \times X]} \left\{ \frac{a_x b_{x'}}{a_{x'} b_x} \right\}. \tag{A.3.18}$$

From setting $x = x'$ in (A.3.18), we see that $d_H(a, b) \ge 0$ for all a and b. d_H is not a distance since $d_H(a, \lambda a) = 0$ for any $\lambda > 0$; however, it satisfies the following modification of the axioms that define a distance:

- $d_H(a, b) = 0$ if and only if $b = \lambda a$ for some $\lambda > 0$,
- symmetry: $d_H(a, b) = d_H(b, a)$,
- triangle inequality: $d_H(a, b) \le d_H(a, c) + d_H(c, b)$.

As a result, d_H becomes a distance when restricted to a subset of $\mathbb{R}^X$ that has no two colinear elements. For instance, one can impose the normalization that a is in the unit simplex, i.e., $\sum_x a_x = 1$.

Given an $X \times X$ matrix A with positive terms A_{xy}, we define $\theta(A)$ as the diameter of the image of A for the metric d_H, that is, $\theta(A) = \sup\{d_H(Au, Av) : u, v \in \mathbb{R}^X_{++}\}$. One can show that this quantity is finite and has the more convenient expression

$$\theta(A) = \max_{x,x',y,y'} \frac{A_{xy}A_{x'y'}}{A_{xy'}A_{x'y}}. \tag{A.3.19}$$

By taking $x = x'$ and $y = y'$ we see that $\theta(A) \geq 1$. Furthermore:

Theorem A.3.3 (Birkhoff). *We have*

$$\sup\left\{\frac{d_H(Au, Av)}{d_H(u, v)} : u, v \in \mathbb{R}^X_{++}, u \neq \lambda v\right\} =: \kappa(A) = \frac{\sqrt{\theta(A)} - 1}{\sqrt{\theta(A)} + 1}. \tag{A.3.20}$$

This result has far-reaching consequences. Suppose that A has two positive eigenvalues λ_1 and λ_2 associated with positive vectors u and v. Then one has $d_H(Au, Av) \leq \kappa(A) d_H(u, v)$, but since $d_H(Au, Av) = d_H(\lambda_1 u, \lambda_2 v) = d_H(u, v)$, this implies that $d_H(u, v) = 0$, and therefore u and v are colinear and $\lambda_1 = \lambda_2$. Further, we can show that for any positive starting point u^0, the sequence defined by $u^k = A^k u^0 / (\mathbf{1}_X^\top A^k u^0)$ converges to a positive vector $u^\star$ such that $d_H(Au^\star, u^\star) = 0$. Hence $u^\star$ is the unique positive eigenvector of A such that $\mathbf{1}_X^\top u^\star = 1$. This is the essence of the Perron–Frobenius theorem, which we discuss next.

The Perron–Frobenius theorem. This theorem is a landmark result of linear algebra, with important implications for Markov chains and dynamic systems. In the version that we shall use, it states that any real square matrix that is positive, in the sense that all its elements are positive, has a unique largest real eigenvalue, known as the *Perron–Frobenius eigenvalue.* This eigenvalue is positive, and is strictly larger than all other eigenvalues of the matrix. There is a unique eigenvector associated with the Perron–Frobenius eigenvalue, called the *right Perron–Frobenius eigenvector,* whose entries are all positive, and which is the only eigenvector of the matrix whose entries are nonnegative. As the transpose of a positive matrix is also positive, it also has a Perron–Frobenius eigenvector, which is called the *left Perron–Frobenius eigenvector* of that matrix. We state:

Theorem A.3.4 (Perron–Frobenius for positive matrices). *Let A be an $X \times X$ matrix with positive terms. Then:*

(i) *There exists a unique positive real eigenvalue λ (the Perron–Frobenius eigenvalue) that is strictly greater than the absolute value of any other eigenvalue of A. Hence λ is the spectral radius of A.*

(ii) *A has a unique (up to scalar multiplication) strictly positive eigenvector u (i.e., such that $u_x > 0$ for all $x \in [X]$) corresponding to the eigenvalue λ, called the right Perron–Frobenius eigenvector.*
(iii) *$A^\top$ has a unique (up to scalar multiplication) strictly positive eigenvector v corresponding to the eigenvalue λ, called the left Perron–Frobenius eigenvector.*
(iv) *The Perron–Frobenius eigenvalue λ is simple, meaning its algebraic multiplicity is 1.*
(v) *The convergence $A^t/\lambda^t \to uv^\top/(v^\top u)$ holds as $t \to +\infty$.*

There is a version of the Perron–Frobenius theorem, not stated here, that applies to a more general class of matrices than the positive matrices, the irreducible nonnegative matrices. We refer the reader to the large literature on the topic.

References

Shapley operators originate with Shapley (1953) in the context of dynamic zero-sum games; see a recent account in Sorin (2004) and also Neyman and Sorin (2003), as well as Filar and Vrieze (2012). What we have called "abstract Shapley operators" have several alternative names (see Gaubert and Gunawardena 2004 as well as Gunawardena and Keane 1995): e.g., dynamic programming operators, topical functions. The link between nonexpansive and order-preserving mappings in theorem A.3.1 was first established by Blackwell (1965), and later refined by Crandall and Tartar (1980). For a proof of theorem A.3.2, see, e.g., Gunawardena and Keane (1995). Our exposition of Hilbert's metric and Birkhoff's theorem is borrowed from the paper Franklin and Lorenz (1989), who credit the main theorem to the paper Birkhoff (1957). Two standard references for the Perron–Frobenius theorem are Berman and Plemmons (1994) and Horn and Johnson (2012).

A.3.5 MONOTONE COMPARATIVE STATICS

Isotonicity. Given X a subset of $\mathbb{R}^n$, and Y a subset of $\mathbb{R}^m$, and $f : X \longmapsto Y$ a function from X to Y, one says that f is *isotone* if $x \leq y$ implies $f(x) \leq f(y)$. One says that f is *antitone* if $x \leq y$ implies $f(x) \geq f(y)$. One says that f is *inverse isotone* if $f(x) \leq f(y)$ implies $x \leq y$. When $n = m = 1$, f is inverse isotone if and only if f is strictly increasing, namely, $x < y$ implies $f(x) < f(y)$. However, this relies on the fact that $\leq$ is a total order on $\mathbb{R}$, and does not extend in general to a higher dimension. However, the following remark, whose proof is immediate, makes inverse isotone functions very useful:

Proposition A.3.1. *Let $X \subseteq \mathbb{R}^n$ and $Y \subseteq \mathbb{R}^m$, and let $f : X \longmapsto Y$ be an inverse isotone function. Then f is injective.*

Sublattices of $\mathbb{R}^n$. We now introduce the important concept of a sublattice of $\mathbb{R}^n$, which we will often simply call a sublattice. For two vectors x and y in $\mathbb{R}^n$, we denote by $x \wedge y$ ("x meet y") the vector in $\mathbb{R}^n$ whose ith entry is $\min\{x_i, y_i\}$, and by $x \vee y$ ("x join y") the vector in $\mathbb{R}^n$ whose ith entry is $\max\{x_i, y_i\}$. A sublattice is a set that is stable with respect to the meet and join operations:

Definition A.3.3 (Sublattice of $\mathbb{R}^n$). *$X \subseteq \mathbb{R}^n$ is a sublattice (of $\mathbb{R}^n$) if for any $x, x' \in X$, $x \wedge x' \in X$, and $x \vee x' \in X$.*

It is important to keep in mind the distinction between sublattices of $\mathbb{R}^n$ and lattice subsets of $\mathbb{R}^n$. A lattice (of $\mathbb{R}^n$) is a subset of $\mathbb{R}^n$ such that for any x and x', (i) the set of z such that $z \leq x$ and $z \leq x'$ hold simultaneously has a smallest maximal element denoted by $x \sqcap x'$, and (ii) the set of z such that $z \geq x$ and $z \geq x'$ hold simultaneously has a smallest element denoted by $x \sqcup x'$. Clearly, a sublattice is a lattice; however, the converse fails to hold. Indeed, consider $X = \{a, b, c, d\}$, where $a = (0,0)$, $b = (2,1)$, $c = (1,2)$, and $d = (3,3)$. Then X is a lattice; however, it is not a sublattice, because one has $b \sqcup c = (3,3) \neq (\max(b_1, c_1), \max(b_2, c_2)) = (2,2)$.

Veinott's lattice order. We next introduce a very important partial order on sublattices called the *lattice order*, also frequently called the *strong set ordering*.

Definition A.3.4 (Lattice order). *Let X and X' be two subsets of $\mathbb{R}^n$. One says that X is smaller than X' in the lattice order, denoted by $X \leq_L X'$, if for any $x \in X$ and $x' \in X'$, $x \wedge x' \in X$ and $x \vee x' \in X'$.*

Therefore, the relation $\leq_L$ is called "order" because it is a partial order on sublattices. A partial order $\preceq$ should be reflexive ($a \preceq a$), antisymmetric ($a \preceq b$ and $b \preceq a$ imply $a = b$), and transitive ($a \preceq b$ and $b \preceq c$ imply $a \preceq c$). It is easy to see that $X \leq_L X$ holds if and only if X is a sublattice. More generally:

Proposition A.3.2. *The relation $\leq_L$ is a partial order on sublattices.*

Increasing differences. Let $X \subseteq \mathbb{R}^n$ and $Y \subseteq \mathbb{R}^m$. Consider a function $f : X \times Y \to \mathbb{R}$. One says $f(x, y)$ has *increasing differences* in (x, y) if for $y \leq y'$, $f(x, y') - f(x, y)$ is nondecreasing in x.

Submodularity. Submodularity is a key property in the theory of monotone comparative statics. One defines it formally as follows:

Definition A.3.5 (Submodular function). *If X is a sublattice of $\mathbb{R}^n$, one says that $f: X \to \mathbb{R}$ is* submodular *if for any x and x' in X, $f(x \wedge x') + f(x \vee x') \leq f(x) + f(x')$.*

Related definitions follow. One says that f is *supermodular* if $-f$ is submodular. One says that f is *modular* if f is both sub- and supermodular. If f is defined on a product of intervals, then f is modular if and only if $f(x) = \sum_{i=1}^{n} f_i(x_i)$. If f is twice continuously differentiable, then it is submodular if and only if $\partial^2 f(x) \partial x_i \partial x_j \leq 0$ for all $i \neq j$, and supermodular if and only if the inequality is reversed. Let us take some examples.

Example A.5. The following are examples of sub- and supermodular functions:

(i) If $F: \mathbb{R}^2 \to \mathbb{R}$ is the c.d.f. of a bivariate random vector, then F is supermodular.
(ii) If $\varphi: \mathbb{R}^n \to \mathbb{R}$ is convex, then $f(x) = \varphi\left(\sum_{i=1}^{n} x_i\right)$ is supermodular.
(iii) If $a_{ij} \geq 0$, then $f(x) = \sum_{1 \leq i,j \leq n} a_{ij} \left|x_i - x_j\right|^2$ is submodular.

Topkis's theorem. Topkis's theorem is an important result in monotone comparative statics.

Theorem A.3.5 (Topkis). *Let X be a sublattice of $\mathbb{R}^n$ and Y a subset of $\mathbb{R}^m$, and f a function $X \times Y \mapsto \mathbb{R} \cup \{-\infty\}$ such that:*

(i) $f(., y)$ is supermodular for each $y \in Y$,
(ii) $f(x, y)$ has increasing differences in (x, y) on $X \times Y$,
(iii) $\arg\max_{x \in X} f(x, y)$ is non-empty for each $y \in Y$.

Then $y \mapsto \arg\max_{x \in X} f(x, y)$ is isotone in the lattice order.

Proof. Conditions (i) and (ii) together imply that for $y \leq y'$, $f(x \wedge x', y) - f(x, y) \geq f(x', y') - f(x \vee x', y')$. Now assume $y \leq y'$ and consider $x \in \arg\max_{x \in X} f(x, y)$ and $x' \in \arg\max_{x \in X} f(x, y')$. We have $0 \geq f(x \wedge x', y) - f(x, y)$ and $f(x', y') - f(x \vee x', y') \geq 0$; but combining this with the previous inequality yields that all these inequalities hold as equalities. Hence $x \wedge x' \in \arg\max_{x \in X} f(x, y)$ and $x \vee x' \in \arg\max_{x \in X} f(x, y')$, which is the desired conclusion. ∎

Submodular convex functions. Consider f a submodular function. The function $(x, y) \mapsto x^\top y - f(x)$ is supermodular, and it is straightforward to show that $f^*(y) = \sup_{x \in \mathbb{R}^n} \{x^\top y - f(x)\}$ is supermodular too. We have therefore:

Proposition A.3.3. *Let X be a sublattice of $\mathbb{R}^n$, and $f : X \to \mathbb{R}$ be a submodular function. Then f^*, the Legendre–Fenchel transform of f, is supermodular.*

When $n = 2$, the converse holds true in some sense: if f is convex and f^* is supermodular, then f is submodular. However, when $n > 2$, the converse no longer holds: supermodularity of f^* does not imply submodularity of f in general.

Example A.6. The following functions f are submodular and convex, and their Legendre–Fenchel transforms f^* are supermodular and convex:

(i) $f(x) = \log\left(1 + \sum_{i=1}^{d} \exp(x_i)\right)$, whose conjugate is $f^*(y) = \sum_{i=0}^{d} y_i \ln y_i$, where $y_0 := 1 - \sum_{i=1}^{d} y_i$.
(ii) $f(x) = \max_{i=1}^{d} \{x_i\}$, whose conjugate is $f^*(y) = 0$ if $y \geq 0$ and $\sum_{i=1}^{d} y_i = 1$, and is $f^*(y) = +\infty$ otherwise.
(iii) Consider the constant elasticity of substitution (CES) function: $f(x) = (\sum_i x_i^r)^{1/r}$. Then f is submodular and convex for $r > 1$, and supermodular and concave for $r < 1$.

Proposition A.3.4. *If $f : \mathbb{R}^d \to \mathbb{R}$ is submodular, then $y \mapsto \partial f^*(y)$ is isotone in the Veinott strong set order: if $x \in \partial f^*(y)$ and $x' \in \partial f^*(y')$ are such that $y \leq y'$, then $x \wedge x' \in \partial f^*(y)$ and $x \vee x' \in \partial f^*(y')$.*

Recall from property A.1.7 that when f is strictly convex, then f^* is C^1. As a result, if $f : \mathbb{R}^d \to \mathbb{R}^d$ is submodular and strictly convex, then $y \mapsto \nabla f^*(y)$ is isotone.

Stieltjes matrices. Take a strictly convex and submodular quadratic function given by $f(x) = x^\intercal Sx/2$, where S is symmetric positive definite and is a Z-matrix. Such a matrix is called a *Stieltjes matrix.*

Definition A.3.6 (Stieltjes matrix). *A Stieltjes matrix is a symmetric positive definite matrix with nonpositive off-diagonal entries.*

We have the following nice characterization. Recall that for $x \in \mathbb{R}^d$, $x^+ = x \vee 0$.

Proposition A.3.5. *Let S be a symmetric $d \times d$ matrix. Then S is a Stieltjes matrix if and only if*

$$x^\top Sx^+ \geq 0$$

holds for all $x \in \mathbb{R}^d$, with a strict inequality unless $x \leq 0$.

As $f^*(y) = y^\intercal S^{-1}y/2$, it follows from proposition A.3.3 that f^* is supermodular. As a result, the off-diagonal terms of S^{-1} are nonnegative. As S^{-1} is a positive definite matrix, its diagonal terms are positive. Hence the inverse of a Stieltjes matrix is symmetric definite positive and entrywise nonnegative.

References

A good reference for this section is Topkis (1976). The definition of the lattice order is due to Veinott (1989).

A.3.6 Z- AND M- FUNCTIONS

One says that a function $f : \mathbb{R}^X \to \mathbb{R}^Y$ is *isotone* as a synonym for its being "order-preserving," i.e., $x \leq y$ implies $f(x) \leq f(y)$. One says that $f : \mathbb{R}^X \to \mathbb{R}^Y$ is *inverse isotone* if $f(x) \leq f(y)$ implies $x \leq y$.

Definition A.3.7 (Z-function). *One says that a function* $\mathbf{q} : \mathbb{R}^X \to \mathbb{R}^X$ *is a Z-function if any* $\mathbf{q}_x$ *is nonincreasing with respect to* p_y *for* $y \neq x$.

The term "Z-function" has many synonyms, including *off-diagonally antitone* function, or function with the *gross substitutes* property. The following are examples of Z-functions:

(i) The function $\mathbf{q}(p) = Ap$ where $A_{xy} \leq 0$ for $x \neq y$. Such a square matrix, whose off-diagonal terms are nonpositive, is called a *Z-matrix*.
(ii) The function $\mathbf{q}$ defined by

$$\mathbf{q}_z(p) = \sum_{x \in [X]} \exp(p_z - p_x - c_{xz}) - \sum_{y \in [X]} \exp(p_y - p_z - c_{zy}), \quad \forall z \in [X].$$

(iii) If $\mathbf{q}$ is a Z-function, then $\tilde{\mathbf{q}}(p) := \mathbf{q}(p^+) - p^-$ is a Z-function.
(iv) If $f : \mathbb{R}^X \to \mathbb{R}$ is a differentiable function, then f is submodular if and only if $p \to \nabla(p)f$ is a Z-function.

The following result offers a characterization of Z-functions in the differentiable case.

Proposition A.3.6. *Let* $\mathbf{q}:\mathbb{R}^X \to \mathbb{R}^X$ *be a differentiable function. Then* $\mathbf{q}$ *is a Z-function if and only if its Jacobian* $D\mathbf{q}(p)$ *is a Z-matrix at each point* $p \in \mathbb{R}^X$.

Definition A.3.8 (Weak nonreversingness). *We say that* $\mathbf{q}:\mathbb{R}^X \to \mathbb{R}^X$ *is weakly nonreversing if*

$$p \leq p' \text{ and } \mathbf{q}(p) \geq \mathbf{q}(p') \quad \text{imply} \quad \mathbf{q}(p) = \mathbf{q}(p').$$

Definition A.3.9 (Strong nonreversingness). *We say that* $\mathbf{q}:\mathbb{R}^X \to \mathbb{R}^X$ *is strongly nonreversing if*

$$p \leq p' \text{ and } \mathbf{q}(p) \geq \mathbf{q}(p') \quad \text{imply} \quad p = p'.$$

For example:

(i) An isotone function is weakly nonreversing;
(ii) $\mathbf{q}(p) = \nabla u(p)$, where $u:\mathbb{R}^X \to \mathbb{R}$ is smooth and convex, then $\mathbf{q}$ is weakly nonreversing; if, in addition, u is strictly convex, then $\mathbf{q}$ is strongly nonreversing;
(iii) $\mathbf{q}(p) = \arg\max_q \varphi(p, q)$, where $\varphi:\mathbb{R}^X \times \mathbb{R}^X \to \mathbb{R}^X$ satisfies increasing differences, and is such that the arg max is a single point, then $\mathbf{q}$ is weakly nonreversing;
(iv) If there is a vector $w \in \mathbb{R}^X_+$ such that $p \mapsto w^\top \mathbf{q}(p)$ is isotone, then $\mathbf{q}$ is weakly nonreversing;
(v) If $\mathbf{q}$ is inverse isotone, then $\mathbf{q}$ is strongly nonreversing;
(vi) If $\mathbf{q}$ is defined on the simplex $\Delta = \{p \in \mathbb{R}^X : \sum_{x\in[X]} p_x = 1\}$, then $\mathbf{q}$ is strongly nonreversing.

Definition A.3.10 (M0-function). *We say that* $\mathbf{q}$ *is an M0-function if it is a Z-function and weakly nonreversing.*

Definition A.3.11 (M-function). *We say that* $\mathbf{q}$ *is an M-function if it is a Z-function and strongly nonreversing.*

As an example, the function $\mathbf{q}(p) = Ap$ where the matrix A is defined by

$$A = \begin{pmatrix} 1 & -\rho \\ -\rho & 1 \end{pmatrix} \tag{A.3.21}$$

is an M0-function if and only if $0 \leq \rho \leq 1$, and an M-function if and only if $0 \leq \rho < 1$. More generally, gradient of convex functions (respectively strictly convex functions) that are submodular provide important examples of M0-functions (respectively M-functions).

Theorem A.3.6. *Let u be a convex and submodular function that is continuously differentiable. Then ∇u is an M0-function. Further, if u is strictly convex, then ∇u is an M-function.*

Proof. Clearly, if u is submodular and differentiable, then ∇u is a Z-function. Assume it is convex. Take p and p' such that $p \leq p'$ and $q := \nabla u(p) \geq q' := \nabla u(p')$. To show that ∇u is weakly reversing, we need to show that $q = q'$. We have

$$u(p) + u^*(q) = p^\top q, \qquad u(p) + u^*(q') \geq p^\top q',$$
$$u(p') + u^*(q') = p'^\top q', \quad u(p') + u^*(q) \geq p'^\top q.$$

By summation, we obtain $p^\top q + p'^\top q' \geq p^\top q' + p'^\top q$, which in turns implies $(p - p')^\top (q - q') \geq 0$, but as $p \geq p'$ and $q \leq q'$, the previous inequality holds as an equality and so do all inequalities that have been summed, which implies in particular that $u(p) + u^*(q') = p^\top q'$, which in turn implies that $q' = \nabla u(p) = q$, which shows that ∇u is weakly reversing. If ∇u is strictly convex, then $\nabla u(p) = \nabla u(p')$ implies $p = p'$, which shows that ∇u is strongly reversing. ■

We now state a series of important characterizations of M0- and M-functions.

Theorem A.3.7 (Weak inverse isotonicity theorem). *Let $\mathbf{q}$ be a Z-function. Then the following statements are equivalent:*

(i) $\mathbf{q}$ is weakly nonreversing (i.e., $\mathbf{q}$ is an M0-function), and
(ii) $\mathbf{q}^{-1}$ is increasing in the strong set order, that is: if $\mathbf{q}(p) \leq \mathbf{q}(p')$, then $\mathbf{q}(p) = \mathbf{q}(p \wedge p')$ and $\mathbf{q}(p') = \mathbf{q}(p \vee p')$.

Proposition A.3.7. *Let $f : \mathbb{R}^X \to \mathbb{R}$ be a convex submodular function that is C^1. Then $\mathbf{q}$ defined by $\mathbf{q}(p) = \nabla f(p)$ is an M0-function. Further, if f is strictly convex, then $\mathbf{q}$ is an M-function.*

Theorem A.3.8 (Strong inverse isotonicity theorem). *Let $\mathbf{q} : \mathbb{R}^X \to \mathbb{R}^X$ be a Z-function. Then the following are equivalent:*

(i) $\mathbf{q}$ is strongly nonreversing (i.e., $\mathbf{q}$ is an M-function), and
(ii) $\mathbf{q}^{-1}$ is an isotone function on its domain, that is: if $\mathbf{q}(p) \leq \mathbf{q}(p')$, then $p \leq p'$.

Let $\mathbf{q} : \mathbb{R}^X \to \mathbb{R}^X$ be a function for which we define the following possible assumptions:

Assumption A.2 (Weak substitutes). *$\mathbf{q}$ is a Z-function, i.e., $\mathbf{q}_x(p)$ is nonincreasing in p_y for each $x \neq y$ in $[X]$.*

Assumption A.3 (Strong substitutes). *$\mathbf{q}_x(p)$ is decreasing in p_y for each $x \neq y$ in $[X]$.*

Assumption A.4 (Weak law of aggregate supply). *The aggregate supply $\sum_{x\in[X]} \mathbf{q}_x(p)$ is nondecreasing in p_x for each $x \in [X]$.*

Assumption A.5 (Strong law of aggregate supply). *The aggregate supply $\sum_{x\in[X]} \mathbf{q}_x(p)$ is increasing in p_x for each $x \in [X]$.*

Defining

$$\mathbf{q}_0(p) = 1 - \sum_{x\in[X]} \mathbf{q}_x(p), \tag{A.3.22}$$

we can restate the weak (respectively, strong) law of aggregate supply by saying that $\mathbf{q}_0(p)$ is nonincreasing (respectively, decreasing) in p_y for each $x \in [X]$.

Assumption A.6 (Connected strong substitutes). *For any $B \subseteq [X]$, and for any p and $p' \in \mathbb{R}^X$ such that $p_x = p'_x$ for all $x \in B$ and $p_x > p'_x$ for $x \notin B$, there exists $x \in B \cup \{0\}$ such that $\mathbf{q}_x(p) < \mathbf{q}_x(p')$.*

The connected strong substitutes assumption is a middle ground between weak substitutes, which assumes that $\mathbf{q}_x$ is weakly decreasing in p_y, and strong substitutes, which assumes that it is strictly decreasing. In many cases, the strong substitutes assumption is stronger than necessary, while the weaker assumption of connected strong substitutes is still satisfied.

The following result provides important implications of weak substitutes and connected strong substitutes:

Theorem A.3.9. *Consider a function $\mathbf{q} : \mathbb{R}^X \to \mathbb{R}^X$. Then:*

(i) If $\mathbf{q}$ satisfies weak substitutes and the weak law of aggregate supply, then it is an M0-function.
(ii) If $\mathbf{q}$ satisfies weak substitutes, the weak law of aggregate supply, and connected strong substitutes, then it is an M-function.
(iii) If $\mathbf{q}$ satisfies weak substitutes and the strong law of aggregate supply, then it is an M-function.

References

McKenzie (2008) attributes the concept of Z-function introduced in definition A.3.7 to Morishima (1964). Z-functions are the natural nonlinear extension of Z-matrices in linear algebra; see Nikaido (2016), section 6.2, where the connection with Leontief's systems in input-output analysis is explained. A matrix whose off-diagonal terms are nonnegative is sometimes known as a *Metzler matrix*, the opposite of a Z-matrix. However, we shall abstain from using the term, as it can induce some confusion with M-matrices introduced below, which are different. The terminology "nonreversing function" is not standard in the literature. This concept was proposed in Galichon, Samuelson, and Vernet (2025), where theorem A.3.8 is proven. A reference for M-functions is Rheinboldt (1998). Theorem A.3.9 as stated here is a close variant of theorem 1 in Berry, Gandhi, and Haile (2012) and of Moré (1972), theorem 4.6.

APPENDIX B

Python Code

B.1 Code for Chapter 1: Random Utility Models

B.1.1 MARKET SHARE SIMULATION

```
# Market share simulation
#
# Importing libraries and generating data:
import numpy as np
np.random.seed(7)
I,Y = 100,3
U_y = np.array([0.0, 0.9,0.7,0.3])
epsilon_i_y = np.random.normal(size=(I,Y+1))
# Computing:
u_i_y = U_y[None,:]+epsilon_i_y
u_i = u_i_y.max(axis = 1)
G = u_i.sum() / I
mu_i_y = 1* ( u_i[:,None] == u_i_y )
mu_y = mu_i_y.sum(axis = 0 ) / I
# Displaying results:
print('G(U)=',G, ';', 'mu_y=', mu_y)
```

B.1.2 DEMAND INVERSION VIA LINEAR PROGRAMMING

```
# Demand inversion via linear programming
#
# Importing libraries and generating data:
import numpy as np, gurobipy as grb
np.random.seed(7)
I,Y = 100,3
mu_y = np.array([27,19,25,29]) / I
epsilon_iy = np.random.normal(size=I*(Y+1) )
```

```
# Computing:
m = grb.Model()
u_i = m.addMVar(I, lb = - grb.GRB.INFINITY)
U_y = m.addMVar(Y+1, lb = - grb.GRB.INFINITY)
m.setObjective( u_i.sum()/I - U_y @ mu_y,
                sense = grb.GRB.MINIMIZE)
m.addConstr(  np.kron(np.eye(I),np.ones((Y+1,1))) @ u_i
            - np.kron(np.ones((I,1)),np.eye(Y+1)) @ U_y
            >= epsilon_iy)
m.addConstr( U_y[0]==0 )
m.optimize()
# Displaying results:
print( 'U_y=' , U_y.x)
```

B.2 Code for Chapter 2: The Logit Model and Its Generalizations

B.2.1 MEV AS A FACTOR MODEL

```
# A max-factor model
#
# Importing libraries and generating the parameters:
import numpy as np
#
np.random.seed(7)
O = 1000
X = 10000000
thelambda = .8
epsilon_o = np.zeros(O)
epsilonprime_o = np.zeros(O)

denom = (np.arange(1,X+1) ** (-thelambda) ).sum()
U_x = np.log( (np.arange(1,X+1) ** (-thelambda) )  / denom )
Uprime_x = np.log(  (np.arange(X,0,-1) ** (-thelambda) ) /
    denom )

rho = 1-thelambda*thelambda

for o in range(O):
    eta_x = - np.log( np.random.uniform(size=X))
    epsilon_o[o] = (U_x + eta_x).max()
    epsilonprime_o[o] = (Uprime_x + eta_x).max()
    if (1+o) % 100 == 0:
        print(np.corrcoef(epsilon_o[:o],
            epsilonprime_o[:o])[0,1] , rho)
```

B.2.2 INVERTING THE NESTED LOGIT MODEL

```
# Nested logit inversion
#
# Importing libraries and generating data:
import numpy as np
#
Y = 10
X = 3
pi_y = \
    np.array([0.05,0.12,0.08,0.21,0.17,0.02,0.14,0.10,0.07,
    0.04])
nest_y = np.array([0,2,1,1,0,2,1,2,2,0])
lambda_x = np.array([0.3, 0.5, 0.8])
# computing the fixed effect matrix
in_x_y = 1* (np.arange(X)[:,None] == nest_y[None,:])
# computing the market share of nests
pix_y = (in_x_y @ pi_y) @ in_x_y
lambdax_y = lambda_x @ in_x_y
# inverse market share map:
U_y = lambdax_y * np.log(pi_y) + (1 - lambdax_y) * np.log(
    pix_y  )
# entropy of choice:
Gstar = lambdax_y * pi_y * np.log(pi_y) + (1 - lambdax_y) * \
    pix_y * np.log( pix_y  )
#
# displaying results
print('G*=',Gstar)
print('U_y=',U_y)
```

B.2.3 CONVERGENCE TO MAX-STABLE DISTRIBUTIONS

```
# Illustration of convergence to max-stable
#
# import libraries and generate date
import numpy as np
import matplotlib.pyplot as plt
#
np.random.seed(7)
O,I = 1000,100000
#
lambda_exp = 0.8
alpha_pareto = 0.6
#
M1_o = np.zeros(O) # max of exponentials
M2_o = np.zeros(O) # max of Paretos
M3_o = np.zeros(O) # max of uniforms
```

```
for o in range(O):
  M1_o[o] = (np.random.exponential( scale = lambda_exp, size
      = I).max() / lambda_exp - np.log(I))
  M2_o[o] = np.random.pareto(a = alpha_pareto, size =
      I).max() / I**(1/alpha_pareto)
  M3_o[o] = I * np.random.uniform(size = I).max() - I
# assess convergence by rescaling to uniform using cdfs
U1tilde_o = np.exp(-np.exp( -  M1_o  )  )
U2tilde_o = np.exp(- M2_o**(-alpha_pareto))
U3tilde_o = np.exp( -(- M3_o) )
#
# plot empirical cdfs
plt.plot(np.sort(U1tilde_o), np.arange(1, O + 1) / O ,
    marker=".", linestyle="none")
plt.gca().set(xlabel="t", ylabel="F(t)", title="Empirical CDF \
    (Exponential => Type I)")
plt.grid(True)
plt.show()
#
plt.plot(np.sort(U2tilde_o), np.arange(1, O + 1) / O ,
    marker=".", linestyle="none")
plt.gca().set(xlabel="t", ylabel="F(t)", title="Empirical CDF \
    (Pareto => Type II)")
plt.grid(True)
plt.show()
#
plt.plot(np.sort(U3tilde_o), np.arange(1, O + 1) / O ,
    marker=".", linestyle="none")
plt.gca().set(xlabel="t", ylabel="F(t)", title="Empirical CDF \
    (Uniform => Type III)")
plt.grid(True)
plt.show()
```

B.3 Code for Chapter 3: Logistic Regression

B.3.1 LOGISTIC REGRESSION VIA GRADIENT DESCENT

```
# Multinomial logistic regression via gradient descent
#
# Importing libraries and generating data :
import numpy as np
#
np.random.seed(7)
I, Y, K = 100, 5, 3
pihat_iy = np.random.multinomial(1,[1/ Y ]* Y, size
    =I).flatten()
phi_iy_k = np.random.normal(size = (I*Y, K))
```

```
# Computing:
alpha = 1 / np.linalg.norm(phi_iy_k @ phi_iy_k.T)
lambda_k = np.zeros(K)
tol,iter = 1e-3, 1000
for i in range(iter):
  expPhi_i_y = np.exp(phi_iy_k @ lambda_k).reshape( (I,Y))
  pi_iy = (expPhi_i_y / expPhi_i_y.sum(axis =
      1)[:,None]).flatten()
  grad_k =  ((pihat_iy - pi_iy) @ phi_iy_k).flatten()
  if np.linalg.norm(grad_k) < tol:
    break
  lambda_k = lambda_k + alpha * grad_k
# Displaying results:
print('Iterations=', i, 'lambda_k=',lambda_k)
```

B.3.2 LOGISTIC REGRESSION AS GLM

```
# Multinomial logistic regression via GLM
#
# Importing libraries and generating data:
import numpy as np, scipy.sparse as sp
from sklearn.linear_model import PoissonRegressor
np.random.seed(7)
I,Y,K = 100,5,3
pihat_iy = np.random.multinomial(1,[1/Y ]*Y,size=I).flatten()
phi_iyk = np.random.normal( size = I*Y*K)
# Computing the parameter:
reg = PoissonRegressor(alpha = 0, fit_intercept=False)
R_omega_p = sp.hstack([phi_iyk.reshape(-1,K),
                -sp.kron(sp.identity(I), np.ones((Y,1)))])
reg.fit(R_omega_p, pihat_iy  )
# Computing the inverse variance-covariance matrix:
pilambda_i_y = np.exp(R_omega_p @ reg.coef_ ).reshape((-1,Y))
phi_i_y_k = phi_iyk.reshape((-1,Y,K))
B1 = (pilambda_i_y[:,:,None,None] * phi_i_y_k[:,:,:,None] *
    phi_i_y_k[:,:,None,:]).sum(axis = (0,1)) / I
B2=  ( pilambda_i_y[:,:,None,None,None] *
    pilambda_i_y[:,None,:,None,None] *
    phi_i_y_k[:,:,None,:,None] * phi_i_y_k[:,None,:,None,:]
    ).sum(axis = (0,1,2)) / I
Bhat_k_l = B1 - B2
# Displaying results:
print('lambda=\n', reg.coef_[:K])
print('Fisher matrix=\n', np.linalg.inv(Bhat_k_l))
```

B.3.3 COERCIVITY DETECTOR

```
# Coercivity detector
#
# Importing libraries and generating data:
import numpy as np, scipy.sparse as sp, gurobipy as grb
np.random.seed(7)
I,Y,K = 100,5,3
muhat_i_y = np.random.multinomial(1, [1/Y]*Y, size=I)
phi_i_y_k = np.random.normal( size = (I,Y,K))
# Computing:
m=grb.Model()
t = m.addMVar(1)
mubar_i_y = m.addMVar( (I,Y))
m.setObjective(t, sense=grb.GRB.MAXIMIZE)
m.addConstr( mubar_i_y >= t )
m.addConstr( mubar_i_y.sum(axis=1) == muhat_i_y.sum(axis=1) )
m.addConstr( (mubar_i_y[:,:,None]*phi_i_y_k).sum(axis=(0,1) )
    == (muhat_i_y[:,:,None]*phi_i_y_k).sum(axis=(0,1) ) )

m.optimize()
# Displaying results:
print('V=', t.X)
```

B.3.4 PROXIMAL GRADIENT DESCENT

```
# Multinomial logistic regression w L1 regularization
#
# Importing libraries and generating data :
import numpy as np, scipy.sparse as sp
np.random.seed (7)
I ,Y , K = 100 ,5 ,3
muhat_iy = np.random.multinomial(1,[1/ Y]* Y, size
    =I).flatten()
phi_iy_k = np.random.normal(size = (I*Y, K))
# Define soft thresholding:
def soft_thresholding(x, eta):
    return np.sign(x) * np.maximum(np.abs(x) - eta, 0)
# Computing :
gamma = 5.0
alpha = 1 / np.linalg.norm(phi_iy_k @ phi_iy_k.T)
lambda_k = np.zeros(K)
tol = 1e-5
iter = 1000
for i in range(iter):
  expPhi_i_y = np.exp(phi_iy_k @ lambda_k).reshape( (I,Y))
  mu_iy = (expPhi_i_y / expPhi_i_y.sum(axis =
      1)[:,None]).flatten()
```

```
  grad_k =  ((muhat_iy - mu_iy) @ phi_iy_k).flatten()
  lambdatilde_k = lambda_k + alpha*grad_k
  new_lambda_k = soft_thresholding(lambdatilde_k ,
      alpha*gamma )
  if np.linalg.norm( lambda_k -new_lambda_k) < alpha*tol:
    iter = i
    break
  lambda_k = new_lambda_k
# Displaying results:
print(iter, lambda_k)
```

B.3.5 MINIMAX REGRET ESTIMATION

```
# Minimax-regret estimation
#
# Importing libraries and generating data:
import numpy as np, scipy.sparse as sp, gurobipy as grb
np.random.seed(7)
I,Y,K = 500,4,3
muhat_iy = np.random.multinomial(1, [1/Y]*Y, size=I).flatten()
phi_iy_k = np.random.normal( size = (I*Y,K))
phi_iy0 = np.random.normal( size = I*Y )
# Computing:
m=grb.Model()
u_i = m.addMVar(I,lb = - grb.GRB.INFINITY)
lambda_k = m.addMVar(K, lb = - grb.GRB.INFINITY)
m.setObjective(u_i.sum()-muhat_iy @ phi_iy_k @ lambda_k,
                sense=grb.GRB.MINIMIZE)
m.addConstr(sp.kron( sp.eye(I),np.ones((Y,1)) ) @ u_i
                >= phi_iy0+phi_iy_k @ lambda_k)
m.optimize()
# Displaying results:
print('lambda=', lambda_k.X, 'val=',m.objval)
```

B.3.6 SMALL NOISE LIMIT

```
# Multinomial logistic regression with small noise
# with the log-sum-exp trick
#
# Importing libraries and generating data :
import numpy as np
from scipy.optimize import minimize

np.random.seed(7)
I, Y, K = 500, 4, 3
pihat_iy = np.random.multinomial(1,[1/ Y ]* Y, size
    =I).flatten()
```

```
phi_iy_k = np.random.normal(size = (I*Y, K))
phi_iy0 = np.random.normal( size = I*Y )
sigma = 0.001
tol,iter = 1e-3, 1000000
# Computing:

def f(lambda_k):
  philambda_iy = phi_iy_k @ lambda_k
  argexpPhi_i_y = (phi_iy0 + philambda_iy).reshape( (I,Y)) / \
      sigma
  maxargexpPhi_i = argexpPhi_i_y.max(axis = 1)
  expPhi_i_y = np.exp( argexpPhi_i_y - maxargexpPhi_i[:,None]
      )
  pi_iy = (expPhi_i_y / expPhi_i_y.sum(axis =
      1)[:,None]).flatten()
  return sigma * (np.log(expPhi_i_y.sum(axis =1)).sum() +
      maxargexpPhi_i.sum()) - pihat_iy.dot(philambda_iy )

def grad_f(lambda_k):
  philambda_iy = phi_iy_k @ lambda_k
  argexpPhi_i_y = (phi_iy0 + philambda_iy).reshape( (I,Y)) / \
      sigma
  maxargexpPhi_i = argexpPhi_i_y.max(axis = 1)
  expPhi_i_y = np.exp( argexpPhi_i_y - maxargexpPhi_i[:,None]
      )
  pi_iy = (expPhi_i_y / expPhi_i_y.sum(axis =
      1)[:,None]).flatten()
  return (pi_iy - pihat_iy ) @ phi_iy_k

res = minimize(f,x0 = np.zeros(K), jac = grad_f,
    method='BFGS', options={'gtol': tol})
# displaying results:
print(sigma, res.x)
```

B.3.7 CODE FOR EXERCISE 3.2

```
import numpy as np
from sklearn.linear_model import LogisticRegression
(I,Y,L) = (1000,5,7)
np.random.seed(1)
Psi_i_l = np.random.uniform(size = (I,L))
pihat_i_y = np.random.multinomial(1, np.ones(Y) / Y,size = I)
yhat_i = pihat_i_y.argmax (axis = 1)
model = LogisticRegression( max_iter=100000,
    fit_intercept=False, penalty = None, tol=1e-8 )
model.fit(Psi_i_l, yhat_i)
theta_y_l = model.coef_
print((theta_y_l - theta_y_l[0,:][None,:]).T)
```

B.3.8 CODE FOR EXERCISE 3.4

```
import numpy as np
(T,X,Y,K) = 5,3,4,2
np.random.seed(7)
phi_t_x_y_k = np.random.uniform(size=(T,X,Y,K))
muhat_t_x_y = np.random.uniform(size=(T,X,Y))
```

B.3.9 LOADING TRAVEL DATA IN PROBLEM 3.1

```
import numpy as np, pandas as pd
#
def load_travel_data():
  thepath ='https://raw.githubusercontent.com/math-econ-code/'
  thepath += 'mec_datasets/main/demand_travelmode/'
  the_data = pd.read_csv(thepath+'travelmodedata.csv')
  pihat_iy = np.where(the_data['choice'] =='yes' , 1, 0)
  Y = the_data['mode'].nunique()
  I = the_data.shape[0] // Y
  covariates = the_data[['travel', 'income', 'gcost']].values
  Phi_iy_k = np.column_stack([ covariates[:,0] , -
      (covariates[:,0]
  * covariates[:,1] ), - covariates[:,2] ])
  _,K = Phi_iy_k.shape
  Phibar_k = Phi_iy_k.mean(axis = 0)
  Phistdev_k = Phi_iy_k.std(axis = 0, ddof = 1)
  Phi_iy_k = ((Phi_iy_k - Phibar_k[None,:]) /
      Phistdev_k[None,:])
  return I,Y,K,pihat_iy,Phi_iy_k

I,Y,K,pihat_iy,Phi_iy_k = load_travel_data()
```

B.4 Code for Chapter 4: Characteristics Models

B.4.1 PURE CHARACTERISTICS MODEL VIA LAGUERRE DIAGRAMS

```
# Laguerre diagrams
#
# Importing libraries and generating data
!pip install -U sdot==2024.12.7.1
from sdot import PowerDiagram, optimal_transport_plan,
    SdotPlan,SumOfDiracs
import numpy as np
```

```
#
np.random.seed( 357 )
#
Y,M = 4,2
np.random.seed(800)
xi_y_m = np.random.uniform(0,2,M*Y).reshape((-1,M))
pwd = PowerDiagram(xi_y_m / 2 )
pwd.boundaries = [[1,0,1],[0,1,1],[-1,0,0],[0,-1,0]]
pwd.plot()
#
# direct problem: given U_y, compute pi_y
def direct(U_y , verbose = True, plot=False):
  pwd.weights =  U_y + ( xi_y_m * xi_y_m).sum(axis=1)
  pi_y = pwd.cell_integrals()
  e_y = pwd.cell_barycenters().T
  if verbose:
    print('U_y =', np.around(U_y,2), '\npi_y=' ,
        np.around(pi_y, 2))
    print('e_y =\n' , np.around(e_y,2))
  if plot:
    pwd.plot()
  return pi_y,e_y
#
# inverse problem: given pi_y, compute U_y
def inverse(pi_y , si0 = None, err=1e-8):
  tp = SdotPlan( source_measure = SumOfDiracs( xi_y_m / 2,
      masses = pi_y  ) )
  tp.adjust_potentials()
  U_y = tp.power_diagram.weights - (xi_y_m *
      xi_y_m).sum(axis=1)
  return U_y - U_y[0]
#
# displaying results
direct(np.array([2,2,0,0.7]), plot=True)
U_y = inverse(np.array([1/4]*4))
direct(U_y,verbose=False, plot=True)
```

B.4.2 COMPARING AR AND GHK SIMULATORS

```
# Geweke, Hajivassiliou and Keane simulator
#
# Importing libraries and generating data:
import numpy as np
from statistics import NormalDist
phi = np.vectorize(NormalDist(mu=0, sigma=1).cdf)
phi_inv = np.vectorize(NormalDist().inv_cdf)
np.random.seed(9)
I,B,J=100000,10,2
```

```
L = np.array([[np.random.uniform()*(i <= j) for i in
    range(J)] for j in range(J)]  )
z_j = np.random.normal(size=J)/100
x_b_i_j = np.random.uniform(size=(B,I,J))
# Coding the simulators:
def iar(L,z_j,x_i_j): # integrated accept-reject simulator
  cond_i_j = (phi_inv(x_i_j)[:,None,:] *
      L[None,:,:]).sum(axis = 2) <= z_j[ None,:]
  argphi_i = (z_j[-1] - (L[-1,:-1][None,:] *
      phi_inv(x_i_j[:,:-1])).sum(axis =1) ) / L[-1,-1]
  return (( cond_i_j[:,:-1]).all(axis = 1) * phi( argphi_i
      )).mean()
#
def ghk(L,z_j,x_i_j): # ghk simulator
    I,J = x_i_j.shape
    xhat_i_j = np.zeros((I,J))
    xhat_i_j[:,0] = x_i_j[:,0] * phi(z_j[0] / L[0,0])
    GHK_i = phi(z_j[0] / L[0,0]) * np.ones(I)
    for j in range(J-1):
        phietc_i = phi((z_j[j+1]-(L[j+1,:(j+1)][None,:]*
            phi_inv(xhat_i_j[:,:(j+1)])).sum(axis= 1) )
            /L[j+1,j+1])
        xhat_i_j[:,j+1]=x_i_j[:,j+1] * phietc_i
        GHK_i *= phietc_i
    return GHK_i.mean()
# Running simulations:
iarSim_b = np.array([iar(L,z_j,x_b_i_j[b,:,:]) for b in
    range(B)])
ghkSim_b = np.array([ghk(L,z_j,x_b_i_j[b,:,:]) for b in
    range(B)])
# Displaying results:
print('IAR, mean=',iarSim_b.mean(),'; std=',iarSim_b.std())
print('GHK, mean=',ghkSim_b.mean(),'; std=',ghkSim_b.std())
```

B.4.3 PROBIT MARKET SHARE MAP USING GHK

This snippet of code assumes that the function ghk has been defined as per appendix B.4.2.

```
# Probit market share
#
# Importing libraries and generating data
import numpy as np
from statistics import NormalDist
phi = np.vectorize(NormalDist(mu=0, sigma=1).cdf)
phi_inv = np.vectorize(NormalDist().inv_cdf)
np.random.seed(9)
I,Y=100000,5
U_y = np.array([ 1.3, 0.9, -0.7, 0.2, 0.6])
x_i_j = np.random.uniform(size=(I,Y-1))
```

```
# Computing:
def ghk(L,z_j,x_i_j): # ghk simulator, as defined above
...
#
def piprobit_y(U_y,sigma_eta,x_i_j):
  Y=U_y.shape[0]
  I = x_i_j.shape[0]
  pi_y = np.zeros(Y)
  for y in range(Y):
      M = np.array( [[(1. if i==j else 0.) - (1. if j==y else
          0.)  for j in \
       range(Y)] for i in range(Y) if i != y])
      L=  np.linalg.cholesky(M @ sigma_eta @ M.T)
      z_j  = - M @ U_y
      pi_y[y] = ghk(L,z_j,x_i_j)
  return (pi_y)
#
sigma_eta = np.eye(Y)
pi_y = piprobit_y(U_y,np.eye(Y),x_i_j)
# Displaying results:
print('pi_y=', pi_y,'\nsum(pi_y)=' ,pi_y.sum())
```

B.4.4 DEMAND ESTIMATION WITH SIMPLE IV

```
# Demand estimation with simple IV regression
#
# importing libraries and generating data
import numpy as np
#
def generate(T= 5000, seed = 77):
  np.random.seed(seed)
  p_t = np.random.uniform(size = T)
  kappa_t = 2 * np.random.normal(size = T)
  zeta_t = np.random.normal(size = T)
  p_t = kappa_t + zeta_t
  pihat_t = np.exp( 2. - 0.5 * p_t +  kappa_t ) / (1 +
      np.exp( 2. - 0.5 * p_t +  kappa_t ))
  return pihat_t, p_t, zeta_t
#
T = 500
pihat_t, p_t, zeta_t = generate(T)
#
# defining instrumental variable regression
def iv(y_i,x_i_k,z_i_k):
  return np.linalg.solve( z_i_k.T @ x_i_k , y_i @ z_i_k )
#
# market share inversion
U_t = np.log(pihat_t / (1 - pihat_t))
regressors_t_k =  np.stack((np.ones(T), p_t), axis = 1 )
instruments_t_k = np.stack((np.ones(T), zeta_t),axis = 1 )
```

```
#
lambdaOLS_k = iv(U_t, regressors_t_k, regressors_t_k )
lambdaIV_k = iv(U_t, regressors_t_k, instruments_t_k )
#
# displaying results
print('OLS estimate without IV:\n', 'lambda_OLS=',lambdaOLS_k)
print('IV estimate:\n', 'lambda_IV =', lambdaIV_k)
```

B.4.5 DEMAND ESTIMATION WITH IV-GMM, LOGIT CASE

```
# Demand estimation with IV-GMM, logit case
#
# importing libraries and generating data
import numpy as np
#
def generate(T= 5000, seed = 77):
  np.random.seed(seed)
  p_t = np.random.uniform(size = T)
  kappa_t = 2 * np.random.normal(size = T)
  zeta_t = np.random.normal(size = T)
  zetaprime_t = np.random.normal(size = T)
  p_t = kappa_t + 0.3* zeta_t + 0.7 * zetaprime_t
  pihat_t = np.exp( 2. - 0.5 * p_t +  kappa_t ) / (1 +
      np.exp( 2. - 0.5 * p_t +  kappa_t ))
  return pihat_t, p_t, zeta_t, zetaprime_t
#
T = 500
pihat_t, p_t, zeta_t, zetaprime_t = generate(T)
#
# defining IV-GMM regression
def lambda_gmm(dep_t, reg_t_k, ins_t_l, W_l_l = None):
  if W_l_l is None:
    W_l_l = np.linalg.inv( ins_t_l.T @ ins_t_l)
  Pi_t_t = ins_t_l @ W_l_l @ ins_t_l.T
  return np.linalg.solve( reg_t_k.T @ Pi_t_t @ reg_t_k, U_t @
      Pi_t_t @ reg_t_k )
#
# market share inversion
U_t = np.log(pihat_t / (1 - pihat_t))
regressors_t_k =  np.stack((np.ones(T), p_t), axis = 1 )
instruments_t_l = np.stack( (np.ones(T), zeta_t,
    zetaprime_t), axis = 1)
#
lambda1S_k = lambda_gmm(U_t,regressors_t_k, instruments_t_l )
kappahat_t = U_t - regressors_t_k @ lambda1S_k
kappaT_t_l = kappahat_t[:,None]*instruments_t_l
W_l_l = T*  np.linalg.inv(kappaT_t_l.T @ kappaT_t_l)
lambda2S_k = lambda_gmm(U_t,regressors_t_k,
    instruments_t_l,W_l_l )
```

```
#
# displaying results
print('IV-GMM 1st stage:\n', 'lambda_1S=',lambda1S_k)
print('IV-GMM 2nd stage:\n', 'lambda_2S =', lambda2S_k)
```

B.4.6 DEMAND ESTIMATION WITH IV-GMM, RANDOM COEFFICIENT LOGIT

```
# Demand estimation with IV-GMM, random coefficient logit case
#
# importing libraries and generating data
import numpy as np
from scipy.optimize import fsolve, minimize

#
def generate(T= 5000, I=10000, seed = 77):
  np.random.seed(seed)
  p_t = np.random.uniform(size = T)
  kappa_t = 2 * np.random.normal(size = T)
  zeta_t = np.random.normal(size = T)
  zetap_t = np.random.normal(size = T)
  zetapp_t = np.random.normal(size = T)
  eta_i = - np.exp( - (1 + np.random.normal(size = I ) ) )
  p_t = kappa_t + 0.3* zeta_t + 0.5 * zetap_t + 0.2 * zetapp_t
  #u_i_t =  2. - 0.5 * p_t[None,:] +  0.1 *
      eta_i[:,None]*p_t[None,:] + kappa_t[None,:]
  #pihat_t = (np.exp( u_i_t) / (1 + np.exp( u_i_t
      ))).mean(axis=0)
  pihat_t = np.exp( 2. - 0.5 * p_t + kappa_t  ) / (1 +
      np.exp( 2. - 0.5 * p_t  + kappa_t ))
  return pihat_t, p_t, zeta_t, zetap_t,zetapp_t, eta_i
#
T,I = 500,1000
pihat_t, p_t, zeta_t, zetap_t,zetapp_t,eta_i = generate(T,I)
#
# inversing the random coefficient logit model
def valUhat_t(tau):
  U_t = np.zeros(T)
  for t in range(T):
    thef = lambda U: \
        (1/(1+np.exp(U+tau*eta_i*p_t[t]))).mean() - 1+ \
        pihat_t[t]
    U_t[t] = fsolve(thef, np.log(pihat_t[t] / \
        (1-pihat_t[t])))[0]
  return U_t

def derUhat_t(tau,Uhat_t = None):
  if Uhat_t is None:
    Uhat_t = valUhat_t(tau)
  epsilonp_i_t = eta_i[:,None]*p_t[None,:]
```

```
  denom_i_t = 1 + np.exp( Uhat_t + tau*epsilonp_i_t )
  return  (epsilonp_i_t / denom_i_t).mean(axis=0) /(1-pihat_t)
#
# defining IV-GMM regression
def nl_iv_gmm(W_l_l = None):
  if W_l_l is None:
    W_l_l = np.linalg.inv( ins_t_l.T @ ins_t_l)
  Pi_t_t = ins_t_l @ W_l_l @ ins_t_l.T
  Gamma_t_t = Pi_t_t - Pi_t_t @ reg_t_k @ \
      np.linalg.inv(reg_t_k.T @ Pi_t_t @ reg_t_k) @ reg_t_k.T \
      @ Pi_t_t
  #
  def F(tau):
    Uhat_t = valUhat_t(tau)
    return Uhat_t @ Gamma_t_t @ Uhat_t
  #
  def derF(tau):
    Uhat_t = valUhat_t(tau)
    dUdtau_t = derUhat_t(tau, Uhat_t)
    return 2* Uhat_t @ Gamma_t_t @ dUdtau_t
  #
  tauhat = minimize(F, 1, jac = derF).x[0]
  Uhat_t = valUhat_t(tauhat)
  lambdahat_k = np.linalg.solve(reg_t_k.T @ Pi_t_t @ reg_t_k,
      Uhat_t @ Pi_t_t @ reg_t_k)
  kappahat_t = Uhat_t - reg_t_k @ lambdahat_k
  return (tauhat, lambdahat_k, kappahat_t)
#
reg_t_k = np.stack((np.ones(T), p_t), axis = 1)
ins_t_l = np.stack ((np.ones(T), zeta_t,zetap_t,zetapp_t),
    axis = 1)
#
tau1, lam1_k, kap1_t = nl_iv_gmm()
kappaT_t_l = kap1_t[:,None]*ins_t_l
W_l_l = T*  np.linalg.inv(kappaT_t_l.T @ kappaT_t_l)
tau2, lam2_k, kap2_t = nl_iv_gmm(W_l_l )
#
# displaying results
print('IV-GMM 1st stage:\n', 'tau_1S', tau1,
    'lambda_1S=',lam1_k)
print('IV-GMM 2nd stage:\n', 'tau_2S', tau2,
    'lambda_2S=',lam2_k)
```

B.4.7 THE BLP METHOD

```
!pip install sparse
import numpy as np, scipy.sparse as sp, sparse as spx
from types import SimpleNamespace
```

```
#
def generate(T=5,Y=3,I=10,M=3,Kx=3,L=3,nbend=2,X=2,seed=77):
    np.random.seed(seed)
    D, S, Ke, K,db = Kx+2+nbend, L, 2, Kx+2,SimpleNamespace()
    x_y = np.arange(Y) if X is None else \
     np.random.choice(np.arange(X),(Y))
    db.p_ty = np.exp( np.random.normal(size = (T*Y)) / 5) / 2
    t2 = (db.p_ty * db.p_ty)[:,None] / 2
    db.phie_ty_k= np.concatenate((db.p_ty[:,None], t2),axis=1)
    db.phieprime_ty_k = np.concatenate( \
     (np.ones((T*Y,1)),db.p_ty[:,None]),axis=1 )
    db.phix_ty_k = np.random.normal( size= (T*Y,Kx)) / 10
    db.phi_ty_k = np.block([[db.phie_ty_k,db.phix_ty_k]])
    kappa_ty=db.p_ty*0.5+np.random.normal(size = (T*Y))/10
    U_ty = db.phi_ty_k @  np.ones(K) + kappa_ty
    t1 = np.random.uniform (size=(T*Y,D-Ke))*db.p_ty[:,None]
        +\
     np.random.normal (size=(T*Y,D-Ke))
    db.zeta_ty_d=np.concatenate((t1,db.phie_ty_k),axis=1)
    eta_i_m = np.random.uniform(size = (I,M)) / 10
    t1 = -0.3 *db.p_ty.reshape((T,Y,1) ) + \
     + 0.95 * np.random.normal(size = (T,Y,1)) / 10
    t2 = np.random.uniform(size = (T,Y,M-1)) / 10
    xi_t_y_m=np.concatenate(( t1,t2), axis=2)
    dxidp_t_y_m = np.concatenate (\
     (- 0.3 * np.ones((T,Y,1)) , np.zeros((T,Y,M-1))),axis=2)
    db.nu_tiy_m=(eta_i_m[None,:,None,:]*xi_t_y_m[:,None,:,:]).\
     reshape((T*I*Y,M))
    nutau_t_i_y=3* (db.nu_tiy_m).sum(axis= 1).reshape((T,I,Y))
    u_t_i = np.log(1 + np.exp(U_ty.reshape((T,Y))[:,None,:] +\
     nutau_t_i_y).sum(axis = 2))
    db.pihat_ty  = np.exp(U_ty.reshape((T,Y))[:,None,:] - \
     u_t_i[:,:,None]).mean(axis = 1).flatten()
    db.psi_ty_l = np.concatenate( (- db.p_ty[:,None], \
     np.random.normal(size = (T*Y,L-1))),axis=1)
    err_ty = -0.5*db.p_ty +  np.random.normal(size =(T*Y))
    db.chi_ty_s = db.psi_ty_l.copy()
    H_t_y_t_y = np.zeros((T,Y,T,Y))
    for t in range(T):
        H_t_y_t_y[t,:,t,:] = 1 * (x_y[:,None] == x_y[None,:])
    db.eta_im,db.xi_tym = eta_i_m.flatten(),xi_t_y_m.flatten()
    db.dxidp_tym=dxidp_t_y_m.flatten()
    db.H_ty_ty = sp.coo_array(H_t_y_t_y.reshape((T*Y,T*Y)))
    for name, value in dict(T=T, Y=Y, I=I, M=M,Ke=Ke, Kx=Kx,\
     K=K, L=L, D=D, S=S, N= Ke+M).items():
        setattr(db, name, value)
    db.SigmaI = sp.kron(sp.kron(eye(T),np.ones((1,I))),eye(Y))
    db.SigmaY = sp.kron(eye(T*I),np.ones((1,Y)))
    db.nuprime_tiy_m = (eta_i_m[None,:,None,:] * \
     dxidp_t_y_m[:,None,:,:]).reshape((T*I*Y,M))
    db.dphieoverdp_ty_tyk = rkron( eye(db.T * db.Y),
```

```
                                     db.phieprime_ty_k)
    db.dnuoverdp_tiy_tym = rkron(sp.kron(eye(db.T) , sp.kron(\
     np.ones((db.I,1)),eye(db.Y))), db.nuprime_tiy_m, )
    return  db
#
# auxiliary functions
def eye(n):
    return sp.coo_array(sp.eye(n))
#
def spx_coo_to_sp(x):
    from scipy.sparse import coo_array
    return coo_array((x.data, x.coords), shape=x.shape)
#
def rkron(A, B):
    A = sp.csr_matrix(A)
    B = sp.csr_matrix(B)
    return sp.coo_array(sp.vstack([sp.kron(A.getrow(i), \
     B.getrow(i)) for i in range( A.shape[0])]))
#
def ipfp(db, nutau_tiy, tol=1e-5, maxiter = 100000):
    nutau_t_i_y=nutau_tiy.reshape((db.T,db.I,db.Y))
    pihat_t_y = db.pihat_ty.reshape((db.T,db.Y))
    U_t_y = np.zeros((db.T,db.Y))
    for iter in range(maxiter):
        max_t_i = np.maximum((U_t_y[:,None,:]
                              + nutau_t_i_y).max(axis = 2),0)
        exp_t_i_y = np.exp(U_t_y[:,None,:]
                           + nutau_t_i_y - max_t_i[:,:,None])
        u_t_i = max_t_i + np.log (np.exp(-max_t_i) +
                                  exp_t_i_y.sum(axis = 2))
        max_t_y = (nutau_t_i_y - u_t_i[:,:,None]).max(axis =
            1)
        exp_t_i_y = np.exp(nutau_t_i_y - u_t_i[:,:,None]
                           - max_t_y[:,None,:])
        Uprime_t_y = np.log(db.I * pihat_t_y) - \
         (max_t_y+ np.log( exp_t_i_y.sum(axis=1) ) )
        if np.abs(U_t_y - Uprime_t_y).max()<tol:
            break
        else:
            U_t_y = Uprime_t_y
    if (iter>=maxiter-1):
        print('ipfp failed')
    return U_t_y.flatten(), u_t_i.flatten()
#
# BLP procedure
#
def F(theta, W = None):
    tau_m = theta[:db.M].reshape((-1,1))
    alpha_ke = theta[db.M:].reshape((-1,1))
    # step 1
    nutau_tiy = (db.nu_tiy_m @ tau_m).flatten()
```

```
Uhat_ty, uhat_ti =  ipfp(db,nutau_tiy)
# step 2
fhat_tiy = np.exp(db.SigmaI.T @ Uhat_ty+ nutau_tiy
                  - db.SigmaY.T @ uhat_ti)
Deltafhat = sp.diags_array(fhat_tiy )
# step 3
Rfhat = eye(db.T*db.I*db.Y) - sp.kron(eye(db.T*db.I), \
 np.ones((db.Y,db.Y))) @ Deltafhat
# step 4
part1 = db.SigmaI.T @ db.dphieoverdp_ty_tyk @ \
 sp.kron( np.eye(db.T*db.Y) ,alpha_ke )
part2=db.dnuoverdp_tiy_tym @ sp.kron(eye(db.T*db.Y),tau_m)
dVoverdp = part1+part2
# step 5
dpioverdp = db.SigmaI @ Deltafhat @ Rfhat @ dVoverdp/db.I
Dhat = db.H_ty_ty * dpioverdp.T / db.I
# step 6
chat_ty = db.p_ty + sp.linalg.spsolve(sp.csc_matrix(Dhat),
                                     db.pihat_ty)
# step 7
b1 = (Uhat_ty.reshape((-1,1)) - db.phie_ty_k @ alpha_ke)
YY = np.block([[ b1],[chat_ty.reshape((-1,1))]])
XX = sp.bmat([[db.phix_ty_k, None],
              [None,db.psi_ty_l]]).todense()
ZZ = sp.bmat([[db.zeta_ty_d, None],
              [None,db.chi_ty_s]]).todense()
# step 8
if W is None:
  W = np.linalg.inv(ZZ.T @ ZZ)
Pi = ZZ @ W @ ZZ.T
Upsilon = Pi - Pi.T @ XX @ np.linalg.inv(XX.T @Pi @XX) @ \
 XX.T @ Pi
# step 9
valF = (YY.T @ Upsilon @ YY / (2*db.T*db.Y)**2).item()
# step 10
mat1 = db.SigmaI @ Deltafhat @ db.SigmaY.T
mat2 = db.SigmaI @ Deltafhat @ Rfhat @ db.nu_tiy_m
dUoverdtau =  sp.linalg.spsolve(sp.csc_matrix(- db.I * \
 sp.diags_array(db.pihat_ty ) +  mat1 @ mat1.T) ,mat2)
# step 11
dVhatoverdtau = db.SigmaI.T @ dUoverdtau + db.nu_tiy_m
dVhatoverdtheta = np.block([[dVhatoverdtau,np.zeros( \
 (db.T*db.I*db.Y,db.Ke))]])
# step 12
transposed1 = spx.COO.from_scipy_sparse( \
 db.dnuoverdp_tiy_tym).reshape((db.T*db.I*db.Y,db.T* \
 db.Y,db.M)).transpose((0,2,1))
d2Voverdtaudp_tiy_mty = transposed1.reshape((db.T*db.I*\
 db.Y,db.M*db.T*db.Y))
d2Voverdpdalpha_tiy_tyk = spx.COO.from_scipy_sparse(\
 db.SigmaI.T @ db.dphieoverdp_ty_tyk)
```

```
    transposed2 = d2Voverdpdalpha_tiy_tyk.reshape((db.T*db.I*\
     db.Y,db.T*db.Y,db.Ke)).transpose((0,2,1))
    d2Voverdalphadp_tiy_kty = transposed2.reshape((db.T*db.I*\
     db.Y,db.Ke*db.T*db.Y))
    # step 13
    b1 = spx_coo_to_sp(d2Voverdtaudp_tiy_mty)
    b2 = spx_coo_to_sp(d2Voverdalphadp_tiy_kty)
    d2Voverdthetadp = sp.bmat([[b1,b2]])
    # step 14
    term1 = sp.kron(np.ones((1,db.N)) ,db.H_ty_ty)
    term2 = db.SigmaI @ Deltafhat @ Rfhat / db.I
    term3 = rkron(Rfhat @ dVhatoverdtheta, Rfhat @ dVoverdp)+\
     d2Voverdthetadp
    dDtopoverdtheta_ty_ntyprime = term1 * (term2 @ term3)
    # step 15
    dDtopoverdtheta_ty_n_typrime=dDtopoverdtheta_ty_ntyprime.\
     reshape((db.T*db.Y,-1,db.T*db.Y))
    dDoverdtheta = dDtopoverdtheta_ty_n_typrime.transpose(\
     (2,1,0)).reshape((db.T*db.Y,-1))
    # step 16
    term = dDoverdtheta @ sp.kron(eye(db.N), \
     (chat_ty-db.p_ty).reshape((-1,1)))
    dcoverdtheta = - sp.linalg.spsolve(sp.csc_matrix(Dhat),
                                       sp.csc_matrix(term))
    # step 17
    mat = np.block([[np.block([[dUoverdtau,-db.phie_ty_k]])],\
     [dcoverdtheta.todense()]])
    gradF = 2* YY.T @ Upsilon @ mat / (2*db.T*db.Y)**2
    return valF, gradF

db = generate()
F(np.ones(db.N))
```

B.4.8 LOADING BLP'S AUTOMOBILE DATA

```
# Loading BLP's automobile data
import pyblp, numpy as np,  pandas as pd
#
product_data = pd.read_csv(pyblp.data.BLP_PRODUCTS_LOCATION)
#
agent_data = pd.read_csv(pyblp.data.BLP_AGENTS_LOCATION)
#
```

B.5 Code for Chapter 5: Applications of GLMs

B.5.1 ENTROPIC OPTIMAL TRANSPORT VIA GLMS

```
# Optimal transport via generalized linear models
#
# Importing libraries and generating data:
import numpy as np, scipy.sparse as sp
from sklearn import linear_model as lm
np.random.seed(7)
X,Y = 5,4
n_x, m_y = np.ones(X)/X, np.ones(Y)/Y
phi_xy = np.random.uniform(size = X*Y)
# Computing:
w_xy = np.exp( phi_xy )
muhat_xy =  np.kron(n_x,m_y) / w_xy
ot_as_glm = lm.PoissonRegressor(fit_intercept=False,alpha=0)
regr = - sp.hstack([ sp.kron(sp.eye(X), np.ones((Y,1))),
                     sp.kron(np.ones((X,1)),sp.eye(Y))])
ot_as_glm.fit( regr, muhat_xy , sample_weight = w_xy )
u_x,v_y  = ot_as_glm.coef_[:X], ot_as_glm.coef_[X:]
# displaying results
print('value=',u_x @ n_x +v_y @ m_y )
```

B.5.2 ENTROPIC OPTIMAL TRANSPORT VIA IPFP

```
# Optimal transport via IPFP
#
# Importing libraries and generating data:
import numpy as np, scipy.sparse as sp
np.random.seed(7)
X,Y = 5,4
n_x, m_y = np.ones(X)/X, np.ones(Y)/Y
phi_xy = np.random.uniform(size = X*Y)
tol,maxit = 1e-5,1000
# Computing:
K_x_y = np.exp(phi_xy).reshape((X,Y))
B_y = np.ones(Y)
for i in range(maxit):
  A_x = n_x / (K_x_y @ B_y)
  Bnew_y = m_y / (A_x @ K_x_y)
  if np.abs(Bnew_y-B_y).max() < tol:
    break
  B_y = Bnew_y
u_x,v_y = - np.log(A_x),- np.log(B_y)
# displaying results
print('Value=',u_x @ n_x +v_y @ m_y , 'Steps=',i)
```

B.5.3 GRAVITY VIA GENERALIZED LINEAR MODELS

```
# Gravity via GLMs
#
# Importing libraries and generating data:
import numpy as np, scipy.sparse as sp
from sklearn import linear_model as lm
np.random.seed(7)
X,Y,K = 5,5,3
muhat_x_y = np.random.uniform(size = (X,Y) )
np.fill_diagonal(muhat_x_y,0)
phi_xy_k = np.random.uniform(size = (X*Y,3) )
# Computing:
regr = sp.hstack([phi_xy_k ,
                    -sp.kron(sp.eye(X), np.ones((Y, 1))),
                    -sp.kron(np.ones((X, 1)),sp.eye(Y))])
w_xy = (np.ones((X,Y)) - np.eye(X)).flatten()
clf = lm.PoissonRegressor(fit_intercept=False, alpha=0)
clf.fit(regr, muhat_x_y.flatten(), sample_weight=w_xy)
# displaying results
print('lambda_k=',clf.coef_[:K])
```

B.5.4 GRAVITY VIA SISTA

```
# Gravity via SISTA
# Importing libraries and generating data:
import numpy as np
np.random.seed(7)
X,Y,K = 5,5,3
muhat_x_y = np.random.uniform(size = (X,Y) )
np.fill_diagonal(muhat_x_y,0)
phi_xy_k = np.random.uniform(size = (X*Y,3) )
def soft_thresholding(x, eta):
    return np.sign(x) * np.maximum(np.abs(x) - eta, 0)
# Computing :
n_x,m_y = muhat_x_y.sum(axis =1), muhat_x_y.sum(axis =0)
gamma = 0.1 # set to zero for no l1 penalty
alpha = 1 / np.linalg.norm(phi_xy_k @ phi_xy_k.T)
lambda_k = np.zeros(K)
tol = 1e-5
iter = 1000
B_y = np.ones(Y)
for i in range(iter):
  K_x_y = np.exp(phi_xy_k @ lambda_k).reshape( (X,Y))
  np.fill_diagonal(K_x_y,0)
  A_x = n_x / (K_x_y @ B_y)
  B_y = m_y / (A_x @ K_x_y)
  mu_x_y =  (A_x[:,None] * B_y[None,:] * K_x_y)
```

```
  grad_k =  (muhat_x_y - mu_x_y).flatten() @ phi_xy_k
  lambdatilde_k = lambda_k + alpha*grad_k
  new_lambda_k = soft_thresholding(lambdatilde_k ,
      alpha*gamma )
  if np.linalg.norm( lambda_k -new_lambda_k) < alpha*tol:
    iter = i
    break
  lambda_k = new_lambda_k
# displaying results
print('lambda_k=',lambda_k,'Steps=',i)
```

B.5.5 MATCHING SURPLUS ESTIMATION VIA GENERALIZED LINEAR MODELS

```
# Cupids via GLM
# Importing libraries and generating data:
import numpy as np, scipy.sparse as sp
from sklearn import linear_model as lm
kr = sp.kron
np.random.seed(7)
X,Y,K = 5,5,3
A = X*Y+X+Y
muhat_a = np.random.uniform(size = A )
phi_xy_k = np.random.uniform(size = (X*Y,K) )
# Computing :
s_a = np.array([2.0]*(X*Y)+[1.0]*(X+Y))
R00 = phi_xy_k/2
R01 = -kr(sp.eye(X),np.ones((Y,1)))/2
R02 = -kr(np.ones((X,1)),sp.eye(Y) )/2
R11 = - sp.eye(X)
R22 = - sp.eye(Y)
R_a_p = sp.bmat([[R00,R01,R02],
                 [None,-sp.eye(X),None],
                 [None,None,-sp.eye(Y)]])
glm = lm.PoissonRegressor(fit_intercept=False,alpha=0)
glm.fit( R_a_p, muhat_a , sample_weight = s_a )
lambda_k  = glm.coef_[:K]
# displaying results
print('lambda_k=',lambda_k)
```

B.5.6 COALITION FORMATION VIA GENERALIZED LINEAR MODELS

```
# Coalition formation via GLM
#
# Importing libraries and generating data:
import numpy as np, scipy.sparse as sp
```

```
from sklearn import linear_model as lm
np.random.seed(7)
x_log_x = np.vectorize(lambda x: x * np.log(x) if x > 0 else
    0)
Z,K = 4,3
B = Z*(Z+1) // 2
# in this example, same-sex marriage: coalitions of size <= 2
count_b_z = np.array([ list(np.eye(Z)[z1] + np.eye(Z)[z2])
          for z2 in range(Z) for z1 in range(Z) if z1<= z2 ])
muhat_b = np.random.uniform(size=B)
phi_b_k = np.random.uniform(size = (B,K) )
# Computing :
s_b = count_b_z.sum(axis = 1)
R_b_p = np.block([phi_b_k , count_b_z ]) /  s_b[:,None]
l_b =  (x_log_x(count_b_z) ).sum(axis = 1) / s_b
glm = lm.PoissonRegressor(fit_intercept=False,alpha=0)
glm.fit( R_b_p, muhat_b * np.exp(l_b) ,
        sample_weight = s_b * np.exp(-l_b) )
# displaying results
print('lambda_k=',glm.coef_[:K])
```

B.5.7 HEDONIC EQUILIBRIUM PRICING VIA GENERALIZED LINEAR MODELS

```
# Hedonic equilibrium pricing via generalized linear models
#
# Importing libraries and generating data:
import numpy as np, scipy.sparse as sp
from sklearn import linear_model
np.random.seed(7)
X,Y,Z = 5,4,10
n_x, m_y = np.ones(X)/X, np.ones(Y)/Y
C_x_z = np.random.uniform(size = (X,Z))
U_y_z = np.random.uniform(size = (Y,Z))
kr=sp.kron
# Computing:
dependent = np.concatenate(\
 [(n_x[:,None] * np.exp(C_x_z)).flatten(), \
 (m_y[:,None] * np.exp(-U_y_z)).flatten()])
weights = np.concatenate( [np.exp(-C_x_z).flatten() ,
                           np.exp(U_y_z).flatten()] )

R00 = kr(np.ones((X,1)),sp.eye(Z))
R01 = - kr(sp.eye(X), np.ones((Z,1) ))
R10 = - kr(np.ones((Y,1)),sp.eye(Z))
R12 = - kr(sp.eye(Y), np.ones((Z,1) ))

regressor = sp.bmat([[ R00, R01 , None],
                      [ R10 ,None , R12]])
```

```
hedonic_as_glm = linear_model.PoissonRegressor( \
 fit_intercept=False,alpha=0)
hedonic_as_glm.fit( regressor,dependent,sample_weight=weights)
p_z  = hedonic_as_glm.coef_[:Z]
# displaying results
print('p_z=',p_z )
```

B.5.8 LOADING MARRIAGE DATA IN PROBLEM 5.1

```
import numpy as np, pandas as pd
# retrieves marriage dataset from:
# Choo and Siow (2006). Who marries whom and why.
# Journal of Political Economy.
def load_marriage_data(nbCateg = 25):
  thepath ='https://raw.githubusercontent.com/math-econ-code/'
  thepath += 'mec_datasets/main/marriage-ChooSiow/'
  n_singles=pd.read_csv(thepath+'n_singles.txt',
    sep='\t',header=None)
  marr = pd.read_csv(thepath+'marr.txt',sep='\t',header=None)
  muhat_x0 = np.array(n_singles[0].iloc[0:nbCateg])
  muhat_0y = np.array(n_singles[1].iloc[0:nbCateg])
  muhat_xy = np.array(marr.iloc[0:nbCateg:,0:nbCateg])
  return muhat_xy, muhat_x0, muhat_xy

muhat_xy,muhat_x0, muhat_0y= load_marriage_data()
```

B.5.9 LOADING INTERNATIONAL TRADE DATA IN PROBLEM 5.2

```
import numpy as np, pandas as pd
# retrieves trade dataset from:
# Yotov et al. (2016). An Advanced Guide to Trade Policy
# Analysis: The Structural Gravity Model. WTO.
def get_trade_data():
  thepath = \
      'https://raw.githubusercontent.com/math-econ-code/'
  thepath += 'mec_datasets/main/gravity_wtodata/'
  tradedata = pd.read_csv(thepath + \
   '1_TraditionalGravity_from_WTO_book.csv')
  tradedata = tradedata[['exporter', 'importer','year', \
   'trade', 'DIST','ln_DIST', 'CNTG', 'LANG', 'CLNY']]
  tradedata.sort_values(['year','exporter','importer'],
                        inplace = True)
  tradedata.reset_index(inplace = True, drop = True)
  nbt = len(tradedata['year'].unique())
  nbi = len(tradedata['importer'].unique())
  nbk = 4
```

```
    years = tradedata['year'].unique()
    distances = np.array(['ln_DIST', 'CNTG', 'LANG', 'CLNY'])
    D_x_y_t_k = np.zeros((nbi,nbi,nbt,nbk))
    tradevol_x_y_t = np.zeros((nbi,nbi,nbt))
    muhat_x_y_t = np.zeros((nbi,nbi,nbt))
    for t, year in enumerate(years):
        tradevol_x_y_t[:, :, t] = np.array(tradedata.loc[ \
         tradedata['year'] == year, 'trade']).reshape((nbi,
             nbi))
        np.fill_diagonal(tradevol_x_y_t[:, :, t], 0)
        for k, distance in enumerate(distances):
            D_x_y_t_k[:, :, t, k] = np.array(tradedata.loc[ \
             tradedata['year'] == year, distance]).reshape( \
             (nbi, nbi))
    muhat_x_y_t=tradevol_x_y_t/(tradevol_x_y_t.sum()/len(years))
    return D_x_y_t_k, muhat_x_y_t

D_x_y_t_k,muhat_x_y_t = get_trade_data()
```

B.5.10 CODE FOR PROBLEM 5.3

```
import numpy as np
(X,Y,Omega) = 5,5,3
np.random.seed(7)
M_y_omega=np.random.uniform(size=(Y,Omega))
n_x = np.ones(X)
P_y = np.ones(Y)
tau_x_y = np.random.uniform(size=(X,Y))
np.fill_diagonal(tau_x_y, 1)
zeta_x_omega = np.random.uniform(size=(X,Omega))
```

B.6 Code for Chapter 6: Dynamic Discrete Choice

B.6.1 NO HETEROGENEITY, FINITE HORIZON VIA LINEAR PROGRAMMING

```
# Dynamic discrete choice via LP
#
# Importing libraries and generating data:
!pip install gurobipy
import numpy as np, gurobipy as grb
T,X,Y = 40,3,2
beta = 0.9
phi_t_x_y = beta**np.arange(T)[:,None,None] \
            * np.array( [[[25,-10], [12,-10 ],[-5,-10]]])
n_x = np.array([10,20,5])
```

```
P_xprime_x_y = np.array([[[1/3, 1],[0, 1],[0, 1]], \
 [[2/3, 0],[1/3,0],[0, 0]],[[0,0],[2/3, 0],[1, 0]]])
c = 9 # size of the workshop
# Computing:
m = grb.Model()
mu_t_x_y = m.addMVar((T,X,Y))
m.setObjective( (mu_t_x_y* phi_t_x_y).sum(),
               sense = grb.GRB.MAXIMIZE)
u_0_x = m.addConstr( n_x == mu_t_x_y[0,:,:].sum(axis = 1))
u_t_x = m.addConstr( (mu_t_x_y[:-1,None,:,:] * P_xprime_x_y[ \
 None,:,:,:]).sum(axis=(2,3))== mu_t_x_y[1:,:,:].sum(axis=2))
lambda_t = m.addConstr( mu_t_x_y[:,:,1].sum(axis = 1) <= c )
m.optimize()
# displaying results
u_0_x.pi
```

B.6.2 NO HETEROGENEITY, FINITE HORIZON VIA BACKWARD INDUCTION

```
# Dynamic discrete choice via backward-forward induction
#
# Importing libraries and generating data:
import numpy as np, gurobipy as grb
T,X,Y = 40,3,2
beta = 0.9
phi_x_y = np.array( [[25,-10], [12,-10 ],[-5,-10]])
phi_t_x_y = np.array([beta**t for t in \
    range(T)])[:,None,None] *  phi_x_y[None,:,:]
n_x = np.array([10,20,5])
P_xprime_x_y = np.array([[[1/3, 1],[0, 1],[0, 1]],[[2/3,
    0],[1/3,0],
                        [0, 0]],[[0,0],[2/3, 0],[1, 0]]])
# Computing -- backward:
u_t_x = np.zeros((T,X))
u_t_x[T-1,:] = phi_t_x_y[T-1,:,:].max(axis = 1)
for t in range(T-2,-1,-1):
  u_t_x[t,:] = ( phi_t_x_y[t,:,:] + (P_xprime_x_y *
      u_t_x[t+1,:][:,None,None]).sum(axis = 0) ).max(axis = 1
      )
# Computing -- forward
mu_t_x_y = np.zeros((T,X,Y))
then_x = n_x
for t in range(T-1):
  y_x = ( phi_t_x_y[t,:,:] +
  (P_xprime_x_y * u_t_x[t+1,:][:,None,None]).sum(axis = 0)
      ).argmax(axis = 1 )
  mu_t_x_y[t,list(range(X)),list(y_x) ]=then_x
  then_x = P_xprime_x_y.reshape((X,X*Y)) @ \
      mu_t_x_y[t,:,:].flatten()
y_x = ( phi_t_x_y[T-1,:,:] ).argmax(axis = 1 )
```

```
mu_t_x_y[T-1,list(range(X)),list(y_x) ]= then_x
# Displaying results
print('u_t_x=',u_t_x[0,:])
print('objective=' ,(mu_t_x_y * phi_t_x_y ).sum() )
```

B.6.3 ESTIMATION OF THE LOGIT MODEL WITH FINITE HORIZON IN NUMPY

```
# Estimation of dynamic discrete choice via scipy.optimize
#
# Importing libraries and generating data:
import numpy as np
from scipy import optimize, sparse as sp
np.random.seed(7)
T,X,Y,K = 2,3,2,3
beta = 0.9
phi_x_y_k = np.random.normal(size=(X,Y,K))
muhat_txy = np.random.uniform(size=(T*X*Y))
P_xprime_xy = np.random.uniform(size= (X,X*Y))
P_xprime_xy = P_xprime_xy / P_xprime_xy.sum(axis=0)[None,:]
# Computing:
phi_t_x_y_k = np.array([beta**t for t in
    range(T)])[:,None,None,None] *  phi_x_y_k[None,:,:,:]
phi_txy_k = phi_t_x_y_k.reshape((-1,K))
SigmaY = sp.kron( sp.eye(T*X),np.ones((1,Y)))
Psi = sp.kron(sp.diags(diagonals=[1]*(T-1), offsets=+1,
    shape=(T,T)), P_xprime_xy.T) - SigmaY.T

def fu_tx(lambda_k):
  u_t_x = np.zeros((T+1,X))
  for t in range(T-1,-1,-1):
    u_t_x[t,:] = np.log( np.exp(
        phi_t_x_y_k[t,:,:,:].reshape((-1,K)) @ lambda_k
        +(u_t_x[t+1,:] @
        P_xprime_xy)).reshape((X,Y)).sum(axis = 1))
  return u_t_x[:-1,:].flatten()

def fdu_tx(lambda_k):
    u_tx = fu_tx(lambda_k).flatten()
    pi_txy = np.exp( phi_txy_k @ lambda_k + Psi @ u_tx)
    return - sp.linalg.spsolve(SigmaY @ sp.diags(pi_txy) @
        Psi , SigmaY @ sp.diags(pi_txy) @ phi_txy_k)

def neg_F(lambda_k):
    return  - muhat_txy.T @ ( phi_txy_k @ lambda_k + Psi @
        fu_tx(lambda_k))

def neg_gradF(lambda_k):
    return muhat_txy.T @ ( - Psi @ fdu_tx(lambda_k) -
        phi_txy_k)
```

```
def neg_hessF(lambda_k):
    u_tx = fu_tx(lambda_k).flatten()
    pi_txy = np.exp( phi_txy_k @ lambda_k + Psi @ u_tx)
    dlogpi = phi_txy_k + Psi @ fdu_tx(lambda_k)
    TXY = T*X*Y
    TTT = sp.csr_matrix( ([1]*TXY,(list(range(TXY)),
        [a*(TXY+1) for a in range(TXY)])),shape =
        (TXY,TXY*TXY) )
    Tdlogpisquare =  TTT @ sp.kron(dlogpi,dlogpi )
    d2u = - sp.linalg.spsolve( SigmaY @ sp.diags(pi_txy) @
        Psi , SigmaY @ sp.diags(pi_txy) @  Tdlogpisquare)
    return (- muhat_txy.T @ Psi @ d2u).reshape((K,K))

result = optimize.minimize(neg_F, np.zeros(K), jac=neg_gradF,
    method='BFGS')

# displaying results
if result.success:
    lambdahat_k = result.x
    print("lambdahat:", lambdahat_k)
    I = muhat_txy.sum()
    V= np.diag(muhat_txy) / I - muhat_txy[:,None] @ \
        muhat_txy[None,:] / (I*I)
    HinvB = np.linalg.solve(neg_hessF(lambdahat_k),
        (phi_txy_k + Psi @ fdu_tx(lambdahat_k)).T)
    print('Cov(lambdahat)=')
    print(HinvB @ V @ HinvB.T)
else:
    print(result.message)
```

B.6.4 ESTIMATION OF THE LOGIT MODEL WITH FINITE HORIZON IN PYTORCH

```
# Estimation of DDC logit model w finite horizon in Torch
#
# Importing libraries and generating data:
import torch, numpy as np, scipy.sparse as sp
np.random.seed(7)
T,X,Y,K = 2,3,2,3
beta = 0.9
phi_x_y_k = np.random.normal(size=(X,Y,K))
muhat_txy = np.random.uniform(size=(T*X*Y))
P_xprime_xy = np.random.uniform(size= (X,X*Y))
P_xprime_xy = P_xprime_xy / P_xprime_xy.sum(axis=0)[None,:]
#####
P_xprime_xy = torch.from_numpy(P_xprime_xy).float()
phi_t_x_y_k = np.array([beta**t for t in
    range(T)])[:,None,None,None] *  phi_x_y_k[None,:,:,:]
```

```
phi_t_x_y_k = torch.from_numpy(phi_t_x_y_k).float()
Ehat_phi_k = torch.from_numpy(muhat_txy.T).float() @ \
    phi_t_x_y_k.reshape((-1,K))
SigmaY = sp.kron( sp.eye(T*X),np.ones((1,Y)))
Psi = sp.kron(sp.diags(diagonals=[1]*(T-1), offsets=+1,
    shape=(T,T)),P_xprime_xy.T) - SigmaY.T
I = muhat_txy.sum()
V= np.diag(muhat_txy) / I - muhat_txy[:,None] @ \
    muhat_txy[None,:] / (I*I)

def F(lambda_k):
    u_t_x = torch.zeros((T+1, X))
    for t in range(T-1, -1, -1):
        u_t_x[t,:] = torch.log(torch.exp(
            (phi_t_x_y_k[t,:,:,:].reshape((-1, K)) @ lambda_k
            + u_t_x[t+1,:] @ P_xprime_xy).reshape((X,
            Y))).sum(axis=1))
    return Ehat_phi_k @ lambda_k + \
        torch.from_numpy(muhat_txy.T @ Psi).float() @ \
        u_t_x[:-1,:].flatten()

lambda_k = torch.ones(K, requires_grad=True)
optimizer = torch.optim.Adam([lambda_k],lr=0.01,
    maximize=True)
num_steps = 1000
for step in range(num_steps):
    optimizer.zero_grad()  # Clear previous gradients
    res = F(lambda_k)      # Compute the function
    res.backward()         # Compute gradients
    optimizer.step()       # Update lambda_k
    if step % 100 == 0:    # Print info every 100 steps
        print(f"Step {step}, Function Value: {res.item()}")

print("lambdahat:", lambda_k) #print("Final Gradient:",
    lambda_k.grad)
lambda_k.grad = None
u_t_x = torch.zeros((T+1, X))
for t in range(T-1, -1, -1):
    u_t_x[t,:] = \
        torch.log(torch.exp((phi_t_x_y_k[t,:,:,:].reshape((-1,
        K)) @ lambda_k + u_t_x[t+1,:] @ \
        P_xprime_xy).reshape((X, Y))).sum(axis=1))

du = torch.zeros((T, X, K))
d2u = torch.zeros((T, X, K, K))
for t in range(T):
    for x in range(X):
         grad1 = torch.autograd.grad(u_t_x[t, x], lambda_k,
             create_graph=True, retain_graph=True,
             allow_unused=True)[0]
         du[t, x] = grad1.detach()
```

```
        for k in range(K):
            grad2 = torch.autograd.grad(grad1[k], lambda_k,
                retain_graph=True, allow_unused=True)[0]
            d2u[t, x, k] = grad2.detach()

neg_hess =  (- muhat_txy.T @ Psi @
    d2u.numpy().reshape((T*X,-1))).reshape((K,K))
HinvB = np.linalg.solve(neg_hess,
    (phi_t_x_y_k.reshape((-1,K)) + Psi @ \
    du.numpy().reshape((T*X,-1) )).T)
print('Cov(lambdahat)=')
print(HinvB @ V @ HinvB.T)
```

B.6.5 ESTIMATION IN INFINITE HORIZON, NESTED FIXED-POINT

```
# Estimation of dynamic discrete choice via nested fixed point
#
# Importing libraries and generating data:
import numpy as np
from scipy import optimize, sparse as sp
np.random.seed(7)
X,Y,K = 10,2,3
beta = 0.9
rel_tol = 1e-5
epsilon = 0.05
max_it = 10000
phi_x_y_k = np.random.normal(size=(X,Y,K))/10
muhat_xy = np.random.uniform(size=(X*Y))
P_xprime_xy = np.random.uniform(size= (X,X*Y))
P_xprime_xy = P_xprime_xy / P_xprime_xy.sum(axis=0)[None,:]
# Computing:
phi_xy_k = phi_x_y_k.reshape((-1,K))
SigmaY = sp.kron( sp.eye(X),np.ones((1,Y)))
Psi = beta* P_xprime_xy.T - np.kron( np.eye(X),np.ones((Y,1)))
tol_lambda = rel_tol* np.abs(phi_x_y_k).max()*muhat_xy.sum()

def fu_x(lambda_k, tol_u=1e-5, max_it = 10000):
    u_x = np.zeros(X)
    phi_x_y = (phi_xy_k @ lambda_k).reshape((X,Y))
    M = 1+ (np.abs(phi_x_y).max() + np.log(Y)) / (1-beta)
    S = int(max ( (np.log(tol_u) - 2* M) / np.log(beta) ,10) )
    for s in range(min(S,max_it)):
        m_x = (phi_x_y + beta * (P_xprime_xy.T @ \
            u_x).reshape((X,Y))).max(axis = 1)
        u_x = m_x + np.log( np.exp(phi_x_y + beta * \
            (P_xprime_xy.T @ u_x).reshape((X,Y)) -m_x[:,None]
            ).sum(axis = 1) )
    return(u_x)
```

```
def fdu_x(lambda_k):
    u_x = fu_x(lambda_k).flatten()
    pi_xy = np.exp( phi_xy_k @ lambda_k + Psi @ u_x)
    return - sp.linalg.spsolve(SigmaY @ sp.diags(pi_xy) @ Psi
        , SigmaY @ sp.diags(pi_xy) @ phi_xy_k)

def neg_F(lambda_k):
    return  - muhat_xy.T @ ( phi_xy_k @ lambda_k + Psi @ \
        fu_x(lambda_k))

def neg_gradF(lambda_k):
    return muhat_xy.T @ ( - Psi @ fdu_x(lambda_k) - phi_xy_k)

def neg_hessF(lambda_k):
    u_x = fu_x(lambda_k).flatten()
    pi_xy = np.exp( phi_xy_k @ lambda_k + Psi @ u_x)
    dlogpi = phi_xy_k + Psi @ fdu_x(lambda_k)
    XY = X*Y
    TTT = sp.csr_matrix( ([1]*XY,(list(range(XY)), [a*(XY+1) \
        for a in range(XY)])),shape = (XY,XY*XY) )
    Tdlogpisquare =  TTT @ sp.kron(dlogpi,dlogpi )
    d2u = - sp.linalg.spsolve( SigmaY @ sp.diags(pi_xy) @ Psi
        , SigmaY @ sp.diags(pi_xy) @  Tdlogpisquare)
    return (- muhat_xy.T @ Psi @ d2u).reshape((K,K))

result = optimize.minimize(neg_F, np.zeros(K), jac=neg_gradF,
    method='BFGS')

# displaying results
if result.success:
    lambdahat_k = result.x
    print("lambdahat:", lambdahat_k)
    I = muhat_xy.sum()
    V = np.diag(muhat_xy) / I - muhat_xy[:,None] @ \
        muhat_xy[None,:] / (I*I)
    HinvB = np.linalg.solve(neg_hessF(lambdahat_k), (phi_xy_k
        + Psi @ fdu_x(lambdahat_k)).T)
    print('Cov(lambdahat)=')
    print(HinvB @ V @ HinvB.T)
else:
    print(result.message)
```

B.6.6 ESTIMATION IN INFINITE HORIZON, AUGMENTED LAGRANGIAN

```
# Estimation of dynamic discrete choice via augmented
    Lagrangian
#
# Importing libraries and generating data:
```

```
import numpy as np
from scipy import optimize, sparse as sp
np.random.seed(7)
X,Y,K = 10,2,3
beta = 0.9
rel_tol = 1e-6
epsilon = 0.02
max_it = 1e6
phi_x_y_k = np.random.normal(size=(X,Y,K))/10
muhat_xy = np.random.uniform(size=(X*Y))
P_xprime_xy = np.random.uniform(size= (X,X*Y))
P_xprime_xy = P_xprime_xy / P_xprime_xy.sum(axis=0)[None,:]
# Computing:
phi_xy_k = phi_x_y_k.reshape((-1,K))
Psi = beta* P_xprime_xy.T - np.kron( np.eye(X),np.ones((Y,1)))
phi_infty = np.abs(phi_x_y_k).max() * muhat_xy.sum()
psi_infty = np.abs(Psi).max() * muhat_xy.sum()
SigmaY = np.kron( np.eye(X),np.ones((1,Y)))

def log_likelihood(lambda_and_u):
    lambda_k, u_x = lambda_and_u[:K], lambda_and_u[K:]
    return - muhat_xy.dot ( phi_xy_k @ lambda_k + Psi @ u_x )

def grad_log_likelihood(lambda_and_u):
    lambda_k, u_x = lambda_and_u[:K], lambda_and_u[K:]
    return - np.concatenate( (muhat_xy.dot( phi_xy_k),
        muhat_xy.dot( Psi)))

def Z_x(lambda_and_u):
    lambda_k, u_x = lambda_and_u[:K], lambda_and_u[K:]
    pi_x_y =  np.exp( phi_xy_k @ lambda_k + Psi @ u_x
        ).reshape((X,Y))
    return (pi_x_y.sum(axis=1)-1 )

def JZ_x(lambda_and_u):
    lambda_k, u_x = lambda_and_u[:K], lambda_and_u[K:]
    pi_xy =  np.exp( phi_xy_k @ lambda_k + Psi @ u_x )
    return  SigmaY @ np.diag(pi_xy) @ np.block([[phi_xy_k,
        Psi]])


def augmented_lagrangian(x, n_x, gamma, obj_val, obj_grad,
    constr_fun, constr_grad):
    f = obj_val(x)
    h = constr_fun(x)
    L = f + np.dot(n_x, h) + 0.5 * gamma * np.sum(h**2)
    grad_f = obj_grad(x)
    J_h = constr_grad(x)
    grad_L = grad_f + np.dot(J_h.T, n_x + gamma * h)
    return L, grad_L
```

```
def solve_augmented_lagrangian(x0, n0_x, gamma0, obj_val,
    obj_grad, constr_fun, constr_grad, coeff=10,
    max_iter=100, tol=1e-6):
    x = np.array(x0)
    n_x = np.array(n0_x)
    gamma = gamma0

    for i in range(max_iter):
        fx = lambda x: augmented_lagrangian(x, n_x, gamma,
            obj_val, obj_grad, constr_fun, constr_grad)[0]
        gx = lambda x: augmented_lagrangian(x, n_x, gamma,
            obj_val, obj_grad, constr_fun, constr_grad)[1]
        result = optimize.minimize(fx, x, method='BFGS', jac
            = gx)
        x = result.x
        h = constr_fun(x)
        n_x += gamma * h
        gamma *= coeff
        if np.linalg.norm(h, np.inf) < tol:
            break
    return x, n_x, gamma


lambda_and_u0 = np.zeros(K+X)
n0_x = muhat_xy.reshape((X,Y)).sum(axis=1)
gamma0 = 1.0

lambda_and_ustar, nstar_x, gammastar =
    solve_augmented_lagrangian(lambda_and_u0, n0_x, gamma0,
    log_likelihood, grad_log_likelihood, Z_x, JZ_x)

lambdastar_k, ustar_x = lambda_and_ustar[:K],
    lambda_and_ustar[K:]
print("lambda_k:", lambdastar_k)
print("n_x:", nstar_x)
```

B.6.7 DYNAMIC DISCRETE CHOICE ESTIMATION VIA CCP ESTIMATOR

```
# CCP estimator for dynamic discrete choice
#
# Importing libraries and generating data:
import numpy as np
from scipy import optimize, sparse as sp
np.random.seed(7)
X,Y,K = 10,2,3
beta = 0.9
rel_tol = 1e-6
epsilon = 0.02
max_it = 1e6
```

```
phi_x_y_k = np.random.normal(size=(X,Y,K))/10
muhat_xy = np.random.uniform(size=(X*Y))
P_xprime_xy = np.random.uniform(size= (X,X*Y))
P_xprime_xy = P_xprime_xy / P_xprime_xy.sum(axis=0)[None,:]
# Computing:
Psi = beta* P_xprime_xy.T - np.kron( np.eye(X),np.ones((Y,1)))
deltaone = np.kron(np.eye(X), np.array([[1.0]+[0.0]*(Y-1)]))
A = np.eye(X*Y) + Psi @ np.linalg.inv( deltaone @ Psi ) @ \
    deltaone
logpihat_xy = np.log(muhat_xy.reshape((X,Y)) /
    muhat_xy.reshape((X,Y)).sum(axis = 1)[:,None]).flatten()
phihat_xy = A @ logpihat_xy
# display results:
print(phihat_xy)
```

B.6.8 LOADING ENGINE REPAIR DATA FOR PROBLEM 6.1

```
# Loading Rust's Harold Zurcher data
import numpy as np
def load_Rust_data():
    pth = 'https://raw.githubusercontent.com/math-econ-code/'
    pth += 'mec_datasets/main/dynamicchoice_Rust/datafiles/'

    def getcleandata(name,nrow):
        filepath = pth + name + '.asc'
        thearray = np.genfromtxt(filepath, delimiter=None,
            dtype=float).reshape((nrow, -1), order='F')
        odometer1 = thearray[5, :]  # mileage at first
            replacement (0 if no replacement)
        odometer2 = thearray[8, :]  # mileage at second
            replacement (0 if no replacement)
        thearray = thearray[11:, :]
        replaced1 = (thearray >= odometer1) * (odometer1 > 0)
             # replaced once
        replaced2 = (thearray >= odometer2) * (odometer2 > 0)
             # replaced twice

        running_odometer = np.floor((thearray- odometer1 *
            replaced1 + (odometer1-odometer2)*replaced2 ) /
            5000).astype(int)
        T,B = thearray.shape
        replact = np.array([[ (1 if (replaced1[t+1,b] and not
            replaced1[t,b]) or (replaced2[t+1,b] and not
            replaced2[t,b])  else 0) for b in range(B)]
                for t in range(T-1)]+
                    [[0 for b in range(B)]])
        increment = np.array([[ 0 for b in range(B)]]+
```

```
                                [[ running_odometer[t+1,b] -
                                    running_odometer[t,b] *
                                    (1-replact[t,b]) for b in
                                    range(B)] for t in
                                    range(T-1)])

        return np.block([[running_odometer.reshape((-1,1)
            ),replact.reshape((-1,1) ),
            increment.reshape((-1,1) )]])

    return np.vstack([getcleandata(name,nrow) for (name,
        nrow) in [
        ('g870',36),('rt50',60),('t8h203',81),('a530875',128)
        ]])
```

B.6.9 SIMULATING DYNAMIC MATCHING DATA FOR PROBLEM 6.2

```
# Data generation for the estimation of dynamic matching
    models
import numpy as np
from scipy import optimize, sparse as sp
from scipy.sparse import kron
np.random.seed(7)
X,Y,K = 15,12,3
B = X*Y+X+Y
beta = 0.9
rel_tol = 1e-5
epsilon = 0.05
max_it = 10000
phi_b_k = np.random.normal(size=(B,K))
muhat_b = np.random.uniform(size=B)
P_xprime_b = np.random.uniform(size= (X,B))
P_xprime_b = P_xprime_b / P_xprime_b.sum(axis=0)[None,:]
Q_yprime_b = np.random.uniform(size= (Y,B))
Q_yprime_b = Q_yprime_b / Q_yprime_b.sum(axis=0)[None,:]
```

B.7 Code for Chapter 7: Constrained Choice

B.7.1 CONSTRAINED LOGIT MODEL

```
# Constrained logit model
#
# Importing libraries and generating data:
import scipy.optimize as opt, numpy as np
U_y = np.array([0.2, -.3, 0.5])
```

```
pibar_y = np.array([0.3, 0.5, 0.6])

# Computing
u = -np.log(opt.brentq(lambda themu : (themu+
    np.minimum(themu *np.exp(U_y),pibar_y).sum() - 1) ,0,1))

# displaying results
print('u=',u)
```

B.7.2 CONSTRAINED CHOICE SIMULATOR

```
# Constrained choice simulation
#
# Importing libraries and generating data:
!pip install gurobipy
import numpy as np, gurobipy as grb
np.random.seed(7)
I,Y = 100,3
U_y = np.array([0.2, -.3, 0.5])
pibar_y = np.array([0.3, 0.2, 0.1])
epsilon_i_y = np.random.normal(size=(I,Y+1) )
# Computing:
m = grb.Model()
lambda_i_y = m.addMVar( (I,Y+1))
m.setObjective(  (lambda_i_y * (
    np.array([0]+list(U_y))[None,:] + epsilon_i_y)).sum(),
                 sense = grb.GRB.MAXIMIZE)
u_i = m.addConstr( lambda_i_y.sum(axis=1) == 1/I )
tau_y = m.addConstr( lambda_i_y.sum(axis=0)[1:] <= pibar_y )
m.optimize()
# Displaying results:
print( 'tau_y=' , tau_y.pi)
```

B.7.3 CLASSES OF CONSTRAINED CHOICE PROBLEMS

```
# Classes of constrained choice problems
#
import numpy as np, scipy.optimize as opt, gurobipy as grb
# define parent class:
class ChoiceProblem:
  def __init__(self,n_x,alpha_x_y):
    self.n_x = n_x
    self.alpha_x_y = alpha_x_y
    self.X,self.Y = alpha_x_y.shape
#
# child class 1: no heterogeneity (individual problem)
```

```
#
class IndividualChoiceProblem(ChoiceProblem):
  def choice( self, constraint = None):
    if constraint is None:
      mubar_x_y = self.n_x[:,None] * np.ones((self.X,self.Y))
    else:
      mubar_x_y = constraint
    alpha_x_y = np.block([[self.alpha_x_y,np.zeros( (self.X,1
        )) ]])
    mu_x_y = np.zeros((self.X,self.Y+1))
    u_x = np.zeros(self.X)
    for x in range(self.X):
        nxres = self.n_x[x]
        for yind in np.argsort(alpha_x_y[x,:])[::-1]:
            if yind == self.Y:
                mu_x_y[x,yind] = nxres
                break
            if mubar_x_y[x,yind] > 0:
                mu_x_y[x,yind] = min(nxres ,
                    mubar_x_y[x,yind])
                nxres -= mu_x_y[x,yind]
            if nxres == 0:
                break
        u_x[x]= alpha_x_y[x,yind]
    return mu_x_y[:,:-1]
#
# child class 2: logit heterogeneity
#
class LogitChoiceProblem(ChoiceProblem):
  def choice( self, constraint = None):
    if constraint is None:
      mubar_x_y = self.n_x[:,None] * np.ones((self.X,self.Y))
    else:
      mubar_x_y = constraint
    mu_x_y = np.zeros((self.X,self.Y))
    for x in range(self.X):
      mu0 = opt.brentq(f=  lambda mu: mu+
          np.minimum(mu*np.exp(self.alpha_x_y[x,:]),
          mubar_x_y[x,:]).sum() -self.n_x[x]
          ,a=0,b=self.n_x[x])
      mu_x_y[x,:] =
          np.minimum(mu0*np.exp(self.alpha_x_y[x,:]),
          mubar_x_y[x,:])
    return mu_x_y
#
# child class 3: simulated heterogeneity
#
class SimulatedChoiceProblem(ChoiceProblem):
  def __init__(self,n_x,alpha_x_y,varepsilon_i_y):
    super().__init__(n_x,alpha_x_y)
    self.varepsilon_i_0 = varepsilon_i_y[:,-1]
```

```
    self.varepsilon_i_y = varepsilon_i_y[:,0:-1]
    self.I,_ = varepsilon_i_y.shape


  def choice( self, constraint = None):
    if constraint is None:
      mubar_x_y = self.n_x[:,None] * np.ones((self.X,self.Y))
    else:
      mubar_x_y = constraint
    m = grb.Model()
    m.setParam('OutputFlag', 0)
    tau_x_y = m.addMVar((self.X,self.Y))
    u_i_x = m.addMVar((self.I,self.X), lb = -
        grb.GRB.INFINITY)
    m.setObjective( (tau_x_y* mubar_x_y).sum() +
        (self.n_x[None,:]* u_i_x).sum()/self.I,
        sense=grb.GRB.MINIMIZE )
    mu_i_x_y = m.addConstr(u_i_x[:,:,None] >=
        self.alpha_x_y[None,:,:] - tau_x_y[None,:,:] +
        self.varepsilon_i_y[:,None,:] )
    m.addConstr(u_i_x >= self.varepsilon_i_0[:,None])
    m.optimize()
    mu_x_y = mu_i_x_y.pi.sum(axis=0)
    return(mu_x_y )
```

B.7.4 THE DEFERRED ACCEPTANCE ALGORITHM

```
# Deferred acceptance under various forms
#
# Before running, need to define the class ChoiceProblem and
# its child classes as above
#
# Importing libraries and generating data:
import numpy as np, scipy.optimize as opt, gurobipy as grb
np.random.seed(7)
X,Y = 15,12
n_x = np.ones(X)
m_y = np.ones(Y)
alpha_x_y = np.random.uniform(size=(X,Y) )
gamma_x_y = np.random.uniform(size=(X,Y) )
muA_x_y = np.minimum(n_x[:,None],m_y[None,:] )
muP_x_y = np.zeros((X,Y))
muK_x_y = np.zeros((X,Y))
proposers = LogitChoiceProblem(n_x,alpha_x_y)
responders = LogitChoiceProblem(m_y,gamma_x_y.T)
# proposers = IndividualChoiceProblem(n_x,alpha_x_y)
# responders = IndividualChoiceProblem(m_y,gamma_x_y.T)
# eps_i_y = np.random.normal(size=(10,Y+1))
# eta_i_x = np.random.normal(size=(10,X+1))
# proposers = SimulatedChoiceProblem(n_x, alpha_x_y,eps_i_y)
```

```
# responders = SimulatedChoiceProblem(m_y,gamma_x_y.T,eta_i_x)
maxiter=int(1e4)
rel_tol = 1e-3
# Main loop:
for i in range(maxiter):
  for x in range(X):
    muP_x_y =  proposers.choice(constraint = muA_x_y )
  for y in range(Y):
    muK_x_y = responders.choice( constraint = muP_x_y.T).T
  muR_x_y = muP_x_y - muK_x_y
  if muR_x_y.sum() < rel_tol:
    break
  muA_x_y = muA_x_y - muR_x_y

print('Converged in '+str(i)+' step.')
print ('Number of matched individuals='+str(muK_x_y.sum()))
```

B.8 Code for Chapter 8: General Heterogeneities

B.8.1 PROBIT MATCHING MODEL BY GRADIENT DESCENT

This code snippet uses the function muprobit defined in appendix B.4.3.

```
# Probit matching
#
# Importing libraries and generating data:
import numpy as np
from statistics import NormalDist
phi = np.vectorize(NormalDist(mu=0, sigma=1).cdf)
phi_inv = np.vectorize(NormalDist().inv_cdf)
def ghk(L,z_j,x_i_j):
    ... # see above
def piprobit_y(U_y,sigma_eps,x_i_j):
    ... # see above
X,Y = 5,3
np.random.seed(7)
phi_x_y = np.random.uniform(size=(X,Y))
n_x = np.ones(X)/X # no unassigned individuals, total masses
    conicide
m_y = np.ones(Y)/Y
I=J= 1000
maxiter= 1000
tol = 1e-3
eps = 0.1
unif1_i_j = np.random.uniform(size=(I,Y-1))
unif2_i_j = np.random.uniform(size=(J,X-1))
# Computing:
def muprobit (U_x_y, n_x,Sigma ,unif_i_j):
    mu_x_y = np.zeros(U_x_y.shape)
```

```
    for x in range(U_x_y.shape[0]):
        mu_x_y[x,:] = n_x[x]*
            piprobit_y(U_x_y[x,:],Sigma,unif_i_j)
    return mu_x_y
#
nablaG = lambda U_x_y:
    muprobit(U_x_y,n_x,np.eye(Y),unif1_i_j)
nablaH = lambda V_x_y:  muprobit(V_x_y.T,m_y, np.eye(X),
    unif2_i_j).T
#
w_x_y = np.zeros((X,Y))
for step  in range(maxiter):
    delta_x_y = nablaG(w_x_y) - nablaH(phi_x_y - w_x_y)
    if np.abs(delta_x_y).max()<tol:
        break
    w_x_y -= eps * delta_x_y
# Displaying results
print('Converged in ',step,' steps')
print('w_x_y=',w_x_y)
```

B.8.2 MATCHING MODELS BY SIMULATION

```
# Estimation of a matching model with probit heterogeneity
# Importing libraries and generating data:
!pip install gurobipy
import numpy as np, scipy.sparse as sp, gurobipy as grb
X,Y,K = 5,4,3
I=J= 10 # with a license, take I=J=1000
np.random.seed(7)
phi_xy_k = np.random.uniform(size=(X*Y,K))
muhat_xy = np.random.uniform(size=(X*Y))
# Simulating heterogeneity:
epsilon_ixy = np.random.normal(size = (I*X*Y))
epsilon_ix0 = np.random.normal(size = I*X)
eta_jxy = np.random.normal(size = (J*X*Y))
eta_jy0 = np.random.normal(size = J*Y )
# Computing:
n_x = muhat_xy.reshape((X,Y)).sum(axis=1)
m_y = muhat_xy.reshape((X,Y)).sum(axis=0)
m=grb.Model()
lambda_k = m.addMVar(K, lb= -grb.GRB.INFINITY)
u_ix = m.addMVar(I*X, lb= -grb.GRB.INFINITY)
v_jy = m.addMVar(J*Y, lb= -grb.GRB.INFINITY)
U_xy =  m.addMVar(X*Y, lb= -grb.GRB.INFINITY)
V_xy =  m.addMVar(X*Y, lb= -grb.GRB.INFINITY)
#theta = m.addMVar(len(b), lb= -grb.GRB.INFINITY)
m.setObjective( (muhat_xy @ phi_xy_k)@ lambda_k
    -(u_ix.reshape((I,X))*n_x[None,:]).sum()/I
               - (v_jy.reshape((I,Y))*m_y[None,:]).sum()/I
                 ,sense=grb.GRB.MAXIMIZE)
```

```
m.addConstr(U_xy+V_xy  >= phi_xy_k @ lambda_k)
m.addConstr(u_ix.reshape((I,X))[:,:,None] >=
    U_xy.reshape((X,Y))[None,:,:] +
    epsilon_ixy.reshape((I,X,Y)) )
m.addConstr(u_ix >= epsilon_ix0)
m.addConstr(v_jy.reshape((J,Y))[:,None,:] >=
    V_xy.reshape((X,Y))[None,:,:] + eta_jxy.reshape((J,X,Y)) )
m.addConstr(v_jy >= eta_jy0)
m.optimize()
# displaying results:
lambda_k = np.array(m.x[:K])
print('lambda_k=',lambda_k)
```

B.8.3 FINITE-HORIZON DYNAMIC CHOICE MODELS BY SIMULATION

```
# Estimation of a model of finite-horizon dynamic discrete
    choice  with probit heterogeneity
# Importing libraries and generating data:
!pip install gurobipy
import numpy as np, gurobipy as grb, scipy.sparse as sp
X,Y,T,K = 5,4,10,3
I=J= 3 # with a license, take I=J=1000
np.random.seed(7)
phi_txy_k = np.random.uniform(size=(T*X*Y,K))
muhat_txy = np.random.uniform(size=(T*X*Y))
muhat_txy = ( muhat_txy.reshape((T,X,Y)) /
    muhat_txy.reshape((T,X,Y)).sum(axis = (1,2)
    )[:,None,None]).flatten()
P_xprime_xy = np.random.uniform(size= (X,X*Y))
P_xprime_xy = P_xprime_xy / P_xprime_xy.sum(axis=0)[None,:]
JT = sp.diags([1], offsets=-1, shape=(T, T), format='csr')
# Simulating heterogeneity:
epsilon_itxy = np.random.normal(size = (I*T*X*Y))
# Computing:
m = grb.Model()
lambda_k = m.addMVar(K,lb = - grb.GRB.INFINITY)
u_tx = m.addMVar((T*X),lb = - grb.GRB.INFINITY)
utilde_itx = m.addMVar((I*T*X),lb = - grb.GRB.INFINITY)
Psi = sp.kron(JT.T,P_xprime_xy.T)
    -sp.kron(sp.eye(T),sp.kron(sp.eye(X),np.ones((Y,1)) ) )

m.setObjective(  ( (muhat_txy @ phi_txy_k) @ lambda_k  +
    (muhat_txy @Psi) @ u_tx ) ,sense = grb.GRB.MAXIMIZE)
m.addConstr(u_tx == sp.kron(np.ones((1,I))/I,sp.eye(T*X)) @
    utilde_itx )
m.addConstr( sp.kron(sp.eye(I*T*X), np.ones((Y,1)))@
    utilde_itx >= sp.kron( np.ones((I,1)),phi_txy_k ) @
    lambda_k
          + sp.kron( np.ones((I,1)) ,sp.kron(JT.T,
              P_xprime_xy.T)) @ u_tx +  epsilon_itxy)
```

```
m.optimize()
# displaying results:
lambda_k = np.array(m.x[:K])
print('lambda_k=',lambda_k)
```

B.8.4 INFINITE-HORIZON DYNAMIC CHOICE MODELS BY SIMULATION

```
# Estimation of a model of infinite-horizon dynamic discrete
    choice  with probit heterogeneity
# Importing libraries and generating data:
import numpy as np, gurobipy as grb, scipy.sparse as sp
X,Y,K = 5,4,3
I=J= 50 # with a license, take I=J=1000
np.random.seed(7)
beta = 0.9
phi_xy_k = np.random.uniform(size=( X*Y,K))
muhat_xy = np.random.uniform(size=(X*Y))
P_xprime_xy = np.random.uniform(size= (X,X*Y))
P_xprime_xy = P_xprime_xy / P_xprime_xy.sum(axis=0)[None,:]
# Simulating heterogeneity:
epsilon_ixy = np.random.normal(size = (I*X*Y))
# Computing:
m = grb.Model()
lambda_k = m.addMVar(K,lb = - grb.GRB.INFINITY)
u_x = m.addMVar((X),lb = - grb.GRB.INFINITY)
utilde_ix = m.addMVar((I*X),lb = - grb.GRB.INFINITY)
Psi = beta* P_xprime_xy.T - sp.kron(sp.eye(X),np.ones((Y,1)) )

m.setObjective(  ( (muhat_xy @ phi_xy_k) @ lambda_k  +
    (muhat_xy @Psi) @ u_x ) ,sense = grb.GRB.MAXIMIZE)
m.addConstr(u_x == sp.kron(np.ones((1,I))/I,sp.eye(X)) @
    utilde_ix )
m.addConstr( sp.kron(sp.eye(I*X), np.ones((Y,1)))@ utilde_ix
    >= sp.kron( np.ones((I,1)),phi_xy_k ) @ lambda_k
          + sp.kron( np.ones((I,1)) ,beta* P_xprime_xy.T ) @
              u_x +  epsilon_ixy)
m.optimize()
# displaying results:
lambda_k = np.array(m.x[:K])
print('lambda_k=',lambda_k)
```

B.9 Code for Appendix A

B.9.1 AUTOMATIC DIFFERENTIATION USING PYTORCH

```
# automatic differentiation using PyTorch
import torch
x = torch.tensor([1.0], requires_grad=True)
y = torch.tensor([2.0], requires_grad=True)
f = x * x + x * y + torch.sin(y)
f.backward()
print(x.grad,y.grad)
```

B.9.2 ROW-WISE KRONECKER PRODUCT

```
# row-wise Kronecker product (sparse and dense)
import scipy.sparse as sp, numpy as np
#
# defines the function:
def rkron(A,B): # dense version
  return np.einsum('ij,ik->ijk',A,B).reshape((A.shape[0],-1))
#
def sp_rkron(A, B): # sparse version
    A, B = sp.csr_matrix(A), sp.csr_matrix(B)
    return sp.coo_array(sp.vstack([sp.kron(A.getrow(i),
         B.getrow(i)) for i in range( A.shape[0])]))
# test it:
np.random.seed(7)
I,J,K=15,23,17
A_i_j = np.random.uniform(size = (I,J))
B_i_k = np.random.uniform(size = (I,K))
rkron(A_i_j,B_i_k).shape, sp_rkron(A_i_j,B_i_k).shape
```

B.9.3 DIAGONAL SELECTION MATRIX

```
# diagonal selection matrix
import numpy as np, scipy.sparse as sp
#
def D_I(I):
  rows = np.arange(I)
  cols = [a*(I+1) for a in range(I)]
  return sp.csr_matrix( ([1]*I,(rows,cols )),shape = (I,I*I) )
# test it:
np.random.seed(7)
I = 15
A_i_i = np.random.uniform(size = (I,I))
D_I(I) @ (A_i_i.flatten()) - A_i_i.diagonal()
```

Bibliography

Abramowitz, Milton and Irene A. Stegun (1965). *Handbook of Mathematical Functions: With Formulas, Graphs, and Mathematical Tables.* Vol. 55. Courier Corporation.

Adachi, Hiroyuki (2000). "On a characterization of stable matchings." In: *Economics Letters* 68.1, pp. 43–49.

Agresti, Alan (2012). *Categorical Data Analysis.* Vol. 792. John Wiley & Sons.

Aguirregabiria, Victor and Pedro Mira (2007). "Sequential estimation of dynamic discrete games." In: *Econometrica* 75.1, pp. 1–53.

— (2010). "Dynamic discrete choice structural models: A survey." In: *Journal of Econometrics* 156.1, pp. 38–67.

Alkan, Ahmet and David Gale (2003). "Stable schedule matching under revealed preference." In: *Journal of Economic Theory* 112.2, pp. 289–306.

Alvarez, Fernando and Robert E. Lucas Jr. (2007). "General equilibrium analysis of the Eaton–Kortum model of international trade." In: *Journal of Monetary Economics* 54.6, pp. 1726–1768.

Anderson, James E. and Eric Van Wincoop (2003). "Gravity with gravitas: A solution to the border puzzle." In: *American Economic Review* 93.1, pp. 170–192.

Anderson, Simon P., Andre De Palma, and Jacques-Francois Thisse (1992). *Discrete Choice Theory of Product Differentiation.* MIT Press.

Andrews, Isaiah, Matthew Gentzkow, and Jesse M. Shapiro (2017). "Measuring the sensitivity of parameter estimates to estimation moments." In: *The Quarterly Journal of Economics* 132.4, pp. 1553–1592.

Arcidiacono, Peter and Robert A. Miller (2011). "Conditional choice probability estimation of dynamic discrete choice models with unobserved heterogeneity." In: *Econometrica* 79.6, pp. 1823–1867.

Arellano, Manuel and Stephen Bond (1991). "Some tests of specification for panel data: Monte Carlo evidence and an application to employment equations." In: *The Review of Economic Studies* 58.2, pp. 277–297.

Azevedo, Eduardo M. and Jacob D. Leshno (2016). "A supply and demand framework for two-sided matching markets." In: *Journal of Political Economy* 124.5, pp. 1235–1268.

Baker, Stuart G. (1994). "The multinomial-Poisson transformation." In: *Journal of the Royal Statistical Society: Series D (The Statistician)* 43.4, pp. 495–504.

Bauschke, Heinz H. and Patrick L. Combettes (2017). *Convex Analysis and Monotone Operator Theory in Hilbert Spaces.* CMS Books in Mathematics. Springer International Publishing.

Baydin, Atilim Gunes, Barak A. Pearlmutter, Alexey Andreyevich Radul, and Jeffrey Mark Siskind (2018). "Automatic differentiation in machine learning: A survey." In: *Journal of Machine Learning Research* 18, pp. 1–43.

Becker, Gary (1973). "A theory of marriage: Part I." In: *Journal of Political Economy* 81.4, pp. 813–846.
Becker, Gordon M., Morris H. DeGroot, and Jacob Marschak (1963). "Stochastic models of choice behavior." In: *Behavioral Science* 8.1, pp. 41–55.
Beggs, Steven, Scott Cardell, and Jerry Hausman (1981). "Assessing the potential demand for electric cars." In: *Journal of Econometrics* 17.1, pp. 1–19.
Ben-Akiva, Moshe E. (1973). "Structure of passenger travel demand models." PhD thesis. Massachusetts Institute of Technology.
Ben-Akiva, Moshe E. and Steven R. Lerman (1985). *Discrete Choice Analysis: Theory and Application to Travel Demand.* Vol. 9.
Ben-Akiva, Moshe E., Nicolaos Litinas, and Koji Tsunekawa (1985). "Continuous spatial choice: The continuous logit model and distributions of trips and urban densities." In: *Transportation Research Part A: General* 19, pp. 83–206.
Ben-Akiva, Moshe E. and Thawat Watanatada (1981). "Application of a continuous spatial choice logit model." In: *Structural Analysis of Discrete Data with Econometric Applications,* pp. 320–343.
Berkson, Joseph (1944). "Application of the logistic function to bio-assay." In: *Journal of the American Statistical Association* 39.227, pp. 357–365.
Berman, Abraham and Robert J. Plemmons (1994). *Nonnegative Matrices in the Mathematical Sciences.* SIAM.
Berry, Steven T. (1994). "Estimating discrete-choice models of product differentiation." In: *The RAND Journal of Economics,* pp. 242–262.
Berry, Steven T., Amit Gandhi, and Philip A. Haile (2013). "Connected substitutes and invertibility of demand." In: *Econometrica* 81.5, pp. 2087–2111.
Berry, Steven T., James A. Levinsohn, and Ariel Pakes (1995). "Automobile prices in market equilibrium." In: *Econometrica* 63.4, pp. 841–890.
— (1999). "Voluntary export restraints on automobiles: Evaluating a trade policy." In: *American Economic Review* 89.3, pp. 400–431.
— (2004). "Differentiated products demand systems from a combination of micro and macro data: The new car market." In: *Journal of Political Economy* 112.1, pp. 68–105.
Berry, Steven T. and Ariel Pakes (2007). "The pure characteristics demand model." In: *International Economic Review* 48.4, pp. 1193–1225.
Bertsekas, Dimitri P. (2012). *Dynamic Programming and Optimal Control.* Vol. 1 and 2. Athena Scientific.
— (2014). *Constrained Optimization and Lagrange Multiplier Methods.* Academic Press.
Bertsimas, Dimitris and John N. Tsitsiklis (1997). *Introduction to Linear Optimization.* Athena Scientific.
Bierlaire, Michel (2006). "A theoretical analysis of the cross-nested logit model." In: *Annals of Operations Research* 144, pp. 287–300.
Birkhoff, Garrett (1957). "Extensions of Jentzsch's theorem." In: *Transactions of the American Mathematical Society* 85.1, pp. 219–227.
Blackwell, David (1965). "Discounted dynamic programming." In: *The Annals of Mathematical Statistics* 36.1, pp. 226–235.
Block, H. D. and Jacob Marschak (1959). "Random orderings and stochastic theories of response." Working Paper, Cowles Foundation for Research in Economics, Yale University.
Bonnet, Odran, Alfred Galichon, Yu-Wei Hsieh, Keith O'Hara, and Matt Shum (2022). "Yogurts choose consumers? Estimation of random-utility models via two-sided matching." In: *The Review of Economic Studies* 89.6, pp. 3085–3114.
Boyd, Stephen P. and Lieven Vandenberghe (2004). *Convex Optimization.* Cambridge University Press. 744 pp.
Brenkers, Randy and Frank Verboven (2006). "Liberalizing a distribution system: The European car market." In: *Journal of the European Economic Association* 4.1, pp. 216–251.

Bubeck, Sébastien and Nicolo Cesa-Bianchi (2012). "Regret analysis of stochastic and nonstochastic multi-armed bandit problems." In: *Foundations and Trends in Machine Learning* 5.1, pp. 1–122.

Cameron, A. Colin and Pravin K. Trivedi (2005). *Microeconometrics: Methods and Applications.* Cambridge University Press.

— (2013). *Regression Analysis of Count Data.* Vol. 53. Cambridge University Press.

Cardell, N. Scott (1997). "Variance components structures for the extreme-value and logistic distributions with application to models of heterogeneity." In: *Econometric Theory* 13.2, pp. 185–213.

Carlier, Guillaume (2022). *Classical and Modern Optimization.* World Scientific.

Carlier, Guillaume, Arnaud Dupuy, Alfred Galichon, and Yifei Sun (2023). "SISTA: Learning optimal transport costs under sparsity constraints." In: *Communications on Pure and Applied Mathematics* 76.9, pp. 1659–1677.

Caron, Stéphane (2023). *pypoman: Python module for polyhedral geometry.* https://github.com/stephane-caron/pypoman. Version 1.0.0.

Chernoff, Herman (1954). "Rational selection of decision functions." In: *Econometrica*, pp. 422–443.

Chernozhukov, Victor, Alfred Galichon, Marc Henry, and Brendan Pass (2021). "Identification of hedonic equilibrium and nonseparable simultaneous equations." In: *Journal of Political Economy* 129.3, pp. 842–870.

Chiappori, Pierre-André, Robert J. McCann, and Lars Nesheim (2010). "Hedonic price equilibria, stable matching, and optimal transport: Equivalence, topology, and uniqueness." In: *Economic Theory* 42, pp. 317–354.

Chiong, Khai Xiang, Alfred Galichon, and Matt Shum (2016). "Duality in dynamic discrete-choice models." In: *Quantitative Economics* 7.1, pp. 83–115.

Choo, Eugene and Aloysius Siow (2006). "Who marries whom and why." In: *Journal of Political Economy* 114.1, pp. 175–201.

Ciscato, Edoardo, Alfred Galichon, and Marion Goussé (2020). "Like attract like? A structural comparison of homogamy across same-sex and different-sex households." In: *Journal of Political Economy* 128.2, pp. 740–781.

Conlon, Christopher and Jeff Gortmaker (2020). "Best practices for differentiated products demand estimation with PyBLP." In: *The RAND Journal of Economics* 51.4, pp. 1108–1161.

Corblet, Pauline (2022a). "Education expansion, sorting, and the decreasing education wage premium." Working Paper.

— (2022b). "Returns to education and experience on the labor market: A matching perspective." PhD thesis. Sciences Po.

Corblet, Pauline, Jeremy Fox, and Alfred Galichon (2024). "Repeated matching games: An empirical framework." Working Paper.

Correia, Sergio, Paulo Guimarães, and Tom Zylkin (2020). "Fast Poisson estimation with high-dimensional fixed effects." In: *The Stata Journal* 20.1, pp. 95–115.

Cosslett, Stephen (1988). "Extreme-value stochastic processes: A model of random utility maximization for a continuous choice set." Preprint, Department of Economics, Ohio State University.

Cox, David R. (1958). "The regression analysis of binary sequences." In: *Journal of the Royal Statistical Society Series B: Statistical Methodology* 20.2, pp. 215–232.

Crandall, Michael G. and Luc Tartar (1980). "Some relations between nonexpansive and order preserving mappings." In: *Proceedings of the American Mathematical Society* 78.3, pp. 385–390.

Cuturi, Marco (2013). "Sinkhorn distances: Lightspeed computation of optimal transport." In: *Advances in Neural Information Processing Systems* 26.

Dagsvik, John K. (1994). "Discrete and continuous choice, max-stable processes, and independence from irrelevant attributes." In: *Econometrica* 62.5, pp. 1179–1205.

Daly, Andrew J. and S. Zachary (1979). "Improved multiple choice models." In: *Identifying and Measuring the Determinants of Mode Choice.* Ed. by D. Hensher and Q. Dalvi. London: Teakfield.

de Haan, Laurens and Ana Ferreira (2006). *Extreme Value Theory: An Introduction.* Springer.

de Haan, Laurens and Sidney I. Resnick (1977). "Limit theory for multivariate sample extremes." In: *Zeitschrift für Wahrscheinlichkeitstheorie und verwandte Gebiete* 40.4, pp. 317–337.

Debreu, Gerard (1960). "Review of *Individual Choice Behavior* by R. D. Luce." In: *American Economic Review* 50.1, pp. 186–188.

Dempster, Arthur P. (1968). "A generalization of Bayesian inference." In: *Journal of the Royal Statistical Society: Series B (Methodological)* 30.2, pp. 205–232.

Dempster, Arthur P., Nan M. Laird, and Donald B. Rubin (1977). "Maximum likelihood from incomplete data via the EM algorithm." In: *Journal of the Royal Statistical Society: Series B (Methodological)* 39.1, pp. 1–22.

Dornbusch, Rudiger, Stanley Fischer, and Paul A. Samuelson (1977). "Comparative advantage, trade, and payments in a Ricardian model with a continuum of goods." In: *The American Economic Review* 67.5, pp. 823–839.

Dubé, Jean-Pierre, Jeremy T. Fox, and Che-Lin Su (2012). "Improving the numerical performance of static and dynamic aggregate discrete choice random coefficients demand estimation." In: *Econometrica* 80.5, pp. 2231–2267.

Dupuy, Arnaud and Alfred Galichon (2014). "Personality traits and the marriage market." In: *Journal of Political Economy* 122.6. The University of Chicago Press, pp. 1271–1319.

Dupuy, Arnaud, Alfred Galichon, and Marc Henry (2014). "Entropy methods for identifying hedonic models." In: *Mathematics and Financial Economics* 8, pp. 405–416.

Dupuy, Arnaud, Alfred Galichon, and Yifei Sun (2019). "Estimating matching affinity matrices under low-rank constraints." In: *Information and Inference: A Journal of the IMA* 8.4, pp. 677–689.

Eaton, Jonathan and Samuel Kortum (2002). "Technology, geography, and trade." In: *Econometrica* 70.5, pp. 1741–1779.

Efron, Bradley (2022). *Exponential Families in Theory and Practice.* Cambridge University Press.

Ekeland, Ivar (2010). "Existence, uniqueness and efficiency of equilibrium in hedonic markets with multidimensional types." In: *Economic Theory* 42, pp. 275–315.

Ekeland, Ivar, James J. Heckman, and Lars Nesheim (2004). "Identification and estimation of hedonic models." In: *Journal of Political Economy* 112.S1, pp. S60–S109.

Ekeland, Ivar and Roger Temam (1976). "Convex analysis and variational problems." In: *Studies in Mathematics and Its Applications* 1.

Fahli, Amine and Antoine Jacquet (2024). "Welfare consequences of rent-control policies: a matching perspective." Working paper.

Fahrmeir, Ludwig, Gerhard Tutz, Wolfgang Hennevogl, and Eliane Salem (1994). *Multivariate Statistical Modelling Based on Generalized Linear Models.* Vol. 425. Springer.

Feller, William (1991). *An Introduction to Probability Theory and Its Applications, Volume 2.* Vol. 81. John Wiley & Sons.

Filar, Jerzy and Koos Vrieze (2012). *Competitive Markov Decision Processes.* Springer Science & Business Media.

Franklin, Joel and Jens Lorenz (1989). "On the scaling of multidimensional matrices." In: *Linear Algebra and Its Applications* 114, pp. 717–735.

Gale, David and Lloyd S. Shapley (1962). "College admissions and the stability of marriage." In: *The American Mathematical Monthly* 69.1, pp. 9–15.

Galichon, Alfred (2016). *Optimal Transport Methods in Economics.* Princeton University Press.

— (2022). "On the representation of the nested logit model." In: *Econometric Theory* 38.2, pp. 370–380.

— (2025a). *'math+econ+code' lecture on dynamic discrete choice.* https://alfredgalichon.github.io/DCM/dynamic-discrete-choice.

— (2025b). *'math+econ+code' lecture on dynamic programming.* https://alfredgalichon.github.io/DCM/dynamic-programming.

— (2025c). *'math+econ+code' lecture on random utility models.* https://alfredgalichon.github.io/DCM/random-utility-models.html.
— (2025d). *'math+econ+code' lecture on the Berry-Levinsohn-Pakes method.* https://alfredgalichon.github.io/DCM/berry-levinsohn-pakes.html.
— (2025e). *'math+econ+code' lecture on the estimation of matching models.* https://alfredgalichon.github.io/DCM/matching-estimation.
— (2025f). *'math+econ+code' lecture on the estimation of the gravity equation.* https://alfredgalichon.github.io/DCM/gravity-equation.
Galichon, Alfred, Yu-Wei Hsieh, and Maxime Sylvestre (2024). "Monotone comparative statics for submodular functions, with an application to aggregated deferred acceptance." arXiv Working Paper No. 2304.12171.
Galichon, Alfred and Bernard Salanié (2009). "Matching with trade-offs: Revealed preferences over competing characteristics." SSRN Working Paper No. 1487307.
— (2022). "Cupid's invisible hand: Social surplus and identification in matching models." In: *The Review of Economic Studies* 89.5, pp. 2600–2629.
— (2024). "Estimating separable matching models." In: *Journal of Applied Econometrics* 39.6, pp. 1021–1044.
Galichon, Alfred, Larry Samuelson, and Lucas Vernet (2025). "Unified gross substitutes and inverse isotonicity for equilibrium problems." In: *Theoretical Economics* 20.3, pp. 941–971.
Gaubert, Stéphane and Jeremy Gunawardena (2004). "The Perron-Frobenius theorem for homogeneous, monotone functions." In: *Transactions of the American Mathematical Society* 356.12, pp. 4931–4950.
Geweke, John (1989). "Bayesian inference in econometric models using Monte Carlo integration." In: *Econometrica*, pp. 1317–1339.
Gittins, John, Kevin Glazebrook, and Richard Weber (2011). *Multi-armed Bandit Allocation Indices.* John Wiley & Sons.
Gorman, William M. (1980). "A possible procedure for analysing quality differentials in the egg market." In: *The Review of Economic Studies* 47.5, pp. 843–856.
Gourieroux, Christian, Alain Monfort, and Alain Trognon (1984). "Pseudo maximum likelihood methods: Theory." In: *Econometrica*, pp. 681–700.
Greene, William H. (2000). *Econometric Analysis.* Fourth edition. Prentice Hall.
Griewank, Andreas and Andrea Walther (2008). *Evaluating Derivatives: Principles and Techniques of Algorithmic Differentiation.* SIAM.
Guimarães, Paulo (2004). "Understanding the multinomial-Poisson transformation." In: *The Stata Journal* 4.3, pp. 265–273.
Guimarães, Paulo and Pedro Portugal (2010). "A simple feasible procedure to fit models with high-dimensional fixed effects." In: *The Stata Journal* 10.4, pp. 628–649.
Gumbel, Emil J. (1960). "Bivariate exponential distributions." In: *Journal of the American Statistical Association* 55.292, pp. 698–707.
Gunawardena, Jeremy and Mike Keane (1995). "On the existence of cycle times for some nonexpansive maps." Technical Report HPL-BRIMS-95-003, Hewlett-Packard Labs.
Gurobi Optimization (2021). *Gurobi Optimizer Reference Manual.*
Hajivassiliou, Vassilis A. and Daniel McFadden (1998). "The method of simulated scores for the estimation of LDV models." In: *Econometrica*, pp. 863–896.
Hansen, Lars Peter (1982). "Large sample properties of generalized method of moments estimators." In: *Econometrica*, pp. 1029–1054.
Hansen, Lars Peter and Thomas J. Sargent (2008). *Robustness.* Princeton University Press.
Hansen, Lars Peter and Kenneth J. Singleton (1982). "Generalized instrumental variables estimation of nonlinear rational expectations models." In: *Econometrica*, pp. 1269–1286.
Hastie, Trevor, Robert Tibshirani, and Jerome Friedman (2009). *The Elements of Statistical Learning: Data Mining, Inference, and Prediction.* Springer Science & Business Media.

Hatfield, John W. and Paul R. Milgrom (2005). "Matching with contracts." In: *American Economic Review* 95.4, pp. 913–935.

Head, Keith and Thierry Mayer (2014). "Gravity equations: Workhorse, toolkit, and cookbook." In: *Handbook of International Economics.* Vol. 4. Elsevier, pp. 131–195.

Heckman, James J., Rosa L. Matzkin, and Lars Nesheim (2010). "Nonparametric identification and estimation of nonadditive hedonic models." In: *Econometrica* 78.5, pp. 1569–1591.

Heckman, James J. and Jose Scheinkman (1987). "The importance of bundling in a Gorman-Lancaster model of earnings." In: *The Review of Economic Studies* 54.2, pp. 243–255.

Hestenes, Magnus R. (1969). "Multiplier and gradient methods." In: *Journal of Optimization Theory and Applications* 4.5, pp. 303–320.

Hiriart-Urruty, Jean-Baptiste, and Claude Lemaréchal (2001). *Fundamentals of Convex Analysis.* Grundlehren Text Editions. Berlin Heidelberg: Springer-Verlag.

Horn, Roger A. and Charles R. Johnson (2012). *Matrix Analysis.* Cambridge University Press.

Hotz, V. Joseph and Robert A. Miller (1993). "Conditional choice probabilities and the estimation of dynamic models." In: *The Review of Economic Studies* 60.3, pp. 497–529.

Idel, Martin (2016). "A review of matrix scaling and Sinkhorn's normal form for matrices and positive maps." arXiv Working Paper No. 1609.06349.

Jacob, Pierre E., Ruobin Gong, Paul T. Edlefsen, and Arthur P. Dempster (2021). "A Gibbs sampler for a class of random convex polytopes." In: *Journal of the American Statistical Association* 116.535, pp. 1181–1192.

James, Willard and Charles M. Stein (1961). "Estimation with quadratic loss." In: *Proceedings of the Fourth Berkeley Symposium on Mathematical Statistics and Probability.* Vol. 4. Univ of California Press, pp. 361.

Jaynes, Edwin T. (1957). "Information theory and statistical mechanics." In: *Physical Review* 106.4, pp. 620.

Joe, Harry (1997). *Multivariate Models and Multivariate Dependence Concepts.* CRC Press.

Judd, Kenneth L. (1998). *Numerical Methods in Economics.* MIT Press.

Kanter, Marek (1975). "Stable densities under change of scale and total variation inequalities." In: *The Annals of Probability,* pp. 697–707.

Keane, Michael P. (1994). "A computationally practical simulation estimator for panel data." In: *Econometrica,* pp. 95–116.

Keane, Michael P., Petra E. Todd, and Kenneth I. Wolpin (2011). "The structural estimation of behavioral models: Discrete choice dynamic programming methods and applications." In: *Handbook of Labor Economics.* Vol. 4. Elsevier, pp. 331–461.

Keane, Michael P. and Kenneth I. Wolpin (1997). "The career decisions of young men." In: *Journal of Political Economy* 105.3, pp. 473–522.

Kitagawa, Toru, Sokbae Lee, and Chen Qiu (2022). "Treatment choice with nonlinear regret." arXiv Working Paper No. 2205.08586.

Knittel, Christopher R. and Konstantinos Metaxoglou (2014). "Estimation of random-coefficient demand models: Two empiricists' perspective." In: *Review of Economics and Statistics* 96.1, pp. 34–59.

Knothe, Herbert (1957). "Contributions to the theory of convex bodies." In: *Michigan Mathematical Journal* 4.1. University of Michigan, Department of Mathematics, pp. 39–52.

Laird, Nan (2014). "Sundberg formulas." In: *Wiley StatsRef: Statistics Reference Online.*

Lancaster, Kelvin J. (1966). "A new approach to consumer theory." In: *Journal of Political Economy* 74.2, pp. 132–157.

Lang, Joseph B. (1996). "On the comparison of multinomial and Poisson log-linear models." In: *Journal of the Royal Statistical Society Series B: Statistical Methodology* 58.1, pp. 253–266.

Leclerc, Hugo and Quentin Mérigot (2024). *pysdot: Python bindings for sdot (semi-discret optimal transportation tools).* https://github.com/sd-ot/pysdot. Version 0.2.18.

Lin, Ranie (2025a). *Berry-Levinsohn-and-Pakes-1995-Replication.* https://github.com/ranielin/Berry-Levinsohn-and-Pakes-1995-Replication.

— (2025b). *Rust-1987-Replication.* https://github.com/ranielin/Rust-1987-Replication.

Luce, R. Duncan (1977). "The choice axiom after twenty years." In: *Journal of Mathematical Psychology* 15.3, pp. 215–233.

— (2005). *Individual Choice Behavior: A Theoretical Analysis.* Courier Corporation.

Luenberger, David G. and Yinyu Ye (2015). *Linear and Nonlinear Programming.* Springer.

Magnac, Thierry and David Thesmar (2002). "Identifying dynamic discrete decision processes." In: *Econometrica* 70.2, pp. 801–816.

Magnus, Jan R. and Heinz Neudecker (2019). *Matrix Differential Calculus with Applications in Statistics and Econometrics.* John Wiley & Sons.

Mauerer, Ingrid and Gerhard Tutz (2023). "Heterogeneity in general multinomial choice models." In: *Statistical Methods & Applications* 32.1, pp. 129–148.

McCullagh, Peter and John Nelder (1989). *Generalized Linear Models.* Second Edition. Chapman and Hall.

McFadden, Daniel (1974). "Conditional logit analysis of qualitative choice behavior." In: *Frontiers in Econometrics.*

— (1976). "The mathematical theory of demand models." In: *Behavioral Travel-Demand Models.* Ed. by P. Stopher and A. Meyburg. Heath and Co., pp. 305–314.

— (1977). "Modelling the choice of residential location." Cowles Foundation Discussion Papers No. 477.

— (1981). "Econometric models of probabilistic choice." In: *Structural Analysis of Discrete Data with Econometric Applications.*

McFadden, Daniel and Kenneth E. Train (2000). "Mixed MNL models for discrete response." In: *Journal of Applied Econometrics* 15.5, pp. 447–470.

McKenzie, Lionel W. (2008). "Gross Substitutes." In: *The New Palgrave Dictionary of Economics.* Ed. by S. Durlauf and L. Blume. 2nd ed. Springer.

Moré, Jorge J. (1972). "Nonlinear generalizations of matrix diagonal dominance with application to Gauss–Seidel iterations." In: *SIAM Journal on Numerical Analysis* 9.2. Society for Industrial and Applied Mathematics, pp. 357–378.

Morishima, Michio (1964). *Equilibrium, Stability, and Growth: A Multi-sectoral Analysis.* Clarendon Press.

Neal, Radford M. and Geoffrey E. Hinton (1998). "A view of the EM algorithm that justifies incremental, sparse, and other variants." In: *Learning in Graphical Models.* Springer, pp. 355–368.

Nevo, Aviv (2000). "A practitioner's guide to estimation of random-coefficients logit models of demand." In: *Journal of Economics & Management Strategy* 9.4, pp. 513–548.

— (2001). "Measuring market power in the ready-to-eat cereal industry." In: *Econometrica* 69.2, pp. 307–342.

Newey, Whitney K. and Kenneth D. West (1987). "Hypothesis testing with efficient method of moments estimation." In: *International Economic Review*, pp. 777–787.

Neyman, Abraham and Sylvain Sorin (2003). *Stochastic Games and Applications.* Vol. 570. Springer Science & Business Media.

Nikaido, Hukukane (2016). *Convex Structures and Economic Theory.* Elsevier.

Pakes, Ariel (1984). "Patents as options: Some estimates of the value of holding European patent stocks." Technical Report, National Bureau of Economic Research.

Palmgren, Juni (1981). "The Fisher information matrix for log linear models arguing conditionally on observed explanatory variable." In: *Biometrika* 68.2, pp. 563–566.

Paszke, Adam et al. (2017). "Automatic differentiation in PyTorch." In: *NIPS 2017 Workshop on Autodiff.* Long Beach, California, USA.

Petrin, Amil (2002). "Quantifying the benefits of new products: The case of the minivan." In: *Journal of Political Economy* 110.4, pp. 705–729.

Peyré, Gabriel and Marco Cuturi (2019). "Computational optimal transport: With applications to data science." In: *Foundations and Trends in Machine Learning* 11.5-6, pp. 355–607.

Pickands, James III (1981). "Multivariate extreme value distributions." In: *Proceedings of the 43rd session of the International Statistical Institute*. Vol. 2, pp. 859–878.

Powell, Michael J. D. (1969). "A method for nonlinear constraints in minimization problems." In: *Optimization*, pp. 283–298.

Puterman, Martin L. (2014). *Markov Decision Processes: Discrete Stochastic Dynamic Programming*. John Wiley & Sons.

Resnick, Sidney I. and Rishin Roy (1991). "Random USC functions, max-stable processes and continuous choice." In: *The Annals of Applied Probability*, pp. 267–292.

Rheinboldt, Werner C. (1998). *Methods for Solving Systems of Nonlinear Equations*. Second Edition. SIAM.

Rockafellar, R. Tyrell and Roger J.-B. Wets (1998). *Variational Analysis*. Grundlehren der mathematischen Wissenschaften. Berlin Heidelberg: Springer-Verlag.

Rosaia, Nicola (2020). "Duality and Estimation of Undiscounted Markov Decision Processes." Mimeo, Harvard University.

Rosen, Sherwin (1974). "Hedonic prices and implicit markets: Product differentiation in pure competition." In: *Journal of Political Economy* 82.1, pp. 34–55.

Rosenblatt, Murray (1952). "Remarks on a multivariate transformation." In: *The Annals of Mathematical Statistics* 23.3. Institute of Mathematical Statistics, pp. 470–472.

Rossi, Peter E., Greg M. Allenby, and Robert McCulloch (2005). *Bayesian Statistics and Marketing*. New York: Wiley.

Ruschendorf, Ludger (1995). "Convergence of the iterative proportional fitting procedure." In: *The Annals of Statistics*, pp. 1160–1174.

Rust, John (1987). "Optimal replacement of GMC bus engines: An empirical model of Harold Zurcher." In: *Econometrica* 55.5, pp. 999–1033.

— (1994). "Structural estimation of Markov decision processes." In: *Handbook of Econometrics* 4, pp. 3081–3143.

Santambrogio, Filippo (2015). *Optimal Transport for Applied Mathematicians: Calculus of Variations, PDEs, and Modeling*. Progress in Nonlinear Differential Equations and Their Applications. Birkhäuser Basel.

Santos Silva, João M. C. and Silvana Tenreyro (2006). "The log of gravity." In: *The Review of Economics and Statistics* 88.4, pp. 641–658.

— (2011). "Poisson: Some convergence issues." In: *The Stata Journal* 11.2, pp. 207–212.

Sargan, John D. (1958). "The estimation of economic relationships using instrumental variables." In: *Econometrica*, pp. 393–415.

Savage, Leonard J. (1951). "The theory of statistical decision." In: *Journal of the American Statistical Association* 46.253, pp. 55–67.

Sen, Amartya K. (1971). "Choice functions and revealed preference." In: *The Review of Economic Studies* 38.3, pp. 307–317.

Shapley, Lloyd S. (1953). "Stochastic games." In: *Proceedings of the National Academy of Sciences* 39.10, pp. 1095–1100.

Shum, Matthew (2016). *Econometric Models for Industrial Organization*. Vol. 3. World Scientific.

Small, Kenneth A. and Harvey S. Rosen (1981). "Applied welfare economics with discrete choice models." In: *Econometrica*, pp. 105–130.

Sorin, Sylvain (2004). "Asymptotic properties of monotonic nonexpansive mappings." In: *Discrete Event Dynamic Systems* 14, pp. 109–122.

Su, Che-Lin and Kenneth L. Judd (2012). "Constrained optimization approaches to estimation of structural models." In: *Econometrica* 80.5, pp. 2213–2230.

Sundberg, Rolf (1974). "Maximum likelihood theory for incomplete data from an exponential family." In: *Scandinavian Journal of Statistics*, pp. 49–58.

— (1976). "An iterative method for solution of the likelihood equations for incomplete data from exponential families." In: *Communication in Statistics-Simulation and Computation* 5.1, pp. 55–64.

Szegő, Gabor (1939). *Orthogonal Polynomials.* Vol. 23. American Mathematical Soc.
Talluri, Kalyan T. and Garrett J. Van Ryzin (2006). *The Theory and Practice of Revenue Management.* Vol. 68. Springer Science & Business Media.
Theil, Henri (1950). "A rank-invariant method of linear and polynomial regression analysis." In: *Indagationes Mathematicae* 12.85, pp. 173.
Tiago de Oliveira, José (1962). "Structure Theory of Bivariate Extremes, Extensions." In: *Estudos de Matemática Estatística e Económica* 7, pp. 165–195.
— ed. (1984). *Statistical Extremes and Applications.* NATO Science Series C: Springer Netherlands.
Tibshirani, Robert (1996). "Regression shrinkage and selection via the lasso." In: *Journal of the Royal Statistical Society Series B: Statistical Methodology* 58.1, pp. 267–288.
Tinbergen, Jan (1956). "On the theory of income distribution." In: *Weltwirtschaftliches archiv*, pp. 155–175.
Topkis, Donald M. (1976). "The structure of sublattices of the product of *n* lattices." In: *Pacific Journal of Mathematics* 65.2, pp. 525–532.
Train, Kenneth E. (2009). *Discrete Choice Methods with Simulation.* Cambridge University Press.
van der Vaart, Aad W. (1998). *Asymptotic Statistics.* Cambridge Series in Statistical and Probabilistic Mathematics. Cambridge: Cambridge University Press.
Veinott, Arthur F. (1989). "Lattice programming." In: Unpublished lectures.
Verbeek, Albert (1989). "The compactification of generalized linear models." In: *Statistical Modelling: Proceedings of GLIM 89 and the 4th International Workshop on Statistical Modelling Held in Trento, Italy, July 17–21, 1989.* Springer, pp. 314–327.
— (1992). "The compactification of generalized linear models." In: *Statistica neerlandica* 46.2-3, pp. 107–142.
Villani, Cédric (2009). *Optimal Transport: Old and New.* Grundlehren der mathematischen Wissenschaften. Berlin, Heidelberg: Springer-Verlag.
Vohra, Rakesh V. (2005). *Advanced Mathematical Economics.* 1st edition. London, New York: Routledge. 208 pp.
Weber, Richard (2016). "Optimization and control." Lecture notes, University of Cambridge.
Weitzman, Martin L. (1979). "Optimal search for the best alternative." In: *Econometrica*, pp. 641–654.
Williams, Huw C. W. L. (1977). "On the formation of travel demand models and economic evaluation measures of user benefit." In: *Environment and Planning A* 9.3, pp. 285–344.
Wilson, Alan G. (1967). "A statistical theory of spatial distribution models." In: *Transportation Research* 1.3, pp. 253–269.
— (2011). *Entropy in Urban and Regional Modelling.* Vol. 1. Routledge.
Winkelmann, Rainer (2008). *Econometric Analysis of Count Data.* Springer.
Wooldridge, Jeffrey M. (2010). *Econometric Analysis of Cross Section and Panel Data.* MIT Press.
Wright, Philip Green (1928). *The Tariff on Animal and Vegetable Oils.* Macmillan.
Yotov, Yoto V., Roberta Piermartini, José-Antonio Monteiro, and Mario Larch (2016). *An Advanced Guide to Trade Policy Analysis: The Structural Gravity Model.* WTO.
Zylkin, Thomas (2019). "Verifying the existence of maximum likelihood estimates for generalized linear models." arXiv Working Paper No. 1903.01633.

Index